T0073351

To my Riemann zeros:
Odile, Julie and Michaël, my muse and offspring

Michel L. Lapidus

To my father and mother,
Ad van Frankenhuysen and Mieke Arts

Machiel van Frankenhuysen

Michel L. Lapidus
Machiel van Frankenhuysen

Fractal Geometry and Number Theory

Complex Dimensions of Fractal Strings and Zeros of Zeta Functions

with 26 illustrations

Birkhäuser
Boston • Basel • Berlin

Michel L. Lapidus
University of California
Department of Mathematics
Sproul Hall
Riverside, CA 92521-0135
USA
lapidus@math.ucr.edu

Machiel van Frankenhuysen
University of California
Department of Mathematics
Sproul Hall
Riverside, CA 92521-0135
USA
machiel@math.ucr.edu

Library of Congress Cataloging-in-Publication Data

Lapidus, Michel L. (Michel Laurent), 1956-
 Fractal geometry and number theory : complex dimensions of fractal strings and zeros
of zeta functions / Michel L. Lapidus, Machiel van Frankenhuysen.
  p. cm.
 Includes bibliographical references and indexes.
 (acid-free paper)
  1. Fractals. 2. Number theory. 3. Functions, Zeta. I. van Frankenhuysen, Machiel,
1967- II. Title.
QA614.86.L36 1999
514'.742–dc21   99-051583
                                                                    CIP

AMS Subject Classifications: Primary–11M26, 11M41, 28A75, 28A80, 35P20, 58G25
                            Secondary–11M06, 11N05, 28A12, 30D35, 58F19, 58F20, 81Q20

Printed on acid-free paper
©2000 Birkhäuser Boston   *Birkhäuser* ℬ®
Softcover reprint of the hardcover 1st edition 2000

ISBN-13: 978-1-4612-5316-7   e-ISBN-13: 978-1-4612-5314-3
DOI: 10.1007/978-1-4612-5314-3

Typeset by the authors in LaTeX.
Cover design by Jeff Cosloy, Newton, MA.
The front cover shows the complex dimensions of the golden string. See page 35, Figure 2.6.

9 8 7 6 5 4 3 2 1

Contents

Appendices

Overview

In this book, we develop a theory of complex dimensions of fractal strings (i.e., one-dimensional drums with fractal boundary). These complex dimensions are defined as the poles of the corresponding (geometric or spectral) zeta function. They describe the oscillations in the geometry or the frequency spectrum of a fractal string by means of an explicit formula.

A long-term objective of this work is to merge aspects of fractal, spectral, and arithmetic geometries. From this perspective, the theory presented in this book enables us to put the theory of Dirichlet series (and of other zeta functions) in the geometric setting of fractal strings. It also allows us to view certain fractal geometries as arithmetic objects by applying number-theoretic methods to the study of the geometry and the spectrum of fractal strings.

In Chapter 1, we first give an introduction to fractal strings and their spectrum, and we precisely define the notion of complex dimension. We then make an extensive study of the complex dimensions of self-similar fractal strings. This study provides a large class of examples to which our theory can be applied fruitfully. In particular, we show in the latter part of Chapter 2 that self-similar strings always have infinitely many complex dimensions with positive real part, and that their complex dimensions are almost periodically distributed. This is established by proving that the lattice strings—the complex dimensions of which are shown to be periodically distributed along finitely many vertical lines—are dense (in a suitable sense) in the set of all self-similar strings.

In Chapter 3, we extend the notion of fractal string to include (possibly virtual) geometries that are needed later on in our work. Then, in Chapter 4, we establish pointwise and distributional explicit formulas (in the sense of number theory, but more general), which should be considered as the basic tools of our theory. In Chapter 5, we apply our explicit formulas to construct the spectral operator, which expresses the spectrum in terms of the geometry of a fractal string. We also illustrate our formulas by studying a number of geometric and direct spectral problems associated with fractal strings.

In Chapter 6, we derive an explicit formula for the volume of the tubular neighborhoods of the boundary of a fractal string. We deduce a new criterion for the Minkowski measurability of a fractal string, in terms of its complex dimensions, extending the earlier criterion obtained by the first author and C. Pomerance (see [LapPo2]). This formula suggests analogies with aspects of Riemannian geometry, thereby giving substance to a geometric interpretation of the complex dimensions.

In the later chapters of this book, Chapters 7, 8 and 9, we analyze the connections between oscillations in the geometry and the spectrum of fractal strings. Thus we place the spectral reformulation of the Riemann Hypothesis, obtained by the first author and H. Maier [LapMa2], in a broader and more conceptual framework, which applies to a large class of zeta functions, including all those for which one expects the generalized Riemann Hypothesis to hold. We also reprove—and extend to a large subclass of the above-mentioned class—Putnam's theorem according to which the Riemann zeta function does not have a sequence of critical zeros in arithmetic progression.

We conclude by proposing as a new definition of fractality the presence of nonreal complex dimensions with positive real part. We also make several suggestions for future research in this area.

This book consists entirely of new research material developed by the authors over the last four years. Some of the main results were announced in [Lap-vF1-4]. The paper [Lap-vF3] combines and supersedes our two IHES preprints [Lap-vF1-2]; it is a slightly expanded version of the IHES preprint M/97/85. The interested reader may wish to consult this paper—in conjunction with the introduction and Chapter 1 below—to have an accessible and relatively self-contained overview of some of the main aspects of this work.

Le plus court chemin entre deux vérités dans le domaine réel

passe par le domaine complexe.

[The shortest path between two truths in the real domain passes through the complex domain.]

Jacques HADAMARD

Introduction

A fractal drum is a bounded open subset of $\mathbb{R}^m$ with a fractal boundary. A difficult problem is to describe the relationship between the shape (geometry) of the drum and its sound (its spectrum). In this book, we restrict ourselves to the one-dimensional case of fractal strings, and their higher dimensional analogues, fractal sprays. We develop a theory of complex dimensions of a fractal string, and we study how these complex dimensions relate the geometry with the spectrum of the fractal string. We refer the reader to [Berr1–2, Lap1–4, LapPol1–3, LapMa1–2, HeLap1–2] and the references therein for further physical and mathematical motivations of this work. (Also see, in particular, Sections 7.1, 10.3 and 10.4, along with Appendix B.)

In Chapter 1, we introduce the basic object of our research, *fractal strings* (see [Lap1–3, LapPol1–3, LapMa1–2, HeLap1–2]). A 'standard fractal string' is a bounded open subset of the real line. Such a set is a disjoint union of open intervals, the lengths of which form a sequence

$$\mathcal{L} = l_1, \, l_2, \, l_3, \dots,$$

which we assume to be infinite. Important information about the geometry of $\mathcal{L}$ is contained in its *geometric zeta function*

$$\zeta_\mathcal{L}(s) = \sum_{j=1}^{\infty} l_j^s.$$

We assume throughout that this function has a suitable meromorphic extension. The central notion of this book, the *complex dimensions* of a fractal string $\mathcal{L}$, is defined as the poles of the meromorphic extension of $\zeta_{\mathcal{L}}$.

The spectrum of a fractal string consists of the sequence of frequencies[1]

$$f = k \cdot l_j^{-1} \qquad (k, j = 1, 2, 3, \dots).$$

The *spectral zeta function* of $\mathcal{L}$ is defined as

$$\zeta_\nu(s) = \sum_f f^{-s}.$$

The geometry and the spectrum of $\mathcal{L}$ are connected by the following formula [Lap2]:

$$\zeta_\nu(s) = \zeta_{\mathcal{L}}(s)\zeta(s), \qquad (*)$$

where $\zeta(s) = 1 + 2^{-s} + 3^{-s} + \dots$ is the classical Riemann zeta function, which in this context can be viewed as the spectral zeta function of the unit interval.

We also define a natural higher-dimensional analogue of fractal strings, the *fractal sprays* [LapPo3], the spectra of which are described by more general zeta functions than $\zeta(s)$. In that case, the counterpart of $(*)$ still holds and can be used to study the spectrum of a fractal spray.[2]

We illustrate these notions throughout Chapter 1 by working out the example of the Cantor string. In this example, we see that the various notions that we have introduced are described by the complex dimensions of the Cantor string. In higher dimensions, a similar example is provided by the Cantor sprays.

This theory of complex dimensions sheds new light on and is partly motivated by the earlier work of the first author in collaboration with C. Pomerance and H. Maier (see [LapPo2] and [LapMa2]). In particular, the heuristic notion of complex dimension suggested by the methods and results of [Lap1–3, LapPo1–3, LapMa1–2, HeLap1–2] is now precisely defined and turned into a useful tool. (Compare, for instance, [LapPo1, LapMa1], [LapPo2, §4.4b], [LapMa2, §3.3], [Lap2, Figure 3.1 and §5], as well as [Lap3, §2.1, §2.2 and p. 150]. Also see Remarks 7.1, 7.2 and Figure 7.1 on pages 164–166 below.)

In Chapter 2, we make an extensive study of the complex dimensions of self-similar strings, which form an important subclass of fractal strings.

---

[1] The eigenvalues of the Dirichlet Laplacian $-d^2/dx^2$ on this set are the numbers $\lambda = \pi^2 k^2 l_j^{-2}$ $(k, j \in \mathbb{N}^*)$. The (normalized) frequencies of $\mathcal{L}$ are the numbers $\sqrt{\lambda}/\pi$.

[2] We refer the interested reader to Appendix B for a brief review of aspects of spectral geometry—including spectral zeta functions and spectral asymptotics—in the classical case of smooth manifolds.

This amounts to a study of the zeros of the function

$$f(s) = 1 - r_1^s - r_2^s - \cdots - r_N^s \qquad (s \in \mathbb{C}),$$

for real numbers $r_j \in (0, 1)$, $j = 1, \ldots, N$, $N \geq 2$. We introduce the subclass of 'lattice self-similar strings', and find a remarkable difference between the complex dimensions of lattice and nonlattice self-similar strings. In the lattice case, each number r_j is a positive integral power of one fixed real number $r \in (0, 1)$. Then f is a polynomial in r^s, and its zeros lie periodically on finitely many vertical lines. The Cantor string is the simplest example of a lattice self-similar string. We refer to Section 2.2 for additional examples. In contrast, the complex dimensions of a nonlattice string are apparently randomly distributed in a vertical strip. We show, however, that these complex dimensions are approximated by those of a sequence of lattice strings. Hence, they exhibit an almost periodic behavior. (See Theorem 2.13 and Section 2.6.) On page 35, Figure 2.6, the reader finds a diagram of the complex dimensions of the golden string, one of the simplest nonlattice self-similar strings. This and other examples are discussed in Section 2.2.

We also show in Chapter 2 that the geometric zeta function of a self-similar string coincides with a suitably defined dynamical (or Ruelle) zeta function, and hence that it admits an appropriate Euler product; see Section 2.1.1.

Chapters 3, 4 and 5 are devoted to the development of the technical tools needed to extract geometric and spectral information from the complex dimensions of a fractal string. In Chapter 3, we introduce the framework in which we will formulate our results, that of 'generalized fractal strings'. These do not in general correspond to a geometric object. Nevertheless, they are not just a gratuitous generalization. They enable us, in particular, to deal with virtual geometries and their associated spectra—suitably defined by means of their zeta functions—as though they arose from actual fractal geometries. In Chapters 7, 8 and 9, the extra flexibility of this framework allows us to study the zeros of several classes of zeta functions.

In Chapter 4, we state and prove our explicit formulas, which can be viewed as our basic tools for obtaining asymptotic expansions of geometric or spectral quantities associated with fractals. Our first explicit formula, which expresses the counting function of the lengths as a sum of oscillatory terms and an error term of smaller order, is only applicable under fairly restrictive assumptions. To obtain a more widely applicable theory, we show in Section 4.4 that this same function, interpreted as a distribution, is given by the same formula, now interpreted distributionally. The resulting distributional formula with error term is applicable under mild assumptions on the analytic continuation of the geometric zeta function. We also obtain a pointwise and a distributional formula without error term, which exists only for the geometry of a small class of fractal strings, including the self-similar strings and the so-called prime string. In Section 4.5, we use this analysis of the prime string to give a proof of the Prime Number Theorem [Da, Edw, In, Pat, Ti].

We note that our explicit formulas are close relatives of—but are significantly more general than—the usual explicit formulas encountered in number theory. (See, for example, [Da, Edw, In, Pat, Wei4–6] along with the discussion and the additional references provided at the end of Section 4.1.) Recall that the original explicit formula was introduced by Riemann [Rie1] in 1859 as an analytical tool to understand the distribution of primes. It was later extended by von Mangoldt [vM1–2] and led in 1896 to the first rigorous proof of the Prime Number Theorem, independently by Hadamard [Had2] and de la Vallée Poussin [dV1] (see [Edw]).

In Chapter 5, we work out the computations that are necessary to find the oscillatory terms in the explicit formulas of a fractal string. We illustrate our results by considering a variety of examples of geometric and direct spectral problems. We also define the spectral operator, which relates the spectrum of a fractal string with its geometry, and we study the geometric and spectral partition functions.

We analyze in detail the special case of self-similar strings. In particular, we deduce from our explicit formulas an analogue of the Prime Number Theorem for the primitive periodic orbits of the dynamical system naturally associated with such strings.

In the subsequent chapters, we investigate the geometric and spectral information contained in the complex dimensions. The main theme of these chapters is that the oscillations in the geometry or in the spectrum of a fractal string are reflected in the presence of oscillatory terms in the explicit formulas.

In Chapter 6, we derive an explicit formula for the volume of the tubular neighborhoods of a fractal string. For example, when the complex dimensions of $\mathcal{L}$ are simple, we obtain the following key formula:

$$V(\varepsilon) = \sum_\omega c_\omega \frac{(2\varepsilon)^{1-\omega}}{\omega(1-\omega)} + R(\varepsilon), \qquad (**)$$

where $V(\varepsilon)$ is the volume of the inner ε-neighborhood of the boundary of $\mathcal{L}$, ω runs over the complex dimensions of $\mathcal{L}$, c_ω denotes the residue of $\zeta_{\mathcal{L}}(s)$ at $s = \omega$, and $R(\varepsilon)$ is an error term of lower order. Formula $(**)$ yields a new criterion for the Minkowski measurability of a fractal string in terms of the absence of nonreal complex dimensions with real part D, the dimension of the string. This extends the joint work of the first author with C. Pomerance [LapPo1; LapPo2, Theorem 2.2], in which a characterization of Minkowski measurability was obtained in terms of the absence of geometric oscillations in the string. A comparison of our formula with Hermann Weyl's formula for tubes in Riemannian geometry [BergGo, p. 235] suggests what kind of geometric information may be associated with the complex dimensions of a fractal string. (See Section 6.1.1.)

In Chapter 7, we study the 'inverse spectral problem', the problem of deducing geometric information from the spectrum of a fractal string: Does the absence of oscillations in the spectrum of a fractal string imply the absence of oscillations in its geometry? In other words, we consider a suitable version of the question (à la Mark Kac [Kac]) *"Can one hear the shape of a fractal string?"* This inverse spectral problem has been considered before by the first author jointly with H. Maier in [LapMa1–2], where it was shown that the audibility of oscillations in the geometry of a fractal string of (Minkowski) dimension $D \in (0,1)$ is equivalent to the absence of zeros of the Riemann zeta function $\zeta(s)$ on the line $\mathrm{Re}\,s = D$. In our framework, this becomes the question of inverting the spectral operator. We deduce, in particular, that the spectral operator is invertible for all fractal strings of dimension $D \neq \frac{1}{2}$ if and only if the Riemann Hypothesis holds, i.e., if and only if the Riemann zeta function $\zeta(s)$ does not vanish when $\mathrm{Re}\,s \neq \frac{1}{2}$, $\mathrm{Re}\,s > 0$.

By considering (generalized) fractal sprays, instead of fractal strings, we extend the above criterion for zeros of $\zeta(s)$ in the critical strip to a large class of zeta functions, including all those for which the analogue of the generalized Riemann Hypothesis is expected to hold. We thus characterize the generalized Riemann Hypothesis as a natural inverse spectral problem for fractal sprays. In addition to the Epstein zeta functions [Te, §1.4], this class includes all Dedekind zeta functions and Dirichlet L-series, and more generally, all Hecke L-series associated with an algebraic number field [ParSh1]. It also includes all zeta functions associated with algebraic varieties over a finite field [ParSh1, Chapter 4, §1]. We refer the interested reader to Appendix A for a brief review of such number-theoretic zeta functions.

In Chapter 8, we make an extensive study of the geometry and the spectrum of 'generalized Cantor strings'. The complex dimensions of such strings $\mathcal{L}$ form an infinite sequence in vertical arithmetic progression, with real part the Minkowski dimension D of $\mathcal{L}$. We show that these strings always have oscillations of order D in their geometry and in their spectrum. In Chapter 9, we deduce from this result that the explicit formulas for the geometry and the spectrum of Cantor strings always contain oscillatory terms of order D. On the other hand, if $\zeta(s)$ had a vertical arithmetic progression of zeros coinciding with the arithmetic progression of complex dimensions of $\mathcal{L}$, then, by formula (∗), the explicit formula for the frequencies would only contain the term corresponding to D, and not any oscillatory term. Thus we prove that $\zeta(s)$ does not have such a sequence of zeros. (See Theorem 9.1. This theorem was already obtained by Putnam [Pu1–2]. However, his methods do not apply to prove the extension to more general zeta functions.)

By considering (generalized) Cantor sprays, we extend this result to a large subclass of the above-mentioned class of zeta functions. This class includes all the Dedekind and Epstein zeta functions, as well as many Dirichlet series not satisfying a functional equation. It does not, however, include the zeta functions associated with varieties over a finite field, for which this result does not hold; see Section 9.4. Indeed, we show that every Dirichlet series with positive coefficients and with only finitely many poles has no infinite sequence of zeros forming a vertical arithmetic progression. (See Theorem 9.5 and see also Section 9.3 for a further extension to all Hecke L-series.)

We conclude with a chapter of a more speculative nature, Chapter 10, in which we make several suggestions for the direction of future research in this area. Our results suggest that important information about the fractality of a string is contained in its complex dimensions. In Section 10.2, we propose as a new definition of fractality the presence of at least one nonreal complex dimension with positive real part. In this new sense, every self-similar set in the real line is fractal. On the other hand, in agreement with geometric intuition, certain compact subsets of $\mathbb{R}^1$, associated with the so-called a-string, are shown here to be 'nonfractal', whereas they are fractal according to the definition of fractality based on the notion of real Minkowski fractal dimension. We suggest one possible way of defining the complex dimensions of higher-dimensional fractals, and we discuss the examples of the Devil's staircase and of the snowflake drum (see Figures 10.1 and 10.6). We note, in particular, that the Devil's staircase is not fractal according to the traditional definition based on the real Hausdorff fractal dimension. However, there is general agreement among fractal geometers that it should be called 'fractal'. (See [Man1, p. 84].) We show that our new definition of fractality does indeed resolve this problem satisfactorily. In spite of this positive outcome, we stress that the theory of the complex dimensions of higher-dimensional fractals still needs to be further developed.

In Sections 10.2 through 10.5, we also briefly discuss several conjectures and open problems regarding possible extensions and geometric, spectral, or dynamical interpretations of the present theory of complex dimensions, both for fractal strings and their higher-dimensional analogue, fractal drums, for which much research remains to be carried out in this context.

1
Complex Dimensions
of Ordinary Fractal Strings

In this chapter, we recall some basic definitions pertaining to the notion of (ordinary) fractal string and introduce several new ones, the most important of which is the notion of complex dimension. We also give a brief overview of some of our results in this context by discussing the simple but illustrative example of the Cantor string. In the last section, we discuss the notion of fractal spray, which is a higher-dimensional analogue of that of fractal string.

1.1   The Geometry of a Fractal String

We begin this chapter by recalling the notion of ordinary fractal string encountered in [Lap1–4, LapPo1–3, LapMa1–2, HeLap1–2].

A *(standard or ordinary) fractal string* $\mathcal{L}$ is a bounded open subset Ω of $\mathbb{R}$. It is well known that such a set consists of countably many open intervals, the lengths of which will be denoted by $l_1, l_2, l_3, \ldots$, called the *lengths* of the string. Note that $\sum_{j=1}^{\infty} l_j$ is finite and equals the Lebesgue measure of Ω. From the point of view of this work, we can and will assume, without loss of generality, that

$$l_1 \geq l_2 \geq \cdots > 0, \tag{1.1}$$

where each length is counted according to its multiplicity. We allow for Ω to be a finite union of open intervals; that is, for the sequence of lengths to be finite.

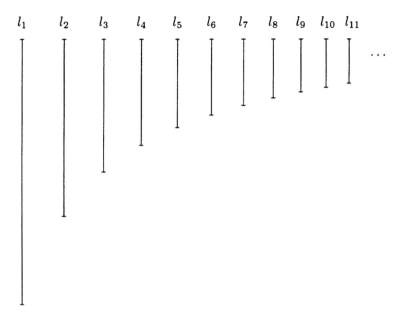

l_1 l_2 l_3 l_4 l_5 l_6 l_7 l_8 l_9 l_{10} l_{11}

Figure 1.1: A fractal harp.

An ordinary fractal string can be thought of as a one-dimensional drum with fractal boundary. Actually, we have given here the usual terminology that is found in the literature. A different terminology may be more suggestive: the open set Ω could be called a 'fractal harp', and each connected interval of Ω could be called a string of the harp; see Figure 1.1.

The *counting function of the lengths*,[1] also called the *geometric counting function* of $\mathcal{L}$, is

$$N_{\mathcal{L}}(x) = \#\{j \geq 1 : l_j^{-1} \leq x\} = \sum_{j \geq 1,\, l_j^{-1} \leq x} 1, \qquad (1.2)$$

for $x > 0$.[2]

The *boundary* of $\mathcal{L}$, denoted $\partial\mathcal{L}$, is defined as the boundary $\partial\Omega$ of Ω. Important geometric information about $\mathcal{L}$ is contained in its *Minkowski dimension* $D = D_{\mathcal{L}}$ and its *Minkowski content* $\mathcal{M} = \mathcal{M}(D; \mathcal{L})$, defined respectively as the inner Minkowski dimension and the inner Minkowski

---

[1] Strictly speaking, this is the counting function of the reciprocal lengths of $\mathcal{L}$.

[2] In agreement with the convention explicitly adopted later on in the book (beginning in Chapter 3, Eq. (3.1)), the integers j such that $l_j^{-1} = x$ must be counted with the weight $1/2$. For notational simplicity, a similar convention is implicitly assumed throughout this chapter for the spectral counting functions (such as, e.g., in Eq. (1.31) below), as well as for the corresponding computations.

content of $\partial\Omega$. To define these quantities, let $d(x, A)$ denote the distance of $x \in \mathbb{R}$ to a subset $A \subset \mathbb{R}$ and let vol_1 denote one-dimensional Lebesgue measure on $\mathbb{R}$. Further, for $\varepsilon > 0$, let $V(\varepsilon)$ be the volume of the ε-neighborhood of $\partial\Omega$ intersected with Ω:

$$V(\varepsilon) = \mathrm{vol}_1 \{x \in \Omega \colon d(x, \partial\Omega) < \varepsilon\}. \tag{1.3}$$

Then

$$D = D_{\mathcal{L}} = \inf\{\alpha \geq 0 \colon V(\varepsilon) = O(\varepsilon^{1-\alpha}) \text{ as } \varepsilon \to 0^+\}, \tag{1.4}$$

and

$$\mathcal{M} = \mathcal{M}(D; \mathcal{L}) = \lim_{\varepsilon \to 0^+} V(\varepsilon)\varepsilon^{-(1-D)}, \tag{1.5}$$

provided this limit exists in $(0, \infty)$, in which case $\mathcal{L}$ is said to be *Minkowski measurable*. We also define the *upper and lower Minkowski content*

$$\mathcal{M}^* = \mathcal{M}^*(D; \mathcal{L}) = \limsup_{\varepsilon \to 0^+} V(\varepsilon)\varepsilon^{-(1-D)} \tag{1.6a}$$

and

$$\mathcal{M}_* = \mathcal{M}_*(D; \mathcal{L}) = \liminf_{\varepsilon \to 0^+} V(\varepsilon)\varepsilon^{-(1-D)}, \tag{1.6b}$$

respectively. Thus, $\mathcal{L}$ is Minkowski measurable if and only if $\mathcal{M}^* = \mathcal{M}_* = \mathcal{M} \in (0, \infty)$.

Remark 1.1. The definitions of Minkowski dimension and content of $\partial\Omega$ extend naturally to the higher-dimensional case when $\Omega \subset \mathbb{R}^d$, with $d \geq 1$, provided that we substitute the exponent $d - \alpha$ for $1 - \alpha$ in (1.4) and $d - D$ for $1 - D$ in (1.5) and (1.6), and that in (1.3), vol_1 is replaced by vol_d, the d-dimensional Lebesgue measure on $\mathbb{R}^d$; see, e.g., [Lap1, Definition 2.1 and §3].

Remark 1.2. The Minkowski dimension is also called, for instance, the 'capacity dimension' or the (upper) 'box dimension' in the literature on fractal geometry. For further information about the notions of Minkowski dimension and content in a related context, we refer, for example, to [BroCa, Lap1–3, LapPol–3, LapMa1–2, LapFl, FlVa, Ger, GerSc1–2, Ca1–2, vB, vB-Le, HuaSl, FlLeVa, LeVa, MolVa, Fa3, vB-Gi, HeLap1–2]. For the notion of Minkowski(–Bouligand) dimension—which was extended by Bouligand [Bou] from integer to real values of D—see also [Bou, Fed2, KahSa, Man1, MarVu, Tr1–3; Fa2, Chapter 3].

Remark 1.3. The more irregular the boundary of $\partial\Omega$, the larger D. Moreover, we always have $d - 1 \leq H \leq D \leq d$, where H denotes the Hausdorff dimension of $\partial\Omega$, and d is the dimension of the ambient space, as in Remark 1.1. Intuitively, D corresponds to coverings of $\partial\Omega$ by d-dimensional

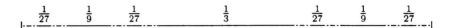

$$\frac{1}{27} \qquad \frac{1}{9} \qquad \frac{1}{27} \qquad\qquad \frac{1}{3} \qquad\qquad \frac{1}{27} \qquad \frac{1}{9} \qquad \frac{1}{27}$$

Figure 1.2: The Cantor string.

Figure 1.3: The .037-tubular neighborhood of the Cantor string.

cubes of size exactly equal to ε, whereas H corresponds to coverings by sets of size at most ε. (See, for example, [Fa2, Chapters 2 and 3], [Lap1, §2.1 and §3], [Mat, Chapter 5], [Rog] and [Tr1–2] or [Tr3, Chapters 2 and 3] for a detailed exposition.) As is discussed in [Lap1], this key difference explains why—from the point of view of harmonic analysis and spectral theory—the Minkowski dimension should be used instead of the more familiar Hausdorff dimension in this setting. (See also the earlier work in [BroCa].) Further arguments are given in [Lap1, Example 5.1, pp. 512–514]; see also [Lap-Po2] and Remark 1.4 below. They exclude similarly another notion of fractal dimension, the packing dimension P [Su, Tr2]. This dimension satisfies the inequality $H \leq P \leq D$. Somewhat paradoxically, as is pointed out in [Lap1, Remark 5.1, p. 514], the Hausdorff and packing dimensions are ruled out in the context of fractal strings (or drums) precisely because they are good mathematical notions. Indeed, they are both associated with countably additive measures, which implies that countable sets have zero dimension. By contrast, the Minkowski content is only finitely subadditive, so that countable sets can have positive Minkowski dimension, like in the example of the a-string in Section 5.5.1 (see especially Equation (5.73)).

1.1.1 The Multiplicity of the Lengths

Another way of representing a fractal string $\mathcal{L}$ is by listing its different lengths l, together with their multiplicity w_l:

$$w_l = \#\{j \geq 1 \colon l_j = l\}. \tag{1.7}$$

Thus, for example,

$$N_{\mathcal{L}}(x) = \sum_{l^{-1} \leq x} w_l. \tag{1.8}$$

In Chapter 3, we will introduce a third way to represent a fractal string, similar to this one, namely, by a measure.

1.1.2 Example: The Cantor String

We consider the ordinary fractal string $\Omega = CS$, the complement in $[0, 1]$ of the usual ternary Cantor set. (See Figure 1.2.) Thus

$$CS = \left(\frac{1}{3}, \frac{2}{3}\right) \cup \left(\frac{1}{9}, \frac{2}{9}\right) \cup \left(\frac{7}{9}, \frac{8}{9}\right) \cup$$

$$\cup \left(\frac{1}{27}, \frac{2}{27}\right) \cup \left(\frac{7}{27}, \frac{8}{27}\right) \cup \left(\frac{19}{27}, \frac{20}{27}\right) \cup \left(\frac{25}{27}, \frac{26}{27}\right) \cup \ldots,$$

so that $l_1 = \frac{1}{3}$, $l_2 = l_3 = \frac{1}{9}$, $l_4 = l_5 = l_6 = l_7 = \frac{1}{27}, \ldots$, or alternatively, the lengths are the numbers 3^{-n-1} with multiplicity $w_{3^{-n-1}} = 2^n$, for $n = 0, 1, 2, \ldots$. This fractal string was studied, in particular, in [LapPo1; Lap-Po2, Example 4.5, pp. 65–67].

We note that by construction, the boundary $\partial\Omega$ of the Cantor string is equal to the ternary Cantor set.

In general, the volume of the tubular neighborhood of the boundary of $\mathcal{L}$ is given by (see [LapPo2, Eq. (3.2), p. 48])

$$V(\varepsilon) = \sum_{j: l_j \geq 2\varepsilon} 2\varepsilon + \sum_{j: l_j < 2\varepsilon} l_j = 2\varepsilon \cdot N_{\mathcal{L}}\left(\frac{1}{2\varepsilon}\right) + \sum_{j: l_j < 2\varepsilon} l_j. \tag{1.9}$$

For the Cantor string (see Figure 1.3), we find

$$V(\varepsilon) = 2\varepsilon \cdot (2^n - 1) + \sum_{k=n}^{\infty} 2^k \cdot 3^{-k-1} = 2\varepsilon \cdot 2^n + \left(\frac{2}{3}\right)^n - 2\varepsilon,$$

where n is such that $3^{-n} \geq 2\varepsilon > 3^{-n-1}$; i.e., $n = [-\log_3(2\varepsilon)]$.[3] To determine the minimal α such that $V(\varepsilon) = O(\varepsilon^{1-\alpha})$ as $\varepsilon \to 0^+$, we write

$$b^{[-\log_3(2\varepsilon)]} = b^{-\log_3(2\varepsilon) - \{-\log_3(2\varepsilon)\}} = (2\varepsilon)^{-\log_3 b} b^{-\{-\log_3(2\varepsilon)\}}$$

for $b = 2$ and for $b = 2/3$. Putting

$$D = \log_3 2 := \frac{\log 2}{\log 3}, \tag{1.10}$$

we find

$$V(\varepsilon) = (2\varepsilon)^{1-D} \left(\left(\frac{1}{2}\right)^{\{-\log_3(2\varepsilon)\}} + \left(\frac{3}{2}\right)^{\{-\log_3(2\varepsilon)\}}\right) - 2\varepsilon. \tag{1.11}$$

The function between parentheses is bounded, and 'multiplicatively periodic': it takes the same value at ε and $\varepsilon/3$. It does not have a limit

---

[3]For $x \in \mathbb{R}$, we write $x = [x] + \{x\}$, where $[x]$ is the integer part and $\{x\}$ the fractional part of x; i.e., $[x] \in \mathbb{Z}$ and $0 \leq \{x\} < 1$.

for $\varepsilon \to 0^+$. We see that the Cantor string has Minkowski dimension $D = \log_3 2$, and that it is not Minkowski measurable. The upper and lower Minkowski content are computed in [LapPo2, Theorem 4.6, p. 65]:

$$\mathcal{M}^* = 2^{2-D}, \qquad \mathcal{M}_* = 2^{1-D}D^{-D}(1-D)^{-(1-D)}. \qquad (1.12)$$

Remark 1.4. Observe that the Minkowski dimension of the Cantor string coincides with its Hausdorff dimension. The basic reason for using the Minkowski dimension is that it is invariant under displacements of the intervals of which the string is composed. This is not the case of the Hausdorff dimension (see [BroCa], [Lap1, Example 5.1, pp. 512–514], [LapPo2]). See also Remark 1.3 above and Remark 2.17 below, as well as the relevant references therein, for further comparison between the various notions of fractal dimensions and for additional justification of the choice of the notion of Minkowski dimension in the context of fractal strings. *Throughout this work, an ordinary fractal string $\mathcal{L}$ is completely determined by the sequence $\{l_j\}_{j=1}^{\infty}$ of its lengths.* Hence, we will often denote such a string by $\mathcal{L} = \{l_j\}_{j=1}^{\infty}$.

We can continue our analysis of the inner tubular neighborhood $V(\varepsilon)$ of the Cantor string by using the Fourier series of the periodic function $u \mapsto b^{-\{u\}}$, for $b > 0$:

$$b^{-\{u\}} = \frac{b-1}{b} \sum_{n \in \mathbb{Z}} \frac{e^{2\pi i n u}}{\log b + 2\pi i n}. \qquad (1.13)$$

Writing $\mathbf{p} = 2\pi/\log 3$ and substituting (1.13) into (1.11), we find

$$V(\varepsilon) = \frac{1}{2\log 3} \sum_{n=-\infty}^{\infty} \frac{(2\varepsilon)^{1-D-in\mathbf{p}}}{(D+in\mathbf{p})(1-D-in\mathbf{p})} - 2\varepsilon. \qquad (1.14)$$

The number $\mathbf{p} = 2\pi/\log 3$ is called the oscillatory period of the Cantor string; see Definition 2.10. A counterpart of formula (1.14) will be derived for a general fractal string in Section 6.1.

Remark 1.5. Note that in view of (1.14), we have

$$\varepsilon^{-(1-D)}V(\varepsilon) = g(\varepsilon) - 2\varepsilon^D, \qquad (1.15)$$

as $\varepsilon \to 0^+$, where g is a multiplicatively periodic function of multiplicative period $3 = e^{2\pi/\mathbf{p}}$ that is bounded away from zero and infinity:

$$0 < \mathcal{M}_* \leq g(\varepsilon) \leq \mathcal{M}^* < \infty,$$

where $\mathcal{M}_*$ and $\mathcal{M}^*$ are given by (1.12) above. Alternatively, g can be viewed as an additively periodic function of $\log \varepsilon^{-1}$ with additive period $\log 3 = 2\pi/\mathbf{p}$. Namely,

$$g(\varepsilon) = f\left(\log \varepsilon^{-1}\right)$$

with

$$f(u) = \frac{1}{2\log 3} \sum_{n=-\infty}^{\infty} \frac{2^{1-D-in\mathbf{p}} e^{in\mathbf{p}u}}{(D+in\mathbf{p})(1-D-in\mathbf{p})}. \tag{1.16}$$

1.2 The Geometric Zeta Function of a Fractal String

Let $\mathcal{L}$ be a fractal string with sequence of lengths $\{l_j\}_{j=1}^{\infty}$. The sum $\sum_{j=1}^{\infty} l_j^{\sigma}$ converges for $\sigma = 1$. It follows that the formula $\zeta_{\mathcal{L}}(s) = \sum_{j=1}^{\infty} l_j^s$ defines a holomorphic function for $\mathrm{Re}\, s > 1$.

Definition 1.6. Let $\mathcal{L}$ be a fractal string. The *dimension* of $\mathcal{L}$, denoted by

$$D = D_{\mathcal{L}}, \tag{1.17}$$

is the infimum of the real numbers σ such that the Dirichlet series $\sum_{j=1}^{\infty} l_j^{\sigma}$ converges.

Further, the *geometric zeta function* of $\mathcal{L}$ is

$$\zeta_{\mathcal{L}}(s) = \sum_{j=1}^{\infty} l_j^s = \sum_l w_l \cdot l^s, \qquad \text{for } \mathrm{Re}\, s > D_{\mathcal{L}}. \tag{1.18}$$

(See Equation (1.7) for the definition of w_l.)

Remark 1.7. By definition, the dimension of $\mathcal{L}$ is the abcissa of convergence of the (generalized) Dirichlet series $\sum_{j=1}^{\infty} l_j^s$; see, e.g., [Ser, §VI.2]. Moreover, it is observed in [Lap2, Eq. (5.4), p. 169] (using a key result of Besicovitch and Taylor [BesTa]) that the dimension of $\mathcal{L}$ coincides with D,[4] the Minkowski dimension of $\mathcal{L}$, as defined earlier in (1.4). Indeed, according to [BesTa], we have

$$D = \inf\left\{\sigma \geq 0 : \sum_{j=1}^{\infty} l_j^{\sigma} < \infty\right\}. \tag{1.19}$$

Note that the Minkowski dimension of an ordinary fractal string always satisfies $0 \leq D \leq 1$.

---

[4]Except in the trivial case when there are only finitely many lengths. Then $D_{\mathcal{L}} = 0$ but the abcissa of convergence is equal to $-\infty$, since the Dirichlet series for $\zeta_{\mathcal{L}}(s)$ converges for all $s \in \mathbb{C}$.

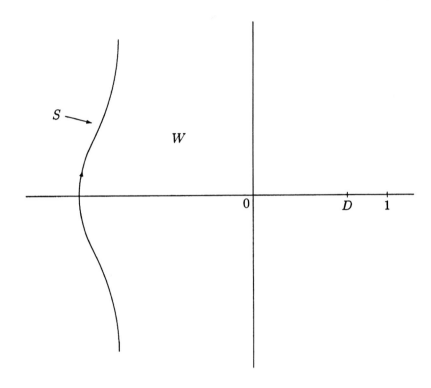

Figure 1.4: The screen S and the window W.

Some values of the geometric zeta function of a string $\mathcal{L}$ have a special interpretation. If there are only finitely many lengths l, then $\zeta_{\mathcal{L}}(0)$ equals the number of lengths of the string. Similarly, the *total length* of the fractal string $\mathcal{L}$ is

$$L := \zeta_{\mathcal{L}}(1) = \sum_{j=1}^{\infty} l_j. \qquad (1.20)$$

1.2.1   The Screen and the Window

In general, $\zeta_{\mathcal{L}}$ may not have an analytic continuation to all of $\mathbb{C}$. We therefore introduce the *screen* S as the contour

$$S(t) = r(t) + it \qquad (t \in \mathbb{R}), \qquad (1.21)$$

for some continuous function $r \colon \mathbb{R} \to [-\infty, D_{\mathcal{L}}]$. (See Figure 1.4.)

The set

$$W = \{s \in \mathbb{C} \colon \operatorname{Re} s \geq r(\operatorname{Im} s)\} \tag{1.22}$$

is called the *window*, and we assume that $\zeta_{\mathcal{L}}$ has a meromorphic extension[5] to a neighborhood of W, with set of poles $\mathcal{D} = \mathcal{D}_{\mathcal{L}}(W) \subset W$, called the (visible) complex dimensions of $\mathcal{L}$. We also require that $\zeta_{\mathcal{L}}$ does not have any pole on the screen S.

Definition 1.8.   (i) The set of *visible complex dimensions* of the fractal string $\mathcal{L}$ is defined as

$$\mathcal{D}_{\mathcal{L}} = \mathcal{D}_{\mathcal{L}}(W) = \{\omega \in W \colon \zeta_{\mathcal{L}} \text{ has a pole at } \omega\}. \tag{1.23}$$

(ii) If $W = \mathbb{C}$ (thus, in particular, if $\zeta_{\mathcal{L}}$ has a meromorphic extension to all of $\mathbb{C}$), we call

$$\mathcal{D}_{\mathcal{L}} = \mathcal{D}_{\mathcal{L}}(\mathbb{C}) = \{\omega \in \mathbb{C} \colon \zeta_{\mathcal{L}} \text{ has a pole at } \omega\} \tag{1.24}$$

the set of *complex dimensions* of $\mathcal{L}$.

Remark 1.9. Since it is contained in the set of poles of a meromorphic function, $\mathcal{D}_{\mathcal{L}}(W)$ is a discrete subset of $\mathbb{C}$. Hence its intersection with any compact subset of $\mathbb{C}$ is finite. When $\mathcal{L}$ consists of finitely many lengths, then $\mathcal{D}_{\mathcal{L}} = \emptyset$.

Remark 1.10. In general, the limit of $\zeta_{\mathcal{L}}(s)$ as $s \to D_{\mathcal{L}}$ from the right is ∞. (This follows, e.g., from [Ser, Proposition 7, p. 67] and is due to the fact that the Dirichlet series (1.18) has nonnegative coefficients.) Thus $s = D_{\mathcal{L}}$ is always a singularity of $\zeta_{\mathcal{L}}(s)$, but not necessarily a pole. If $D = D_{\mathcal{L}}$ is a pole and $D_{\mathcal{L}} \in W$, then

$$D_{\mathcal{L}} = \max \{\operatorname{Re} \omega \colon \omega \in \mathcal{D}_{\mathcal{L}}\}, \tag{1.25}$$

since $\zeta_{\mathcal{L}}$ is holomorphic for $\operatorname{Re} s > D$.

Remark 1.11. Assume that W is symmetric with respect to the real axis. Since $\zeta_{\mathcal{L}}(\bar{s}) = \overline{\zeta_{\mathcal{L}}(s)}$, it then follows that the nonreal complex dimensions of a (standard) fractal string always come in complex conjugate pairs $\omega, \bar{\omega}$; that is, $\omega \in \mathcal{D}_{\mathcal{L}}$ if and only if $\bar{\omega} \in \mathcal{D}_{\mathcal{L}}$.

The general theme of this monograph is that *the complex dimensions describe oscillations in the geometry (and the spectrum) of a fractal string.*

---

[5]Following traditional usage, we will continue to denote by $\zeta_{\mathcal{L}}$ the meromorphic extension of the Dirichlet series given by (1.18).

1.2.2   The Cantor String (Continued)

The geometric zeta function of the Cantor string is

$$\zeta_{CS}(s) = \sum_{n=0}^{\infty} 2^n \cdot 3^{-(n+1)s} = \frac{3^{-s}}{1 - 2 \cdot 3^{-s}}. \tag{1.26}$$

We choose $W = \mathbb{C}$. The complex dimensions of the Cantor string are found by solving the equation $1 - 2 \cdot 3^{-\omega} = 0$, with $\omega \in \mathbb{C}$. (Note that they are all simple poles of ζ_{CS}.) Thus

$$\mathcal{D}_{CS} = \{D + in\mathbf{p} : n \in \mathbb{Z}\}, \tag{1.27}$$

where $D = \log_3 2$ is the dimension of CS and $\mathbf{p} = 2\pi/\log 3$ is its oscillatory period.[6] (See Figure 2.2 in Section 2.2.1.) Formula (1.14) above expresses $V(\varepsilon)$ as a sum of terms proportional to $\varepsilon^{1-\omega}$, where $\omega = D + in\mathbf{p}$ runs over the complex dimensions of the Cantor string. (See Sections 6.1 and 6.3.)

In Section 6.2, we will extend and reinterpret in terms of complex dimensions the criterion for Minkowski measurability obtained in [LapPo1; LapPo2, Theorem 2.2, p. 46]. In particular, under mild growth conditions on $\zeta_{\mathcal{L}}$, we will show in Theorem 6.12 that an ordinary fractal string is Minkowski measurable if and only if D is a simple pole of $\zeta_{\mathcal{L}}$ and the only complex dimension of $\mathcal{L}$ on the vertical line $\operatorname{Re} s = D$ is D itself. Heuristically, nonreal complex dimensions above D would create oscillations in the geometry of $\mathcal{L}$, and therefore in the volume of the tubular neighborhood, $V(\varepsilon)$; see (1.28) below and (1.14)–(1.16) above, and compare with the intuition expressed in [LapMa1] or [LapMa2, esp. §3.3].

In view of (1.26) and (1.27), this makes particularly transparent the non-Minkowski measurability of the Cantor string (deduced above and established earlier in [LapPo2, Theorem 4.6]). As we will see in Chapter 6, a similar argument can be used in many other situations.

As another example, we compute the counting function of the lengths: There are $1+2+4+\cdots+2^{n-1}$ lengths greater than x^{-1}, where $n = [\log_3 x]$. Thus $N_{CS}(x) = 2^n - 1$. Using the Fourier series (1.13), we obtain

$$N_{CS}(x) = \frac{1}{2\log 3} \sum_{n \in \mathbb{Z}} \frac{x^{D+in\mathbf{p}}}{D + in\mathbf{p}} - 1. \tag{1.28}$$

Again, we find a sum extended over the complex dimensions of the Cantor string. In Chapter 4, we will formulate and prove our explicit formulas, which allow us to derive such formulas in much greater generality.

---

[6]Compare [Lap2, Example 5.2(i), p. 170] and [Lap3, p. 150], where our present terminology was not used.

1.3 The Frequencies of a Fractal String and the Spectral Zeta Function

Given a fractal string $\mathcal{L}$, we can listen to its sound. In mathematical terms, we consider the bounded open set $\Omega \subset \mathbb{R}$, together with the (positive) Dirichlet Laplacian $\Delta = -d^2/dx^2$ on Ω. An eigenvalue λ of Δ corresponds to the (normalized) frequency $f = \sqrt{\lambda}/\pi$ of the fractal string.

The frequencies of the unit interval are $1, 2, 3, \ldots$ (each counted with multiplicity one), and the frequencies of an interval of length l are l^{-1}, $2l^{-1}, 3l^{-1}, \ldots$ (also counted with multiplicity one). Thus the frequencies of $\mathcal{L}$ are the numbers

$$f = k \cdot l_j^{-1}, \qquad (1.29)$$

where $k, j = 1, 2, 3, \ldots$; that is, they are the integer multiples of the reciprocal lengths of $\mathcal{L}$. The total multiplicity of the frequency f is equal to

$$w_f^{(\nu)} = \sum_{j:\, f \cdot l_j \in \mathbb{N}^*} 1 = \sum_{l:\, f \cdot l \in \mathbb{N}^*} w_l = w_{1/f} + w_{2/f} + w_{3/f} + \cdots. \qquad (1.30)$$

To study the frequencies, we introduce the spectral counting function and the spectral zeta function.

Definition 1.12. The *counting function of the frequencies*, also called the *spectral counting function* of $\mathcal{L}$, is

$$N_\nu(x) = \#\{f \le x : \text{frequency of } \mathcal{L}, \text{ counted with multiplicity}\}$$
$$= \sum_{f \le x} w_f^{(\nu)}, \qquad (1.31)$$

for $x > 0$.[7]

The *spectral zeta function* of $\mathcal{L}$ is

$$\zeta_\nu(s) = \sum_{k,j=1}^{\infty} \left(k \cdot l_j^{-1} \right)^{-s} = \sum_f w_f^{(\nu)} f^{-s}, \qquad (1.32)$$

for Re s sufficiently large.[8]

Note that N_ν and ζ_ν depend on $\mathcal{L}$. However, for simplicity, we do not indicate this explicitly in our notation.

---

[7]See footnote 2 above (according to which the frequencies f such that $f = x$ should be counted with the weight $1/2$). Also note that a frequency f is proportional to an inverse (or reciprocal) length.

[8]In the second equality of (1.31) and (1.32), the sum is extended over all distinct frequencies of $\mathcal{L}$.

Let $\zeta(s)$ be the Riemann zeta function defined by $\zeta(s) = \sum_{n=1}^{\infty} n^{-s}$ for $\mathrm{Re}\, s > 1$. It is well known that $\zeta(s)$ has an extension to the whole complex plane as a meromorphic function, with one simple pole at $s = 1$, with residue 1. We refer, for example, to [Da, Edw, In, Ivi, Pat, Ti] for the classical theory of the Riemann zeta function.

The following theorem relates the spectrum of an ordinary fractal string with its geometry.

Theorem 1.13. *The spectral counting function of $\mathcal{L}$ is given by*

$$N_\nu(x) = N_{\mathcal{L}}(x) + N_{\mathcal{L}}\left(\frac{x}{2}\right) + N_{\mathcal{L}}\left(\frac{x}{3}\right) + \dots \tag{1.33}$$

$$= \sum_{j=1}^{\infty} [l_j x], \tag{1.34}$$

and the spectral zeta function of $\mathcal{L}$ is given by

$$\zeta_\nu(s) = \zeta_{\mathcal{L}}(s)\zeta(s), \tag{1.35}$$

where $\zeta(s)$ is the Riemann zeta function. Thus $\zeta_\nu(s)$ is holomorphic for $\mathrm{Re}\, s > 1$. It has a pole at $s = 1$, with residue L, the total length of $\mathcal{L}$. (Compare (1.20).) Moreover, it has a meromorphic extension to a neighborhood of the window W.

Proof. For the spectral counting function, this follows from the following computation:

$$N_\nu(x) = \sum_{k=1}^{\infty} \sum_{j:\, k \cdot l_j^{-1} \leq x} 1 = \sum_{k=1}^{\infty} \#\{j : l_j^{-1} \leq x/k\} = \sum_{k=1}^{\infty} N_{\mathcal{L}}\left(\frac{x}{k}\right).$$

Observe that this sum is finite, since $N_{\mathcal{L}}(y) = 0$ for $y < l_1^{-1}$. The second expression is derived similarly:

$$N_\nu(x) = \sum_{j=1}^{\infty} \sum_{k \leq l_j x} 1 = \sum_{j=1}^{\infty} [l_j x].$$

For the spectral zeta function, we have successively

$$\zeta_\nu(s) = \sum_{k,j=1}^{\infty} k^{-s} l_j^s = \sum_{j=1}^{\infty} l_j^s \sum_{k=1}^{\infty} k^{-s} = \zeta_{\mathcal{L}}(s)\zeta(s).$$

This completes the proof of the theorem.   □

Remark 1.14. Formula (1.35) of Theorem 1.13 above was established in [Lap2, Eqs. (5.2) and (5.3), p. 169] and used, in particular, in [Lap2–3, LapPo1–3, LapMa1–2, HeLap1–2].

One of the problems we are interested in is the inverse spectral problem for fractal strings (as considered in [LapMa1–2] in connection with the Riemann Hypothesis); i.e., to derive information about the geometry of an ordinary fractal string from certain information (for example, asymptotics) about its spectrum. We study this problem in Chapters 7 and 9, with the help of explicit formulas for the frequency counting function, which will be established in Chapters 5 and 8, using the results of Chapter 4. These explicit formulas express the various functions associated with the geometry or the spectrum of a fractal string as a sum of oscillatory terms of the form a constant times x^ω, where ω runs over the complex dimensions of $\mathcal{L}$. In particular, in the case of the 'direct spectral problem' where $N_\nu(x)$ is expressed in terms of the geometry of $\mathcal{L}$, the spectral counting function $N_\nu(x)$ is given by (5.19). We stress that if ω is simple, then the coefficient of x^ω is proportional to $\zeta(\omega)$, the value of the Riemann zeta function at the complex dimension ω; see formula (5.20b).

We note here that if the Riemann zeta function had a zero at one of the complex dimensions of a fractal string, then the expansion of the counting function of the frequencies would no longer have the corresponding term. Thus, for example, the Cantor string would sound similar to a string without, say, the complex dimensions $D \pm 37i\mathbf{p}$, if $\zeta(D \pm 37i\mathbf{p}) = \zeta(\log_3 2 \pm 74\pi i/\log 3)$ happened to vanish. This line of reasoning—based on the now rigorous notion of complex dimensions and the use of our explicit formulas—enables us to reformulate (and extend to many other zeta functions) the characterization of the Riemann Hypothesis obtained in [LapMa1–2]. (See Chapter 7.)

Observe that in particular, if $\zeta(s)$ happened to vanish at all the points $D + in\mathbf{p}$ $(n \in \mathbb{Z}\backslash\{0\})$, the Cantor string would sound the same as a Minkowski measurable fractal string of the same dimension $D = \log_3 2$. (See, in particular, Theorem 6.12.) In Chapter 9, we will show, however, that this is not the case for a rather general class of zeta functions, and for any arithmetic sequence of points $D + in\mathbf{p}$ $(n \in \mathbb{Z})$. These zeta functions can be thought of as being associated with fractal sprays, as discussed in the next section. In Chapter 8, we will study generalized Cantor strings, which can have any sequence $\{D + in\mathbf{p}\}_{n\in\mathbb{Z}}$ (for arbitrary $D \in (0,1)$ and $\mathbf{p} > 0$) as their complex dimensions. We note that such strings can no longer be realized geometrically as subsets of Euclidean space.

1.4   Higher-Dimensional Analogue: Fractal Sprays

Fractal sprays were introduced in [LapPo3] (see also [Lap2, §4] announcing some of the results in [LapPo3]) as a natural higher-dimensional analogue of fractal strings and as a tool to explore various conjectures about the

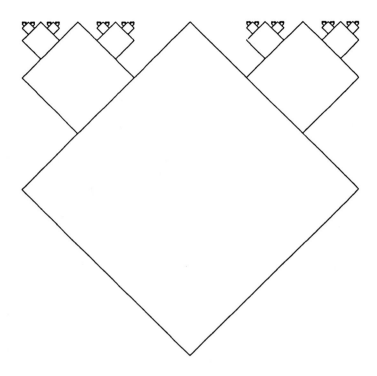

Figure 1.5: A Cantor spray: a fractal spray Ω in the plane, with basic shape B, the unit square, scaled by the Cantor string.

spectrum (and the geometry) of drums with fractal boundary[9] in $\mathbb{R}^d$. In the present book, fractal sprays and their generalizations (to be introduced later on) will continue to be a useful exploratory tool and will enable us to extend several of our results to zeta functions other than the Riemann zeta function. (See especially Chapters 7 and 9.)[10]

For example, we will consider the spray of Figure 1.5, obtained by scaling an open square B of size 1 by the lengths of the Cantor string $\mathcal{L} = \mathrm{CS}$. Thus Ω is a bounded open subset of $\mathbb{R}^2$ consisting of one open square of size 1/3, two open squares of size 1/9, four open squares of size 1/27, and so on. The spectral zeta function for the Dirichlet Laplacian on the square

---

[9]Of course, when $d \geq 2$, a drum with fractal boundary in $\mathbb{R}^d$ can be much more complicated than a fractal spray.

[10]With the present terminology introduced in [LapPo3], we see that the main examples studied, e.g., in [BroCa], [Lap2, §4 and §5.2], [Lap3, §4.4.1b], [FlVa, Ger, GerSc1–2, LapPo3, LeVa], are special cases of ordinary fractal sprays.

is

$$\zeta_B(s) = \sum_{n_1,n_2=1}^{\infty} \left(n_1^2 + n_2^2\right)^{-s/2}, \tag{1.36}$$

and hence the spectral zeta function of this spray is given by

$$\zeta_\nu(s) = \zeta_{CS}(s) \cdot \zeta_B(s). \tag{1.37}$$

In general, a (standard) fractal spray Ω in $\mathbb{R}^d$ ($d \geq 1$) is given by a (nonempty) bounded open set $B \subset \mathbb{R}^d$ (called here the *basic shape*), scaled by a (standard) fractal string $\mathcal{L} = \{l_j\}_{j=1}^{\infty}$. More precisely, we call a *fractal spray of $\mathcal{L}$ on B* (or with basic shape B) any bounded open set Ω in $\mathbb{R}^d$ which is the disjoint union of open sets Ω_j for $j = 1, 2, \ldots$, where Ω_j is congruent to $l_j B$ (the homothetic of Ω by the ratio l_j) for each j. (See [LapPo3, §2].) Note that a fractal string $\mathcal{L} = (l_j)_{j=1}^{\infty}$ can be viewed as a fractal spray of $\mathcal{L}$ with basic shape $B = (0, 1)$, the unit interval.

Let $\{\lambda_k(B)\}_{k=1}^{\infty}$ be the sequence of nonzero eigenvalues (counted with multiplicity and written in nondecreasing order) of the (positive) Laplacian (with Dirichlet or other suitable boundary conditions) $\Delta = -\sum_{q=1}^{d} \partial^2/\partial x_q^2$ on B. That is,

$$0 < \lambda_1(B) \leq \lambda_2(B) \leq \cdots \leq \lambda_k(B) \leq \cdots \to \infty, \quad \text{as } k \to \infty.$$

Then the (normalized) frequencies of the Laplacian Δ on B are the numbers $f_k = f_k(B) = \pi^{-1}\sqrt{\lambda_k(B)}$ (for $k = 1, 2, 3, \ldots$),[11] and the spectral zeta function of B is[12]

$$\zeta_B(s) := \sum_{k=1}^{\infty} (f_k(B))^{-s} = \pi^s \sum_{k=1}^{\infty} (\lambda_k(B))^{-s/2}. \tag{1.38}$$

As in Equation (1.29), a simple calculation (see, e.g., [LapPo3, §3])—relying on the invariance of the Laplacian under isometries—shows that the frequencies of the spray (that is, of the Laplacian Δ on Ω) are the numbers

$$f = f_k(B) \cdot l_j^{-1}, \tag{1.39}$$

---

[11]The spectrum of the Dirichlet Laplacian Δ is always discrete if B is bounded (or more generally, if B has finite volume); further, 0 is never an eigenvalue. For the Neumann Laplacian, we assume that B has a locally Lipschitz boundary or more generally, that B satisfies the so-called extension property, to insure that the spectrum is discrete. (See, e.g., [BiSo, Met, EdEv], [Maz, §1.5.1], or [Lap1, esp. Chapter 2 and pp. 510–511], as well as the relevant references therein.)

[12]We adopt here the convention of [Lap2–3] and of Appendix B rather than the traditional convention used, e.g., in [Se1] or [Gi], according to which $(\lambda_k(B))^{-s/2}$ would be replaced by $(\lambda_k(B))^{-s}$.

where $k, j = 1, 2, 3, \ldots$. The total multiplicity of the frequency f is given by the analogue of (1.30) in this more general context:

$$
\begin{aligned}
w_f^{(\nu)} &= \#\left\{(k, j) \colon f = f_k(B) \cdot l_j^{-1}\right\} \\
&= \sum_{k=1}^{\infty} w_{f_k(B)/f} \\
&= w_{f_1(B)/f} + w_{f_2(B)/f} + w_{f_3(B)/f} + \cdots .
\end{aligned}
\tag{1.40}
$$

It follows, exactly as in the proof of formula (1.35), that the spectral zeta function $\zeta_\nu(s)$ of the fractal spray—defined by (1.32), with $w_f^{(\nu)}$ as in (1.40)—is given by

$$
\zeta_\nu(s) = \zeta_{\mathcal{L}}(s) \cdot \zeta_B(s).
\tag{1.41}
$$

For example, if $B \subset \mathbb{R}^d$ is the d-dimensional unit cube $(0, 1)^d$, with its opposite sides identified (so that Δ is the Laplacian with periodic boundary conditions on B), then[13]

$$
\zeta_B(s) = \zeta_d(s) = \sum_{(n_1, \ldots, n_d) \in \mathbb{Z}^d \setminus \{0\}} \left(n_1^2 + \cdots + n_d^2\right)^{-s/2} .
\tag{1.42}
$$

Thus $\zeta_d(s)$ is the classical Epstein zeta function; see [Te, §1.4, p. 58] or Section A.4 in Appendix A. (Note that $\zeta_1(s) = 2\zeta(s)$.) It follows that $\zeta_B(s)$ admits a meromorphic continuation to all of $\mathbb{C}$, with a single simple pole at $s = d$.

Hence, for instance, in light of (1.26), (1.41) and (1.42), for the fractal spray of the Cantor string CS on the unit square B (as represented in Figure 1.5), the corresponding spectral zeta function is given by

$$
\zeta_\nu(s) = \frac{3^{-s}}{1 - 2 \cdot 3^{-s}} \cdot \zeta_2(s) .
\tag{1.43}
$$

It is meromorphic in all of $\mathbb{C}$. Further, it has a simple pole at $s = 2$, and at each point $s = \log_3 2 + 2\pi i n / \log 3$ $(n \in \mathbb{Z})$ where $\zeta_2(s)$ does not vanish, it has a simple pole as well.

Later on in this book, we will use generalized fractal sprays to extend several of our results to number-theoretic zeta functions and many other Dirichlet series. (See especially Sections 7.3 and 9.2.)

---

[13]Recall that we exclude the eigenvalue 0.

2
Complex Dimensions
of Self-Similar Fractal Strings

An important class of examples of ordinary fractal strings is provided by the so-called self-similar strings, which we will use throughout this book to illustrate our theory. These are constructed in the usual way with the aid of contraction mappings. In this chapter, we give a detailed analysis of the structure of the complex dimensions of such fractal strings.

2.1 The Geometric Zeta Function of a Self-Similar String

Let N scaling factors $r_1, r_2, \ldots, r_N$ be given ($N \geq 2$), with

$$1 > r_1 \geq r_2 \geq \ldots \geq r_N > 0.$$

Assume that

$$R := \sum_{j=1}^{N} r_j < 1. \tag{2.1}$$

Given an open interval of length L, we construct a *self-similar string* $\mathcal{L}$ with scaling ratios $r_1, r_2, \ldots, r_N$ by a procedure reminiscent of the construction of the Cantor set. Subdivide I into intervals of length $r_1 L, \ldots, r_N L$. The remaining piece of length $(1 - R)L$ is the first member of the string, also called the first length in Remark 2.5 below. Repeat this process with the remaining intervals. (See Figure 2.1.)

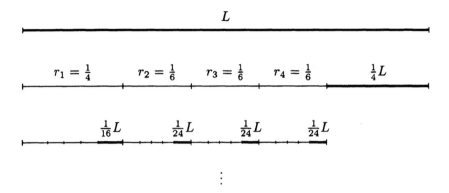

Figure 2.1: An example of a self-similar string (with $N = 4$ and scaling ratios $r_1 = \frac{1}{4}$, $r_2 = r_3 = r_4 = \frac{1}{6}$).

As a result, we obtain a string $\mathcal{L}$ consisting of intervals of length

$$L(1 - R)r_1^{k_1} \ldots r_N^{k_N} \qquad (k_1, \ldots, k_N \in \mathbb{N}),$$

with multiplicity the number of ways to write $r_1^{k_1} \ldots r_N^{k_N}$ as a product of the form $r_{\nu_1} \ldots r_{\nu_q}$, where $q = \sum_{j=1}^{N} k_j$ and $\nu_j \in \{1, \ldots, N\}$. Thus the multiplicity of the length $L(1 - R)r_1^{k_1} \ldots r_N^{k_N}$ is equal to the multinomial coefficient

$$\binom{q}{k_1 \ldots k_N} = \frac{q!}{k_1! \ldots k_N!} \qquad (q = \sum_{j=1}^{N} k_j), \tag{2.2}$$

and the total multiplicity of a length l is the sum of all the multinomial coefficients $\binom{q}{k_1 \ldots k_N}$ for which $L(1 - R)r_1^{k_1} \ldots r_N^{k_N} = l$.

Remark 2.1. Throughout this book, *we will always assume that a self-similar string is nontrivial; that is, we exclude the trivial case when $\mathcal{L}$ is composed of a single interval.* This will permit us to avoid having to consider separately this obvious exception to some of our theorems.

Remark 2.2. It may be helpful for some readers to recall that a self-similar set in $\mathbb{R}^d$ can be thought of as the union of N scaled copies of itself. More precisely, a compact subset F of $\mathbb{R}^d$ is said to be a *self-similar set* if there exist N similarity transformations W_j ($j = 1, \ldots, N$, $N \geq 2$) of $\mathbb{R}^d$ with scaling ratios $r_j \in (0, 1)$ such that $F = \bigcup_{j=1}^{N} W_j(F)$. (See, for example, [Mor, Hut] or [Fa2, Sections 9.1 and 9.2].) As can be easily checked, the boundary of a self-similar string is a self-similar set in $\mathbb{R}$. Conversely, every self-similar set in $\mathbb{R}$ (satisfying the classical 'open set condition' [Hut], [Fa2, Sections 9.1 and 9.2]) can be obtained in this way.

For the expert reader, we mention that, in this work, we always consider self-similar sets satisfying the open set condition. This condition states that there exists a nonempty bounded open set $V \subset \mathbb{R}^d$ such that the sets $W_j(V)$ are pairwise disjoint (for $j = 1, \ldots, N$) and $\bigcup_{j=1}^N W_j(V) \subset V$; see [Hut] or [Fa2, p. 118]. Roughly speaking, this means that the scaled copies $W_j(F)$ of F do not overlap too much.

Theorem 2.3. *Let* $\mathcal{L}$ *be a self-similar string, constructed as above with scaling ratios* $r_1, \ldots, r_N$. *Then the geometric zeta function of this string has a meromorphic continuation to the whole complex plane, given by*

$$\zeta_{\mathcal{L}}(s) = \frac{(L(1-R))^s}{1 - \sum_{j=1}^N r_j^s}, \quad \text{for } s \in \mathbb{C}. \tag{2.3}$$

Here, L *is the total length of* $\mathcal{L}$ *(as defined by (1.20)), which is also the length of* I, *the initial interval, and* R *is given by (2.1).*

Proof. Indeed, we have

$$\sum_{\nu_1=1}^N \cdots \sum_{\nu_q=1}^N \left(r_{\nu_1} \ldots r_{\nu_q} \right)^s = \sum_{\nu_1=1}^N \cdots \sum_{\nu_q=1}^N r_{\nu_1}^s \ldots r_{\nu_q}^s = \left(\sum_{j=1}^N r_j^s \right)^q.$$

Hence, in view of (1.18) and of the above discussion, we deduce that

$$\zeta_{\mathcal{L}}(s) = \sum_{q=0}^\infty \left(\sum_{\nu_1=1}^N \cdots \sum_{\nu_q=1}^N \left(L(1-R) r_{\nu_1} \ldots r_{\nu_q} \right)^s \right)$$

$$= (L(1-R))^s \sum_{q=0}^\infty \left(\sum_{j=1}^N r_j^s \right)^q.$$

Let D be the unique real solution of $\sum_{j=1}^N r_j^D = 1$ (see also the first part of the proof of Theorem 2.13, in particular Equations (2.31) and (2.32) below). For $\operatorname{Re} s > D$, we have $\left| \sum_{j=1}^N r_j^s \right| < 1$, so that the above sum converges (hence D is the dimension of $\mathcal{L}$). We obtain that

$$\zeta_{\mathcal{L}}(s) = \frac{(L(1-R))^s}{1 - \sum_{j=1}^N r_j^s}.$$

This computation is valid for $\operatorname{Re} s > D$, but now the meromorphic continuation of $\zeta_{\mathcal{L}}$ is given by the last formula, as desired. □

Throughout this chapter, we choose W, the window of $\mathcal{L}$ (as defined in Section 1.2.1), to be the entire complex plane, so that $W = \mathbb{C}$ and $\mathcal{D}_{\mathcal{L}} = \mathcal{D}_{\mathcal{L}}(\mathbb{C})$ can be called (without any ambiguity) the set of complex dimensions of $\mathcal{L}$ (as in Definition 1.8(ii)). This choice of W is justified by Theorem 2.3.

Note that in view of Remark 1.11, the nonreal complex dimensions of a self-similar string come in complex conjugate pairs ω, $\bar{\omega}$. This is also clear in view of the following corollary.

Corollary 2.4. *The set of complex dimensions $\mathcal{D}_{\mathcal{L}}$ of the self-similar string $\mathcal{L}$ is the set of solutions of the equation*

$$\sum_{j=1}^{N} r_j^{\omega} = 1, \qquad \omega \in \mathbb{C}. \tag{2.4}$$

Moreover, the multiplicity of the complex dimension ω in $\mathcal{D}_{\mathcal{L}}$ (that is, the multiplicity of ω as a pole of $\zeta_{\mathcal{L}}$) is equal to the multiplicity of ω as a solution of (2.4).

Remark 2.5. For a self-similar string, the total length of $\mathcal{L}$ is also the length of the initial interval I in the above construction. Note that we can always normalize a self-similar string in such a way that the numerator in formula (2.3) equals 1 (equivalently, that the first length of $\mathcal{L}$ is 1), by choosing

$$L = \frac{1}{1 - R}. \tag{2.5}$$

This will not affect the complex dimensions of the string.

2.1.1 Dynamical Interpretation, Euler Product

Let $\mathcal{L}$ be a self-similar string with scaling ratios $r_1, \ldots, r_N$ and normalized as in Remark 2.5. Consider the associated dynamical system of finite sequences on N symbols with the action of the left shift. That is, we consider finite sequences $\mathfrak{x} = a_1, a_2, \ldots, a_l$, with $a_1, a_2, \ldots, a_l \in \{1, \ldots, N\}$, equipped with the left shift $\sigma(\mathfrak{x}) = a_2, a_3, \ldots, a_l, a_1$. (See [PaPol].) Let $l(\mathfrak{x}) = l$ denote the length of $\mathfrak{x}$, and let $\iota(\mathfrak{x})$ denote the length of the shortest period of $\mathfrak{x}$. Thus $\iota(\mathfrak{x})$ is the length of the periodic orbit of σ associated with $\mathfrak{x}$. Note that $\iota(\mathfrak{x}) \mid l(\mathfrak{x})$. We say that $\mathfrak{x}$ is *primitive* if $\iota(\mathfrak{x}) = l(\mathfrak{x})$. Furthermore, we define $r(\mathfrak{x}) = r_{a_1}$. In the following, we will refer to arbitrary finite sequences as *periodic orbits* and to the primitive ones as *primitive periodic orbits*.

Theorem 2.6. *The dynamical zeta function associated with this system is given by*

$$\exp\left(\sum_{\mathfrak{x}} \frac{1}{l(\mathfrak{x})} \left(r(\mathfrak{x})r(\sigma\mathfrak{x})\ldots r(\sigma^{l(\mathfrak{x})-1}\mathfrak{x})\right)^s\right) = \frac{1}{1 - \sum_{j=1}^{N} r_j^s} = \zeta_{\mathcal{L}}(s), \tag{2.6}$$

the geometric zeta function of $\mathcal{L}$. The sum on the left-hand side of (2.6) is taken over all periodic orbits $\mathfrak{x}$ and converges for $\operatorname{Re} s > D$.

Proof. The sum over periodic orbits of fixed length n can be computed, just as in the proof of Theorem 2.3:

$$\sum_{\mathfrak{x}:\, l(\mathfrak{x})=n} \left(r(\mathfrak{x})r(\sigma\mathfrak{x})\dots r(\sigma^{n-1}\mathfrak{x})\right)^s = \sum_{a_1=1}^{N}\sum_{a_2=1}^{N}\cdots\sum_{a_n=1}^{N}\left(r_{a_1}\dots r_{a_n}\right)^s$$

$$= (r_1^s + \cdots + r_N^s)^n.$$

Hence, for $\mathrm{Re}\, s > D$, the sum over all periodic orbits is equal to

$$\sum_{n=1}^{\infty}\frac{1}{n}\sum_{\mathfrak{x}:\, l(\mathfrak{x})=n}\left(r(\mathfrak{x})r(\sigma\mathfrak{x})\dots r(\sigma^{n-1}\mathfrak{x})\right)^s = -\log\left(1 - r_1^s - \cdots - r_N^s\right),$$

and the theorem follows upon exponentiation. Note that the last equality in (2.6) is a consequence of Theorem 2.3 and Remark 2.5.   □

Theorem 2.7 (Euler product). *The function $\zeta_{\mathcal{L}}(s)$ has the following expansion as a product:*

$$\zeta_{\mathcal{L}}(s) = \prod_{\mathfrak{p}}\left(\frac{1}{1 - \left(r(\mathfrak{p})r(\sigma\mathfrak{p})\dots r(\sigma^{l(\mathfrak{p})-1}\mathfrak{p})\right)^s}\right)^{\frac{1}{l(\mathfrak{p})}}, \qquad (2.7)$$

where $\mathfrak{p}$ runs over all primitive periodic orbits. The infinite product in (2.7) converges for $\mathrm{Re}\, s > D$.

Proof. We write the sum over the periodic orbits $\mathfrak{x}$ as a sum over the primitive periodic orbits $\mathfrak{p}$ and repetitions of these, to obtain

$$\sum_{\mathfrak{x}}\frac{1}{l(\mathfrak{x})}\left(r(\mathfrak{x})r(\sigma\mathfrak{x})\dots r(\sigma^{l(\mathfrak{x})-1}\mathfrak{x})\right)^s =$$

$$= \sum_{\mathfrak{p}}\sum_{n=1}^{\infty}\frac{1}{nl(\mathfrak{p})}\left(r(\mathfrak{p})r(\sigma\mathfrak{p})\dots r(\sigma^{nl(\mathfrak{p})-1}\mathfrak{p})\right)^s$$

$$= \sum_{\mathfrak{p}}\frac{1}{l(\mathfrak{p})}\sum_{n=1}^{\infty}\frac{1}{n}\left(r(\mathfrak{p})r(\sigma\mathfrak{p})\dots r(\sigma^{l(\mathfrak{p})-1}\mathfrak{p})\right)^{ns}$$

$$= \sum_{\mathfrak{p}}-\frac{1}{l(\mathfrak{p})}\log\left(1 - \left(r(\mathfrak{p})\dots r(\sigma^{l(\mathfrak{p})-1}\mathfrak{p})\right)^s\right),$$

for $\mathrm{Re}\, s > D$. The theorem then follows upon taking exponentials.   □

In Section 5.4.4, we will apply our explicit formulas of Chapter 4 to derive a Prime Number Theorem for primitive periodic orbits.

Remark 2.8. The above dynamical zeta function (with weight) is often called in the literature the Ruelle (or Bowen–Ruelle) zeta function of the suspended flow [Bow, Rue]. We refer the interested reader to the monograph by Parry and Pollicott [PaPol] for a number of relevant references on this subject and for a detailed account of the theory of such zeta functions. Also see, for example, [BedKS] and [Lal2–3].

2.2 Examples of Complex Dimensions of Self-Similar Strings

In each of these examples, we normalize the string so that its first length is 1, as was explained in Remark 2.5.

2.2.1 The Cantor String

We take two equal scaling factors $r_1 = r_2 = 1/3$. The self-similar string CS with total length 3 and these scaling factors is the *Cantor string*.[1] It consists of lengths 3^{-n} with multiplicity 2^n, $n = 0, 1, 2, \ldots$. The geometric zeta function of this string is

$$\zeta_{CS}(s) = \frac{1}{1 - 2 \cdot 3^{-s}}. \tag{2.8}$$

The complex dimensions are found by solving the equation

$$2 \cdot 3^{-\omega} = 1 \qquad (\omega \in \mathbb{C}). \tag{2.9}$$

We find

$$\mathcal{D}_{CS} = \{D + in\mathbf{p} : n \in \mathbb{Z}\}, \tag{2.10}$$

with $D = \log_3 2$ and $\mathbf{p} = 2\pi/\log 3$. (See Figure 2.2.) All poles are simple. Further, the residue at each pole is $1/\log 3$.

Remark 2.9. This example will be studied in much more detail and extended to so-called generalized Cantor strings in Chapter 8. Also see Chapter 1, Section 5.4.1 and Example 6.23 below.

2.2.2 The Fibonacci String

Next we consider a self-similar string with two lines of complex dimensions. The *Fibonacci string* is the string Fib with total length 4 and scaling factors $r_1 = 1/2$, $r_2 = 1/4$. This string does not seem to have been considered previously in the literature. Its lengths are $1, 1/2, 1/4, 1/8, \ldots, 1/2^n, \ldots$, with multiplicity respectively $1, 1, 2, 3, \ldots, F_{n+1}, \ldots$, the Fibonacci numbers. Recall that these numbers are defined by the following recursive equation:

$$F_{n+1} = F_n + F_{n-1}, \text{ with } F_0 = 0, \ F_1 = 1. \tag{2.11}$$

---

[1] It differs from the Cantor string discussed in Chapter 1 in the normalization of the first length. Alternatively, one could say that we have used another unit to measure the lengths.

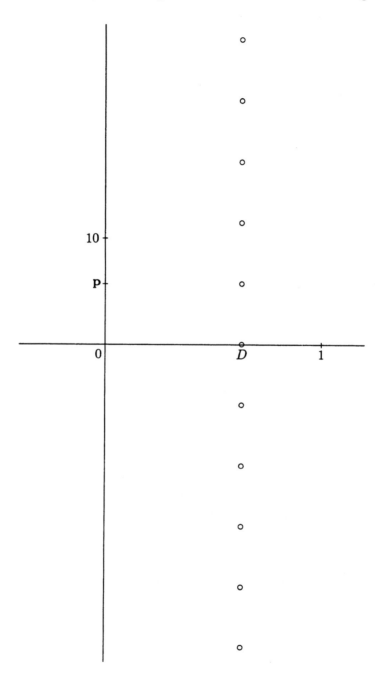

Figure 2.2: The complex dimensions of the Cantor string. $D = \log_3 2$ and $\mathbf{p} = 2\pi/\log 3$.

The geometric zeta function of the Fibonacci string is

$$\zeta_{\text{Fib}}(s) = \frac{1}{1 - 2^{-s} - 4^{-s}}. \tag{2.12}$$

The complex dimensions are found by solving the quadratic equation

$$(2^{-\omega})^2 + 2^{-\omega} = 1 \qquad (\omega \in \mathbb{C}). \tag{2.13}$$

We find $2^{-\omega} = \left(-1 + \sqrt{5}\right)/2 = \phi^{-1}$ and $2^{-\omega} = -\phi$, where

$$\phi = \frac{1 + \sqrt{5}}{2} \tag{2.14}$$

is the golden ratio. Hence

$$\mathcal{D}_{\text{Fib}} = \{D + in\mathbf{p} \colon n \in \mathbb{Z}\} \cup \{-D + i(n + 1/2)\mathbf{p} \colon n \in \mathbb{Z}\}, \tag{2.15}$$

with $D = \log_2 \phi$ and $\mathbf{p} = 2\pi/\log 2$. (See Figure 2.3.) Again, all the poles are simple. Further, for all $n \in \mathbb{Z}$, the residue at the poles $D + in\mathbf{p}$ is $\frac{\phi+2}{5\log 2}$ and the residue at $-D + i(n + 1/2)\mathbf{p}$ is $\frac{3-\phi}{5\log 2}$.

2.2.3 A String with Multiple Poles

Let $\mathcal{L}$ be the self-similar string with scaling factors $r_1 = r_2 = r_3 = 1/9$ and $r_4 = r_5 = 1/27$ and with total length $L = \frac{27}{16}$ (so that its first length is 1). Then the geometric zeta function of $\mathcal{L}$ is given by

$$\zeta_{\mathcal{L}}(s) = \frac{1}{1 - 3 \cdot 9^{-s} - 2 \cdot 27^{-s}}. \tag{2.16}$$

The complex dimensions are found by solving the cubic equation

$$2z^3 + 3z^2 = 1, \quad z = 3^{-\omega} \qquad \text{(with } z, \omega \in \mathbb{C}\text{)}. \tag{2.17}$$

This factors as $(2z - 1)(z + 1)^2 = 0$. Thus we see that there is one line of simple poles $\omega = D + in\mathbf{p}$ ($D = \log_3 2$ and $\mathbf{p} = 2\pi/\log 3$, $n \in \mathbb{Z}$), with residue $\frac{4}{9\log 3}$, corresponding to the solution $z = 1/2$, and another line[2] of double poles $\omega = \frac{1}{2}i\mathbf{p} + in\mathbf{p}$ ($n \in \mathbb{Z}$) corresponding to the double solution $z = -1$. (See Figure 2.4.) The Laurent series at a double pole is

$$\frac{1}{3(\log 3)^2} \left(s - \tfrac{1}{2}i\mathbf{p} - in\mathbf{p}\right)^{-2} + \frac{5}{9\log 3} \left(s - \tfrac{1}{2}i\mathbf{p} - in\mathbf{p}\right)^{-1} + O(1),$$

as $s \to \frac{1}{2}i\mathbf{p} + in\mathbf{p}$.

---

[2]When we talk about a 'line of poles' in this context, we mean a discrete line, as depicted in Figure 2.4. In the following, for convenience, we will continue using this abuse of language.

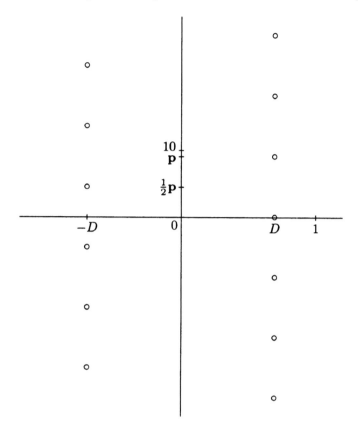

Figure 2.3: The complex dimensions of the Fibonacci string. $D = \log_2 \phi$ and $\mathbf{p} = 2\pi / \log 2$.

2.2.4   Two Nonlattice Examples

The above examples are all self-similar lattice strings, as will be defined in Section 2.3. The reader may get the mistaken impression that in general, it is easy to find the complex dimensions of a self-similar string. However, in the nonlattice case, in the sense of Definition 2.10 below, it is practically impossible to obtain complete information about the complex dimensions. Nevertheless, in Sections 2.4–2.6, we will obtain some partial information about the location and the density of the complex dimensions of a nonlattice string. (See, in particular, Theorem 2.13 below.)

We now give our first explicit example of a nonlattice string. (See Remark 2.12 below for some further information.) We take two scaling factors $r_1 = 1/2$, $r_2 = 1/3$. The self-similar string $\mathcal{L}$ with total length 6 and these scaling factors is nonlattice. It consists of lengths $2^{-m}3^{-n}$ with multiplicity the binomial coefficient $\binom{m+n}{m}$ for $m, n = 0, 1, \ldots$ (see Section 2.1). The

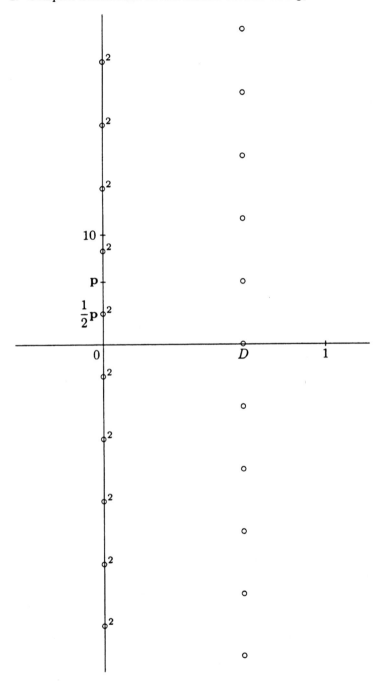

Figure 2.4: The complex dimensions of a string with multiple poles. $D = \log_3 2$ and $\mathbf{p} = 2\pi/\log 3$; here, the symbol $\circ^2$ denotes a multiple pole of order two.

geometric zeta function of this string is

$$\zeta_{\mathcal{L}}(s) = \frac{1}{1 - 2^{-s} - 3^{-s}}. \tag{2.18}$$

The complex dimensions are found by solving the *transcendental* equation

$$2^{-\omega} + 3^{-\omega} = 1 \qquad (\omega \in \mathbb{C}). \tag{2.19}$$

We cannot solve this equation explicitly. We do not even know the exact value of the dimension D of this string; i.e., the precise value of the unique real solution of Equation (2.19). We only know that all complex dimensions have real part $\geq -1 = \sigma_{\text{left}}$ (see Theorem 2.13), and that $D \approx .78788\ldots$. Naturally, with the help of a computer (and of Theorems 2.26 and 2.29 below), we could also sketch an approximate plot of the set of complex dimensions $\mathcal{D} = \mathcal{D}_{\mathcal{L}}(\mathbb{C})$.

The Golden String

Next, we consider the nonlattice string GS with scaling factors $r_1 = 2^{-1}$ and $r_2 = 2^{-\phi}$, where ϕ is the golden ratio given by (2.14). We call this string the *golden string*. Its geometric zeta function is

$$\zeta_{\text{GS}}(s) = \frac{1}{1 - 2^{-s} - 2^{-\phi s}}, \tag{2.20}$$

and its complex dimensions are the solutions of the transcendental equation

$$2^{-\omega} + 2^{-\phi\omega} = 1 \qquad (\omega \in \mathbb{C}). \tag{2.21}$$

A diagram of the complex dimensions of GS is given in Figure 2.5. We have not obtained it by directly solving (2.21) numerically. Instead, we have obtained it by applying Theorems 2.26 and 2.29 below, in which the complex dimensions of a nonlattice string are approximated by those of a lattice string with a large oscillatory period: We chose the approximation $\phi \approx 987/610$ to approximate $\mathcal{L}$ by the lattice string with scaling factors $r_1 = r^{610}$, $r_2 = r^{987}$, where $r = 2^{-1/610}$, and hence with multiplicative generator $2^{-1/610}$. We note that the proof of Theorem 2.29 (and of Lemma 2.28) provides a concrete method for obtaining more and more accurate approximations of $\mathcal{D} = \mathcal{D}_{\text{GS}}(\mathbb{C})$. In particular, the dimension D of the golden string is approximately equal to $D = .77921\ldots$.

Our numerical investigations—and the theoretical information contained in Lemma 2.28 and Theorem 2.29—indicate that the complex dimensions of the golden string have a very interesting and beautiful structure; see Figure 2.6. We intend to discuss this in more detail in a later work.

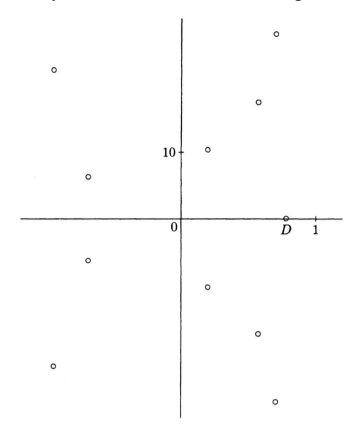

Figure 2.5: The complex dimensions of the golden string (the nonlattice string with scaling ratios $r_1 = 2^{-1}$ and $r_2 = 2^{-\phi}$).

2.3   The Lattice and Nonlattice Case

There is an important dichotomy regarding the scaling ratios r_j ($j = 1, \ldots, N$) with which a self-similar string is constructed. Indeed, recall that an additive subgroup of the real numbers is either dense in $\mathbb{R}$ or else discrete. In the latter case, there exists a number $\tau > 0$, called the *additive generator*, such that the subgroup is equal to $\tau\mathbb{Z}$.

We now apply this basic fact to the additive group

$$\sum_{j=1}^{N} (\log r_j)\mathbb{Z},$$

generated by the logarithms $\log r_j$ ($j = 1, \ldots, N$) of the scaling factors. Alternatively, in light of the isomorphism $y \mapsto e^y$ between the groups $(\mathbb{R}, +)$ and $(\mathbb{R}_+^*, \cdot)$, we formulate the next definition in terms of the multiplicative

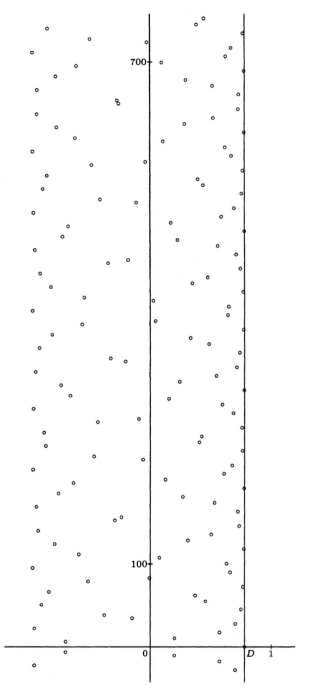

Figure 2.6: The almost periodic behavior of the complex dimensions of the golden string.

subgroup

$$G = \prod_{j=1}^{N} r_j^{\mathbb{Z}} \qquad (2.22)$$

of $\mathbb{R}_+^*$, the positive real line.

Definition 2.10. The case when G is dense in $\mathbb{R}_+^*$ is called the *nonlattice case*. We then say that $\mathcal{L}$ is a *nonlattice string*.

The case when G is not dense (and hence discrete) in $\mathbb{R}_+^*$ is called the *lattice case*. We then say that $\mathcal{L}$ is a *lattice string*. In this situation there exists a unique real number r, $0 < r < 1$, called the *multiplicative generator* of the string, and positive integers $k_1, \ldots, k_N$ without common divisor, such that $1 \le k_1 \le \cdots \le k_N$ and

$$r_j = r^{k_j}, \qquad (2.23)$$

for $j = 1, \ldots, N$.

We note that r is the generator of the multiplicative group $\prod_{j=1}^{N} r_j^{\mathbb{Z}}$ such that $r < 1$. Similarly, by taking logarithms, we see that $\log r^{-1}$ is the positive generator of the additive group $\sum_{j=1}^{N} (\log r_j)\mathbb{Z}$.

Remark 2.11. The dichotomy lattice vs. nonlattice comes from (probabilistic) renewal theory [Fel, Chapter XI] and was used by Lalley in [Lal1–3]. It was then introduced by the first author in [Lap3, §4] in the present setting of self-similar strings (or, more generally, of self-similar drums). (In [Fel], the lattice and nonlattice case are called the arithmetic and nonarithmetic case, respectively.) See also, for example, [Lap3, §5] and [KiLap, Lap5–6] for the case of self-similar drums with fractal membrane (rather than with fractal boundary); also see, for instance, [Str1–2] for a different but related context.

2.3.1 Generic Nonlattice Strings

The explicit examples discussed in the literature are mostly lattice strings. However, the generic self-similar string is a nonlattice string. We refine this further with the remark that, generically, the $\mathbb{Q}$-vector space spanned by the numbers $\log r_j$ $(j = 1, \ldots, N)$ is N-dimensional. (This is the case if and only if the logarithms of the scaling ratios, $\log r_1, \ldots, \log r_N$, are rationally independent.) In this case, we call the string *generic nonlattice*. Otherwise, if this dimension is less than N, but at least 2, the string is *nongeneric nonlattice*. The lattice case corresponds to the case when this dimension is equal to 1.

Remark 2.12. We note that, with our present terminology, the two nonlattice self-similar strings discussed in Section 2.2.4 are generic nonlattice.

Indeed, for the first string, this follows since $\log 2/\log 3$ is irrational, and for the second string, since the golden ratio ϕ is irrational.

2.4 The Structure of the Complex Dimensions

The following key theorem summarizes many of the results that we will obtain about the complex dimensions of self-similar strings. It provides, in particular, a useful criterion for distinguishing between a lattice and a non-lattice string, by looking at the right-most elements in its set of complex dimensions $\mathcal{D}$. It also gives the basic structure of $\mathcal{D}$ in the lattice case. We will establish all the facts stated in Theorem 2.13 (with the exception of those regarding Minkowski measurability, which will be proved in Chapter 6) in the present section and in Sections 2.5 and 2.6 below. More precise and additional information can be found in the latter sections as well as in Theorems 6.20 and 6.21 of Section 6.3.1.

Theorem 2.13. *Let $\mathcal{L}$ be a self-similar string of dimension D and with scaling ratios $r_1,\ldots,r_N$. Then all the complex dimensions of $\mathcal{L}$ lie to the left of or on the line* $\operatorname{Re} s = D$:

$$\mathcal{D} = \mathcal{D}_{\mathcal{L}} \subset \{s \in \mathbb{C}:\ \operatorname{Re} s \le D\}. \tag{2.24}$$

The value $s = D$ is the only pole of $\zeta_{\mathcal{L}}$ on the real line. (In particular, Equation (1.25) holds, with $D = D_{\mathcal{L}}$.) Moreover, $0 < D < 1$ and D is equal to the Minkowski dimension of the (boundary of the) string.
    The set $\mathcal{D} = \mathcal{D}_{\mathcal{L}}(\mathbb{C})$ of complex dimensions of $\mathcal{L}$ is symmetric with respect to the real axis and is contained in a bounded strip $\sigma_{\text{left}} \le \operatorname{Re} s \le D$, for some real number σ_{left}. It is infinite, with density[3]

$$\#(\mathcal{D} \cap \{\omega \in \mathbb{C}:\ |\operatorname{Im}\omega| \le T\}) = \frac{\log r_N^{-1}}{\pi}T + O\left(\sqrt{T}\right), \tag{2.25}$$

as $T \to \infty$.

    In the lattice *case, $\zeta_{\mathcal{L}}(s)$ is a rational function of r^s, where r is the multiplicative generator of $\mathcal{L}$, as given in Definition 2.10. The complex dimensions ω are obtained by finding the complex solutions z of the polynomial equation (of degree k_N)*

$$\sum_{j=1}^{N} z^{k_j} = 1, \quad \text{with } r^\omega = z. \tag{2.26}$$

---

[3]In (2.25), the elements of $\mathcal{D}$ are counted according to their multiplicity as poles of $\zeta_{\mathcal{L}}$.

Hence there exist finitely many poles $\omega_1(=D), \omega_2, \ldots, \omega_q$, and a positive real number $\mathbf{p}$, called the oscillatory period *of the string, and given by*

$$\mathbf{p} = \frac{2\pi}{\log r^{-1}}, \tag{2.27}$$

such that

$$\mathcal{D} = \{\omega_u + in\mathbf{p} : n \in \mathbb{Z}, u = 1, \ldots, q\}. \tag{2.28}$$

In other words, the poles lie periodically on finitely many vertical lines, and on each line they are separated by $\mathbf{p}$. The multiplicity of the complex dimensions corresponding to one value of $z = r^\omega$ is the same as that of z. (In particular, the poles on the line above D are all simple.) It follows that a lattice string is not Minkowski measurable; its geometry has oscillations of order D.

Finally, as a function of r^s, $\zeta_{\mathcal{L}}(s)$ is a rational function. Moreover, if the lattice string is normalized as in Remark 2.5 above (so that its first length is equal to 1), then the residue of $\zeta_{\mathcal{L}}(s)$ at $s = D + in\mathbf{p}$ is independent of $n \in \mathbb{Z}$ and equal to

$$\frac{1}{\log r^{-1} \sum_{j=1}^N k_j r^{k_j D}}. \tag{2.29}$$

More generally, for $u = 1, \ldots, q$, the principal part of the Laurent series of the function $\zeta_{\mathcal{L}}(s + \omega_u + in\mathbf{p})$ at $s = 0$ does not depend on $n \in \mathbb{Z}$.

In the *nonlattice case, D is simple and is the unique pole of $\zeta_{\mathcal{L}}$ on the line $\mathrm{Re}\, s = D$.[4] It follows that a nonlattice string is Minkowski measurable; its geometry does not have oscillations of order D. Further, there is an infinite sequence of simple complex dimensions of $\mathcal{L}$ coming arbitrarily close (from the left) to the line $\mathrm{Re}\, s = D$.*

Finally, the complex dimensions of $\mathcal{L}$ can be approximated (via an explicit algorithm) by the complex dimensions of a sequence of lattice strings, with larger and larger oscillatory period. Hence the complex dimensions of a nonlattice string have an almost periodic structure.

Proof of part of Theorem 2.13. Recall from Corollary 2.4 that the complex dimensions of $\mathcal{L}$ are given by the complex solutions of Equation (2.4):

$$\sum_{j=1}^N r_j^s = 1. \tag{2.30}$$

---

[4]Then the residue of $\zeta_{\mathcal{L}}(s)$ at $s = D$ is equal to $\left(\sum_{j=1}^N \left(\log r_j^{-1} \right) r_j^D \right)^{-1}$.

Consider the function

$$f(s) = \sum_{j=1}^{N} r_j^s \tag{2.31}$$

for real values of s. Because $0 < r_j < 1$, f is strictly decreasing. Since $f(0) = N > 1$ and $f(1) = R < 1$, there exists a unique value D strictly between 0 and 1 such that $f(D) = 1$. This is the only real value of s where $\zeta_{\mathcal{L}}$ has a pole.

Consider now a complex number s with real part $\sigma > D$. For this number,

$$|f(s)| \leq \sum_{j=1}^{N} r_j^{\sigma} < \sum_{j=1}^{N} r_j^{D} = 1. \tag{2.32}$$

It follows that all the poles of $\zeta_{\mathcal{L}}$ lie to the left of or on the line $\operatorname{Re} s = D$.

Suppose there is another pole on the line $\operatorname{Re} s = D$, say at $s = D + it$. Without loss of generality, we may assume that $t > 0$. The first inequality in (2.32) is an equality if and only if all numbers r_j^s have the same argument (i.e., point in the same direction), and then $f(s)$ itself is equal to 1 if and only if all the numbers r_j^s are real and positive. This means that $r_j^{it} > 0$, and hence is equal to 1, for all $j = 1, \ldots, N$. Thus there exist a positive integer l and positive integers k_j without common divisor, such that

$$t \log r_j^{-1} = 2\pi k_j l \qquad (j = 1, \ldots, N).$$

This means that we are in the lattice case. Note that $1 \leq k_1 \leq \ldots \leq k_N$ and that $s = D + it/l$ is also a pole of $\zeta_{\mathcal{L}}$.

In this case, we proceed as follows. Write $z = r^s$, so that $r_j^s = r^{k_j s} = z^{k_j}$, for $j = 1, \ldots, N$. Then, by (2.31), $f(s) = 1$ is equivalent to

$$\sum_{j=1}^{N} z^{k_j} = 1. \tag{2.33}$$

This is a polynomial equation of degree k_N. Therefore, it has k_N complex solutions, counted with multiplicity. To every solution $z = |z|e^{i\theta}$ ($-\pi < \theta \leq \pi$) of this equation there corresponds a unique solution

$$s_0 = -\frac{\log |z|}{\log r^{-1}} - \frac{i\theta}{\log r^{-1}}$$

with imaginary part $-\pi/\log r^{-1} \leq \operatorname{Im} s_0 < \pi/\log r^{-1}$, and a line of solutions

$$s = -\frac{\log |z|}{\log r^{-1}} - \frac{i(\theta - 2n\pi)}{\log r^{-1}} = s_0 + in\mathbf{p},$$

with $n \in \mathbb{Z}$. These points are all poles of $\zeta_{\mathcal{L}}$; i.e., complex dimensions of the string. Note that the orders of these poles are all equal, and coincide with the multiplicity of the corresponding solution z of the algebraic equation (2.33).

If we normalize the string by $L(1-R) = 1$, as in Remark 2.5 above, then $\zeta_{\mathcal{L}}(s) = 1/f(s)$ is periodic, with period $i\mathbf{p}$. It follows that for $u = 1, \ldots, q$, the principal part of the Laurent series of this function at $s = \omega_u + in\mathbf{p}$ does not depend on $n \in \mathbb{Z}$. In particular, the residue at $s = D + in\mathbf{p}$ is equal to the residue at $s = D$. Since D is simple, we find by a direct computation, analogous to that carried out for the Cantor and the Fibonacci strings in Sections 2.2.1 and 2.2.2,

$$\operatorname{res}\left(\zeta_{\mathcal{L}}(s); D\right) = \lim_{s \to D} \frac{s - D}{1 - \sum_{j=1}^{N} r_j^s} = \frac{1}{\sum_{j=1}^{N} (\log r_j^{-1}) r_j^D}$$
$$= \frac{1}{\log r^{-1} \sum_{j=1}^{N} k_j r^{k_j D}}. \tag{2.34}$$

If we do not normalize $\mathcal{L}$, the residue at $s = D + in\mathbf{p}$ is equal to

$$\frac{(L(1-R))^{D+in\mathbf{p}}}{\log r^{-1} \sum_{j=1}^{N} k_j r^{k_j D}} = (L(1-R))^{in\mathbf{p}} \operatorname{res}\left(\zeta_{\mathcal{L}}(s); D\right).$$

Finally, the statement about the density of the poles is proved in Section 2.5 (see Theorem 2.22), where we will also show that the poles of $\zeta_{\mathcal{L}}$ lie to the right of some vertical line (see Remark 2.23). In the nonlattice case, the statement about poles of $\zeta_{\mathcal{L}}$ close to the line $\operatorname{Re} s = D$ is proved in Theorem 2.31. Further, that about the Minkowski measurability is proved in Section 6.3.1, Theorems 6.20 and 6.21. (In the nonlattice case, it was proved by different means in [Lap3, §4.4] and later on, independently, in [Fa3]; see Remark 6.27 below.)

The last statement of the nonlattice case of Theorem 2.13—concerning the approximation of the complex dimensions of nonlattice strings by those of a sequence of lattice strings—follows from Theorems 2.26 and 2.29 in Section 2.6. The explicit algorithm is provided by the method of proof of Theorem 2.29 and Lemma 2.28.   □

Remark 2.14. In the lattice case, if ω_u is a simple pole of $\zeta_{\mathcal{L}}(s)$, then a computation similar to that leading to (2.34) shows that

$$\operatorname{res}\left(\zeta_{\mathcal{L}}(s); \omega_u + in\mathbf{p}\right) = \frac{(L(1-R))^{in\mathbf{p}}}{\log r^{-1} \sum_{j=1}^{N} k_j r^{k_j \omega_u}}, \quad \text{for } n \in \mathbb{Z}, \tag{2.35}$$

where $L(1-R)$ is the first length of $\mathcal{L}$ (which is equal to 1 when $\mathcal{L}$ is normalized).

Remark 2.15. For a general lattice string, the degree k_N of the polynomial equation (2.26) can be any positive integer. Indeed, given any integer

$k \geq 1$, the lattice string with $N = 2$ and scaling ratios $r_1 = r$ and $r_2 = r^k$, with $0 < r < 1$, gives rise to the algebraic equation of degree k

$$z^k + z = 1 \qquad (2.36)$$

for the complex dimensions.

The following corollary of Theorem 2.13 will show that every (nontrivial) self-similar string is 'fractal', in the new sense that we will introduce in Section 10.2. Namely, it has at least one (and hence at least two complex conjugate) nonreal complex dimensions with positive real part.

Corollary 2.16. *Every self-similar string has infinitely many complex dimensions with positive real part.*[5]

Proof. In the lattice case, this follows from the fact that $\mathcal{D}$ always contains $\{D + inp \colon n \in \mathbb{Z}\}$ (see formula (2.28) in Theorem 2.13).

In the nonlattice case, this follows from the fact that $D > 0$, combined with the statement about the complex dimensions approaching $\operatorname{Re} s = D$ from the left in the nonlattice part of Theorem 2.13 (which will be established and made more precise in Theorem 2.31). $\qquad \square$

We first comment on some aspects of Corollary 2.16.

Remark 2.17. It is well known that for a self-similar set satisfying the open set condition—as is the case of the boundary of a self-similar string $\mathcal{L}$ considered here, by Remark 2.2 above—the Minkowski dimension is the unique real solution of Equation (2.4). (See, for example, [Fa2, Theorems 9.1 and 9.3, pp. 114 and 118].)[6] In other words, D is equal to the 'similarity dimension' s [Man1], defined by

$$\sum_{j=1}^{N} r_j^s = 1 \qquad (s > 0). \qquad (2.37)$$

In our terminology, this is the unique complex dimension of $\mathcal{L}$ located on the real axis. In view of Theorem 2.3 and Corollary 2.4, what is new and much less easy to establish in the statement of Corollary 2.16 is that Equation (2.4) (or equivalently, (2.37)) has infinitely many complex solutions with positive real part.

Next, we briefly comment on some aspects of the lattice vs. nonlattice dichotomy exhibited by Theorem 2.13.

---

[5]Recall from Remark 2.1 that we exclude the trivial case when $\mathcal{L}$ consists of a single interval.

[6]In the present one-dimensional situation, this is essentially due to Moran [Mor] and in higher dimension, it follows from the results of Hutchinson in [Hut], along with the fact that for a self-similar set satisfying the open set condition, the Hausdorff and Minkowski dimensions coincide (see, e.g., [Fa2, Theorem 9.3, p. 118]).

Remark 2.18. As was noted earlier, the Cantor string, the Fibonacci string and the example of a self-similar string with multiple poles given in Sections 2.2.1, 2.2.2 and 2.2.3, respectively, are all lattice strings. Their multiplicative generator r is equal to $1/3$, $1/2$ and $1/3$, respectively. Further, the algebraic equation (2.26) giving rise to their complex dimensions is provided, respectively, by (2.9), (2.13) and (2.17), which are of degree 1, 2 and 3.

Remark 2.19. As was mentioned in Section 2.2.4, in the nonlattice case, the poles of the geometric zeta function $\zeta_\mathcal{L} = \zeta_\mathcal{L}(s)$ are usually intractable. For example, we guess that in general all, or almost all, poles are simple. We hope to study this question in subsequent work. As for the real parts of the poles, they could be dense in the 'critical interval' (i.e., the intersection of the real line with the narrowest strip that contains all the poles). Or there could be dense pieces (or a discrete set of points) with pole-free regions in between; i.e., substrips of the critical strip that contain no poles. Regarding these questions, we do not yet have a reasonable conjecture to propose. However, thanks to the algebraic equation (2.26), we can say a lot more in the lattice case.

2.5   The Density of the Poles in the Nonlattice Case

In the lattice case, the algebraic equation (2.26) allows one to gain a complete understanding of the structure of the poles. To study the poles in the nonlattice case, we shall use Nevanlinna theory to obtain information about the solutions of the equation $\sum_{j=1}^{N} r_j^s = 1$. We will prove that the solutions of this equation lie in a bounded strip, and that they have the same asymptotic density as in the lattice case; that is, a linear density given by (2.25). On a first reading, the reader may wish to note the main result of this section, Theorem 2.22 (along with the remark following it), and skip the details of the proof.

We refer, for example, to the monograph [LanCh] for a clear exposition of Nevanlinna theory.

2.5.1   Nevanlinna Theory

Recall that $\mathbb{P}^1(\mathbb{C})$ denotes the complex projective line; i.e., the complex line $\mathbb{C}$, completed by a point at infinity, denoted ∞. Alternatively, $\mathbb{P}^1(\mathbb{C})$ could be realized as the Riemann sphere. The *distance* between two points a, a' in $\mathbb{P}^1(\mathbb{C})$ is defined by

$$\|a, a'\| = \frac{|a - a'|}{\sqrt{1 + |a|^2}\sqrt{1 + |a'|^2}}, \quad \text{if } a, a' \neq \infty, \tag{2.38a}$$

and by

$$\|a, \infty\| = \frac{1}{\sqrt{1 + |a|^2}}, \quad \text{if } a \neq \infty. \tag{2.38b}$$

Here, $|z|$ denotes the absolute value of the complex number z. (See, e.g., [Bea, §2.1].) When one views $\mathbb{P}^1(\mathbb{C})$ as a sphere of diameter 1 in three-dimensional Euclidean space, the distance is simply the chordal distance between the inverse images of a and a' under stereographic projection.

Let f be a nonconstant meromorphic function and let $a \in \mathbb{P}^1(\mathbb{C})$. The *mean proximity* function of f is the function of the positive real variable ϱ given by

$$m_f(a, \varrho) = \int_{|z|=\varrho} -\log \|f(z), a\| \frac{dz}{2\pi i z}. \tag{2.39}$$

The *counting function* of f is defined as[7]

$$n_f(a, \varrho) = \#\{z \in \mathbb{C} : |z| \leq \varrho, \; f(z) = a\}, \tag{2.40}$$

and, for $a \neq f(0)$, we set

$$N_f(a, \varrho) = \int_0^\varrho n_f(a, t) \frac{dt}{t}. \tag{2.41}$$

Finally, the *height* of f is defined, for $a \neq f(0)$, by

$$T_f(\varrho) = N_f(a, \varrho) + m_f(a, \varrho) + \log \|f(0), a\|, \tag{2.42}$$

which is independent of a (cf. [LanCh, Theorem 1.6, p. 19]). It is this independence that we will exploit, for $a = \infty$ and for $a = 0$.

2.5.2   *Complex Zeros of Dirichlet Polynomials*

We are ready to study the zeros of the Dirichlet polynomial $1 - \sum_{j=1}^N r_j^s$. For this purpose, we will investigate the equation

$$\sum_{j=0}^n a_j^{s+p_j} = 0, \quad \text{where } a_0 = 1 > a_1 > a_2 > \ldots > a_n > 0.$$

We are interested in the application where, as sets,

$$\{a_1, \ldots, a_n\} = \{r_1, \ldots, r_N\},$$

---

[7]This function takes finite values since the zeros of a nonconstant meromorphic function form a discrete subset of $\mathbb{C}$.

$a_0 = 1$, $p_0 = 0$, and p_j $(j = 1, \ldots, n)$ is such that

$$-a_j^{p_j} = \#\{k \colon k = 1, \ldots, N, \; r_k = a_j\}.$$

Our analysis is partly similar to that of Jorgenson and Lang in [JorLan3]. In particular, the result that the zeros of a Dirichlet polynomial lie in a bounded strip can be found in [JorLan3, p. 58]. On the other hand, in the present situation, we obtain more precise results than in [JorLan3]. Similar results were also obtained by B. Jessen. (See [Bohr, Appendix II].)

Lemma 2.20. *Let $a_0 = 1 > a_1 > \ldots > a_n > 0$ and let $p_0, \ldots, p_n$ be arbitrary complex numbers. Assume that $\sum_{j=0}^{n} a_j^{p_j} \neq 0$. Then the number of complex zeros s of $\sum_{j=0}^{n} a_j^{s+p_j}$ in the closed disc with radius ϱ is*

$$\frac{\log a_n^{-1}}{\pi} \varrho + O(\sqrt{\varrho}), \quad as \; \varrho \to \infty.$$

Moreover, there exist real numbers σ_0 and σ_1 (given respectively by formulas (2.44) and (2.46) below) such that these zeros lie in the bounded strip $\sigma_0 \leq \operatorname{Re} s \leq \sigma_1$.

Proof. Let $f(s) := \sum_{j=0}^{n} a_j^{s+p_j}$. We will obtain information about the zeros of this function (i.e., about $N_f(0, \varrho)$) by first computing the height

$$T_f(\varrho) = m_f(\infty, \varrho) + \log \|f(0), \infty\|$$

and then combining the relation

$$T_f(\varrho) = N_f(0, \varrho) + m_f(0, \varrho) + \log \|f(0), 0\|$$

with estimates for $m_f(0, \varrho)$. Thus we shall obtain estimates for $N_f(0, \varrho)$, and from this we shall deduce estimates for $n_f(0, \varrho)$.

Ignoring the $\log \|f(0), \infty\|$ term, the height of f is

$$m_f(\infty, \varrho) = \int_{|s|=\varrho} \log \sqrt{1 + |f(s)|^2} \, \frac{ds}{2\pi i s}.$$

We need to estimate $|f(s)|$ in the above integral. Let

$$p = \max_{j=1,\ldots,n} |\operatorname{Re} p_j| \tag{2.43}$$

and let $\sigma = \operatorname{Re} s$. Then $|f(s)|$ is bounded from above by $a_n^\sigma (n+1) a_n^{-p}$. On the other hand, let

$$\sigma_0 = -p + \frac{2p \log a_{n-1} - \log 2n}{\log a_{n-1} - \log a_n}; \tag{2.44a}$$

i.e., σ_0 is the unique real solution of the equation

$$a_n^p = 2na_{n-1}^{-p}\left(\frac{a_{n-1}}{a_n}\right)^{\sigma_0}.$$ (2.44b)

For $\sigma = \operatorname{Re} s \leq \sigma_0$, we have

$$|f(s)| = \left|\sum_{j=0}^{n} a_j^{s+p_j}\right| \geq a_n^{\sigma+\operatorname{Re} p_n} - \sum_{j=0}^{n-1} a_j^{\sigma+\operatorname{Re} p_j}$$

$$\geq a_n^{\sigma+p} - \sum_{j=0}^{n-1} a_j^{\sigma-p}$$

$$\geq a_n^{\sigma+p} - na_{n-1}^{\sigma-p}$$

$$= a_n^{\sigma}\left(a_n^p - na_{n-1}^{-p}\left(\frac{a_{n-1}}{a_n}\right)^{\sigma}\right).$$

Hence $|f(s)|$ is bounded from below as follows:

$$|f(s)| \geq a_n^{\sigma} \cdot \frac{1}{2}a_n^p.$$ (2.45)

Putting these estimates together, we find

$$\log\sqrt{1+|f(s)|^2} = \begin{cases} \sigma \log a_n + O(1) & (\sigma \leq \sigma_0), \\ O(1) & (\sigma \geq \sigma_0), \end{cases}$$

$$= \begin{cases} \sigma \log a_n + O(1) & (\sigma \leq 0), \\ O(1) & (\sigma \geq 0). \end{cases}$$

Here and in the rest of this proof, the asymptotic estimates hold as $|s| = \varrho \to \infty$. On the circle with radius ϱ, the real part of $s = \varrho e^{i\theta}$ is $\varrho\cos\theta$. For the height, we thus find

$$T_f(\varrho) = \log a_n \int_{\pi/2}^{3\pi/2} \varrho \cos\theta \, \frac{d\theta}{2\pi} + O(1) = \frac{\log a_n^{-1}}{\pi}\varrho + O(1),$$

where the implied constant depends only on n and the numbers a_j and p_j.

Now, clearly, $m_f(0, \varrho)$ is positive. Hence $N_f(0, \varrho) \leq \frac{\log a_n^{-1}}{\pi}\varrho + O(1)$. To show that this is the correct asymptotic order for $N_f(0, \varrho)$, we have to bound $m_f(0, \varrho)$ from above. In view of (2.38), inequality (2.45) shows that

$$\|f(s), 0\|^{-1} = \frac{\sqrt{1+|f(s)|^2}}{|f(s)|}$$

is uniformly bounded for $\sigma \leq \sigma_0$. Let

$$\sigma_1 = p - \frac{\log 2n}{\log a_1};$$ (2.46a)

i.e., σ_1 is the unique real solution of the equation

$$na_1^{\sigma-p} = \frac{1}{2}, \qquad (2.46b)$$

where p is still defined by (2.43). Then we have, for $\sigma \geq \sigma_1$:

$$|f(s)| \geq 1 - \sum_{j=1}^{n} a_j^{\sigma + \operatorname{Re} p_j} \geq 1 - na_1^{\sigma-p} \geq \frac{1}{2}.$$

Thus $\log \| f(s), 0 \|^{-1}$ is uniformly bounded for $\sigma \geq \sigma_1$ and $\sigma \leq \sigma_0$. Observe that we have now shown that the complex zeros of f lie in the strip $\sigma_0 \leq \operatorname{Re} s \leq \sigma_1$. The integral for $m_f(0, \varrho)$ over the parts of the circle $|s| = \varrho$ between $\operatorname{Re} s = \sigma_0$ and $\operatorname{Re} s = \sigma_1$ is bounded since $x \mapsto \log |x|$ is an integrable function around $x = 0$. This shows that

$$N_f(0, \varrho) = \frac{\log a_n^{-1}}{\pi} \varrho + O(1),$$

as $\varrho \to \infty$. Finally, in view of (2.40) and (2.41), the statement for $n_f(0, \varrho)$ is a consequence of the following general calculus lemma, applied to $n(t) := n_f(0, t)$ and $N(\varrho) := N_f(0, \varrho)$. $\qquad \square$

Lemma 2.21. *Let $n(t)$ be a nondecreasing, nonnegative function on $[0, \infty)$ for which there exists $t_0 > 0$ such that $n(t) = 0$ for $t \leq t_0$. Let*

$$N(\varrho) = \int_0^\varrho n(t) \frac{dt}{t}$$

and suppose that there exist positive constants c and C such that

$$|N(\varrho) - c\varrho| \leq C \qquad \text{for all } \varrho > 0.$$

Then

$$|n(\varrho) - c\varrho| \leq \sqrt{8Cc\varrho}$$

for all sufficiently large positive values of ϱ.

Proof. Consider a value of ϱ for which $n(\varrho) > c\varrho$. For this value, we have

$$n(\varrho) + C \geq N(n(\varrho)/c) = N(\varrho) + \int_\varrho^{n(\varrho)/c} n(t) \frac{dt}{t} \geq c\varrho - C + n(\varrho) \log \frac{n(\varrho)}{c\varrho}.$$

Hence, writing $x = \frac{n(\varrho)}{c\varrho}$, we deduce that

$$2C \geq c\varrho - n(\varrho) + n(\varrho) \log \frac{n(\varrho)}{c\varrho} = c\varrho \left(1 - x + x \log x\right).$$

Consider now a value of ϱ for which $n(\varrho) < c\varrho$. In the same way as above, we find

$$c\varrho - C \leq N(\varrho) = N(n(\varrho)/c) + \int_{n(\varrho)/c}^{\varrho} n(t)\,\frac{dt}{t} \leq n(\varrho) + C + n(\varrho)\log\frac{c\varrho}{n(\varrho)}.$$

Again, we find that $c\varrho\,(1 - x + x\log x) \leq 2C$, with $x = \frac{n(\varrho)}{c\varrho}$.

The function

$$1 - x + x\log x = \frac{(x-1)^2}{2} + O\left((x-1)^3\right) \qquad (\text{as } x \to 1)$$

is nonnegative and vanishes at $x = 1$. It follows that for ϱ sufficiently large, x is close to 1. Around $x = 1$, this function takes values larger than $(x-1)^2/4$. Hence for large positive ϱ, $(x-1)^2 \leq \frac{8C}{c\varrho}$. This is equivalent to $|n(\varrho) - c\varrho| \leq \sqrt{8Cc\varrho}$, as was to be proved. $\square$

We apply Lemma 2.20, choosing for $a_1, \ldots, a_n$ the sequence of distinct numbers among $r_1, \ldots, r_N$, and letting p_j be such that $-a_j^{p_j}$ is the number of integers k in $\{1, \ldots, N\}$ such that $r_k = a_j$. Note that Lemma 2.20 applies since $N \geq 2$ ensures that $\sum_{j=1}^{N} r_j^0 = N \neq 1$. We then obtain the following result:

Theorem 2.22. *The number of complex zeros (counted with multiplicity) of the Dirichlet polynomial $1 - \sum_{j=1}^{N} r_j^s$ in the disc $|s| \leq \varrho$ is asymptotically given by*

$$\frac{\log r_N^{-1}}{\pi}\varrho + O(\sqrt{\varrho}), \qquad \text{as } \varrho \to +\infty.$$

Remark 2.23. Let $\mathcal{L}$ be a self-similar string with scaling ratios $r_1, \ldots, r_N$. Then, in view of Corollary 2.4, Theorem 2.22 yields the asymptotic density (2.25) of the set of complex dimensions $\mathcal{D}_{\mathcal{L}}$, as stated in Theorem 2.13. Moreover, the fact (also stated in Theorem 2.13) that the poles of $\zeta_{\mathcal{L}}(s)$ (i.e., the complex zeros of $1 - \sum_{j=1}^{N} r_j^s$) lie in a bounded strip $\sigma_{\text{left}} \leq \operatorname{Re} s \leq D$, also follows from Lemma 2.20, with the abovementioned choice of the numbers p_j and a_j. (Recall that we already know that $\mathcal{D}_{\mathcal{L}} \subset \{s : \operatorname{Re} s \leq D\}$.) Observe that Lemma 2.20 implies that $\sigma_{\text{left}} \geq \sigma_0$, where σ_0 is the real number given by (2.44a) or, equivalently, (2.44b). Even though, in general, this inequality does not yield the optimal choice for σ_{left}, it provides an effective method for estimating σ_{left}.

2.6 Approximating a Fractal String and Its Complex Dimensions

The generic self-similar string is nonlattice. Such a string has a slightly more regular behavior in its geometry: It is Minkowski measurable, and this fact

will be explained in Section 6.3.1 and in Section 6.2, Theorem 6.12, by the
absence of nonreal complex dimensions with real part D. On the other hand,
a lattice string is never Minkowski measurable and always has periodic
oscillations of order D in its geometry because its complex dimensions with
real part D form an infinite, vertical arithmetic progression. In Section 6.3.1
(see especially Theorem 6.20), we establish a version of Theorem 6.12 valid
for all nonlattice strings, to be contrasted with Theorem 6.21 for lattice
strings.

In the present section, we show, in particular, that the set of lattice
strings is dense (in a suitable sense) in the set of all self-similar strings:
Every nonlattice string can be approximated by lattice strings. This approximation is such that it results in an approximation of the complex
dimensions: Fixing $T > 0$, there exists a lattice string such that the complex dimensions of the nonlattice string with imaginary part less than T
(in absolute value) are approximated by those of the lattice string. Moreover, the oscillatory period of this lattice string is much smaller than T.
This means that the complex dimensions of all self-similar strings exhibit
an almost periodic behavior.

We begin by stating several definitions and a result regarding the convergence of a sequence of fractal strings and of the associated complex
dimensions. In Section 2.6.1 below, we will study in more detail the particular situation of self-similar strings.

In the following, it will be helpful to keep in mind that a meromorphic
function with values in $\mathbb{C}$ can be viewed as a holomorphic function with
values in $\mathbb{P}^1(\mathbb{C})$, by defining the value at a pole to be $\infty \in \mathbb{P}^1(\mathbb{C})$. More-
over, $\mathbb{P}^1(\mathbb{C})$ will be equipped with the metric $\|\cdot, \cdot\|$ defined in Section 2.5.1,
Equation (2.38).

Definition 2.24. Let $\mathcal{L}$ be a string with window W, and let $\left\{\mathcal{L}^{(n)}\right\}_{n=1}^{\infty}$ be
a sequence of strings with windows $W^{(n)}$. We say that the *sequence* $\mathcal{L}^{(n)}$
converges to $\mathcal{L}$ (*and write* $\mathcal{L}^{(n)} \to \mathcal{L}$) if for every compact set $K \subset W$, we
have that $K \subset W^{(n)}$ for all sufficiently large n, and $\zeta_{\mathcal{L}^{(n)}}(s) \to \zeta_{\mathcal{L}}(s)$ (in
the topology of $\mathbb{P}^1(\mathbb{C})$) uniformly on K.

Definition 2.25. Let $\mathcal{L}$ be a string with complex dimensions $\mathcal{D} = \mathcal{D}(W)$
and let $\mathcal{L}^{(n)}$ be a sequence of fractal strings, with complex dimensions
$\mathcal{D}^{(n)} = \mathcal{D}(W^{(n)})$. We say that the *complex dimensions of* $\mathcal{L}^{(n)}$ *converge
locally to those of* $\mathcal{L}$ (*and write* $\mathcal{D}^{(n)} \to \mathcal{D}$), if for every compact set $K \subset W$
and every $\varepsilon > 0$, there is an integer n_0 such that for all integers $n \geq n_0$,
there exists a bijection

$$b_n \colon K \cap \mathcal{D} \to \mathcal{D}^{(n)} \tag{2.47}$$

that respects multiplicities and such that

$$|\omega - b_n(\omega)| < \varepsilon \text{ for all } \omega \in K \cap \mathcal{D}.$$

More precisely, b_n is a set-valued map from $K \cap \mathcal{D}$ to finite subsets of $\mathcal{D}^{(n)}$, such that the multiplicities of the elements of $b_n(\omega)$ add up to the multiplicity of ω and the distance from ω to each of the elements of $b_n(\omega)$ is bounded by ε.

Theorem 2.26. *Let $\mathcal{L}$ be a fractal string and let $\mathcal{L}^{(n)}$ be a sequence of fractal strings. Then (with the notation of Definitions 2.24 and 2.25), $\mathcal{L}^{(n)} \to \mathcal{L}$ implies $\mathcal{D}^{(n)} \to \mathcal{D}$.*

Proof. Let $\mathcal{L}^{(n)} \to \mathcal{L}$. Let $K \subset W$ and choose a circle T around $\infty \in \mathbb{P}^1(\mathbb{C})$, so small in the metric $\|\cdot, \cdot\|$ that the pre-image $\zeta_{\mathcal{L}}^{-1}(T) \cap K$ is the union of disjoint small 'circles' (really, Jordan curves) around each point $\omega \in K \cap \mathcal{D}$. Let $\varepsilon > 0$ and assume that ε is smaller than the distance from T to 0 and ∞. Let n_0 be such that

$$\|\zeta_{\mathcal{L}^{(n)}}(s), \zeta_{\mathcal{L}}(s)\| \le \frac{\varepsilon}{2},$$

for all $n \ge n_0$ and all $s \in K$. On the 'circle' around ω, $\zeta_{\mathcal{L}^{(n)}}(s)$ is away from infinity by at least $\varepsilon/2$, and $\zeta_{\mathcal{L}^{(n)}}(\omega)$ is away from infinity by at most $\varepsilon/2$. By the Maximum Modulus Principle [Ahl, Theorem 12′, p. 134], $\zeta_{\mathcal{L}^{(n)}}$ has a pole inside T. Refining this argument, by comparing the arguments of $\zeta_{\mathcal{L}^{(n)}}$ and $\zeta_{\mathcal{L}}$, we see that $\zeta_{\mathcal{L}^{(n)}}$ has the same number of poles inside the 'circle' around ω as the multiplicity of ω. We define $b_n(\omega)$ as the set of poles of $\zeta_{\mathcal{L}^{(n)}}$ inside the 'circle' around ω. Then one checks that b_n satisfies the requirements of Definition 2.25. Hence, we conclude that $\mathcal{D}^{(n)} \to \mathcal{D}$. □

Remark 2.27. Definitions 2.24, 2.25 and Theorem 2.26 extend in the obvious way to generalized (rather than ordinary) fractal strings, which will be introduced in Section 3.1 below.

2.6.1 Approximating a Nonlattice String by Lattice Strings

We now focus our attention on the case of self-similar strings. Our main objective in the present section is to show that in the sense of Definition 2.24, every nonlattice string can be approximated by a sequence of lattice strings (Theorem 2.29). In view of Theorem 2.26 above, this will show that the complex dimensions of a nonlattice string can be approximated by those of a sequence of lattice strings. It will imply, in particular, that the complex dimensions of a nonlattice string have an almost periodic structure, and that they come arbitrarily close to the line $\operatorname{Re} s = D$ (Theorem 2.31), as was stated in Theorem 2.13.

Let scaling ratios $r_1, \ldots, r_N$ be given that generate a nonlattice string $\mathcal{L}$; i.e., the dimension of the $\mathbb{Q}$-vector space generated by the numbers $\log r_j$ $(j = 1, \ldots, N)$ is at least 2. (See Section 2.3.1 above.)

Lemma 2.28. *Given scaling ratios $1 > r_1 \geq \ldots \geq r_N > 0$ generating a nonlattice string $\mathcal{L}$, there exists a sequence of scaling ratios $1 > r_1^{(n)} \geq \ldots \geq r_N^{(n)} > 0$, generating lattice strings $\mathcal{L}^{(n)}$, such that for each $j = 1, \ldots, N$, $r_j^{(n)} \to r_j$ as $n \to \infty$.*

Proof. The following lemma about Diophantine approximation can be found in [In, Theorem J, p. 94]:

> *Let N real numbers $\theta_1, \ldots, \theta_N$ be given. Then, for every $T > 0$, there exist a real number $q > T$ and integers $k_1, \ldots, k_N$ such that*
>
> $$|q\theta_j - k_j| < \frac{1}{q^{1/N}},$$
>
> *for $j = 1, \ldots, N$.*

We apply this lemma with $\theta_j = \log r_j^{-1}$, for $j = 1, \ldots, N$. Writing $\log r^{-1} = q^{-1}$, we find positive integers $k_N \geq \cdots \geq k_1$ such that

$$|\log r_j - k_j \log r| < q^{-1-1/N}, \tag{2.48}$$

for each $j = 1, \ldots, N$. The numbers r^{k_j} $(j = 1, \ldots, N)$ generate a lattice string with period

$$\mathbf{p} = \frac{2\pi}{\log r^{-1}} = 2\pi q.$$

Let

$$f(s) = 1 - \sum_{j=1}^{N} r_j^s \tag{2.49a}$$

and

$$g(s) = 1 - \sum_{j=1}^{N} r^{k_j s}. \tag{2.49b}$$

Determine, as in Section 2.5, the real value $\sigma_0 \geq 1$ such that $|f(s)| \geq r_N^{-\sigma_0}/2$ for $\text{Re}\, s \leq -\sigma_0$. (See especially Equation (2.44).) To show that f is well approximated by g, we consider the expression

$$r_j^s - r^{k_j s} = s \int_{k_j \log r}^{\log r_j} e^{st}\, dt.$$

Taking absolute values, and restricting $s = \sigma + it$ to $-\sigma_0 \leq \text{Re}\, s \leq 1$, we find the bound

$$|r_j^s - r^{k_j s}| \leq |s| \int_{k_j \log r}^{\log r_j} e^{\sigma t}\, |dt| \leq 2\sigma_0 q^{-1-1/N} r_N^{-\sigma_0 t}, \quad \text{for } j = 1, \ldots, N.$$

Then, by (2.49), we have

$$|f(s) - g(s)| \leq \sum_{j=1}^{N} \left| r_j^s - r^{k_j s} \right| \leq 2\sigma_0 N r_N^{-\sigma_0} q^{-1-1/N} t.$$

Letting $q \to \infty$, we find functions g arbitrarily close to f, as desired. $\qquad \square$

Theorem 2.29. *Given a nonlattice string $\mathcal{L}$, there exists a sequence of lattice strings $\mathcal{L}^{(n)}$ approximating it; that is, such that $\mathcal{L}^{(n)} \to \mathcal{L}$ (in the sense of Definition 2.24).*

Note that by Theorem 2.26, this in turn implies that $\mathcal{D}^{(n)} \to \mathcal{D}$ (in the sense of Definition 2.25). Here, we take the full sets of complex dimensions; i.e., we choose $W = \mathbb{C}$ and $W^{(n)} = \mathbb{C}$ for each $n \geq 1$.

Proof. Construct a sequence of scaling ratios as in Lemma 2.28. Since the scaling ratios converge, it follows that $\zeta_{\mathcal{L}^{(n)}}$ converges to $\zeta_{\mathcal{L}}$ uniformly on compact subsets of $W = \mathbb{C}$. $\qquad \square$

Note that the sequence $\mathcal{L}^{(n)}$ is obtained by constructing approximations of the scaling ratios of $\mathcal{L}$. The following theorem shows that this is the only way to approximate a given self-similar string by a sequence of self-similar strings. It is of independent interest even though we will not use it in this book.

Theorem 2.30. *Let $\mathcal{L}$ be a self-similar string, with scaling ratios $1 > r_1 \geq \ldots \geq r_N > 0$, and let $\mathcal{L}^{(n)}$ be a sequence of self-similar strings, with scaling ratios $1 > r_1^{(n)} \geq \ldots \geq r_{N^{(n)}}^{(n)} > 0$. Let $W = W^{(n)} = \mathbb{C}$ for all n. If $\mathcal{L}^{(n)} \to \mathcal{L}$, then $N^{(n)} = N$ for all sufficiently large n, and the scaling ratios of $\mathcal{L}^{(n)}$ converge to those of $\mathcal{L}$.*

Proof. The convergence of $\mathcal{L}^{(n)}$ to $\mathcal{L}$ means that $\sum_{j=1}^{N^{(n)}} \left(r_j^{(n)} \right)^s$ converges to $\sum_{j=1}^{N} r_j^s$ on compact subsets of $\mathbb{C}$. By evaluating these functions at $s = 0$, we see first of all that $N^{(n)} = N$ for all sufficiently large n. For notational simplicity, we assume from now on that $N^{(n)} = N$ for all $n \geq 1$.

Suppose first that $\mathcal{L}^{(n)}$ and $\mathcal{L}$ are lattice strings with the same multiplicative generator r. Since, then, $\sum_{j=1}^{N} \left(r_j^{(n)} \right)^s$ and $\sum_{j=1}^{N} r_j^s$ are polynomials in r^s, the theorem follows from [Ahl, §5.5]. In general, we choose lattice strings approximating everything on the level of the scaling ratios. Then the lattice strings are close, from which we deduce the convergence of the scaling ratios. It follows that the original scaling ratios are close. $\qquad \square$

Next, in Theorems 2.31 and 2.34 below, we apply the construction of the proof of Lemma 2.28 in the following way. Let the functions f and g be

defined as in (2.49). We choose $\varepsilon > 0$ and we first determine T so large
that we have

$$2\pi\sigma_0 N r_N^{-\sigma_0} q^{-1/2N} < \varepsilon \qquad \text{for } q > T.$$

Then f is approximated by g to within an error of ε on the region

$$\left\{ s = \sigma + it: \ -\sigma_0 \le \sigma \le 1, \ |t| \le \pi q^{1+1/2N} \right\}.$$

That is, g approximates f well on $\left[q^{1/2N}\right]$ (positive and negative) periods
of g.

The next theorem completes the proof of the nonlattice part of Theorem 2.13.

Theorem 2.31. *Let $\mathcal{L}$ be a nonlattice self-similar string. Then there exists a sequence of simple complex dimensions of $\mathcal{L}$ approaching the line $\operatorname{Re} s = D$ from the left.*

Proof. Note that $\lim_{\sigma \to +\infty} f(\sigma + it) = 1$ and $\lim_{\sigma \to -\infty} |f(\sigma + it)| = \infty$, for any given real value of t. Given $t \in \mathbb{R}$, let

$$m(t) = \inf_{\sigma \in \mathbb{R}} |f(\sigma + it)| \tag{2.50}$$

denote the infimum of $|f(s)|$ on the horizontal line $s = \sigma + it$ ($\sigma \in \mathbb{R}$). Thus $0 \le m(t) \le 1$, $m(0) = 0$, and in general, $m(t) = 0$ if and only if $f(\sigma + it) = 0$ for some $\sigma \in \mathbb{R}$. In particular, the function m does not vanish identically. Choose t_0 such that $m(t) \ne 0$ for $0 < t \le t_0$. Take an approximation g of f to within $\varepsilon \le \frac{1}{3} m(t_0)$. This means in particular that the real value D' for which g vanishes is very close to D, and that D and D' are the unique zeros of f and g, respectively, in the region

$$-\sigma_0 \le \sigma \le 1, \ |t| \le t_0.$$

By Rouché's Theorem [Ahl, Corollary to Theorem 18, p. 153], D and D' actually lie within the 'circle' $\{s \in \mathbb{C}: |g(s)| = \varepsilon\}$ in this region. Again by Rouché's Theorem, the translate of this 'circle' by a period of g also contains a zero of f, at least for the first $\left[q^{1/2N}\right]$ (positive and negative) periods of g. Now we can make sure that this 'circle' is arbitrarily small around D, by choosing ε small. This shows that f has a sequence of simple zeros approaching $\operatorname{Re} s = D$, and this proves the theorem, since (by (2.49a) and Corollary 2.4) zeros of f are complex dimensions of $\mathcal{L}$ (with the same multiplicity). $\qquad\square$

Remark 2.32. Note that the plot of the complex dimensions of the golden string $\mathcal{L}$ given in Figure 2.6 on page 35 is in agreement with Theorem 2.31. As was mentioned earlier, further application of the method of proof of Theorems 2.26, 2.29 and of Lemma 2.28 would provide increasingly more accurate plots of $\mathcal{D}$; it would yield, in particular, two sequences of complex dimensions of $\mathcal{L}$ (symmetric with respect to the real axis) converging from the left to the vertical line $\operatorname{Re} s = D$.

The last two theorems of this section are of a technical nature and verify conditions that will be needed (in Sections 5.4 and 6.3.1) in order to apply the explicit formulas of Chapter 4 to the case of self-similar strings. (See Remark 5.9.)

The following theorem can be applied when one cannot choose a screen passing between $\operatorname{Re} s = D$ and the complex dimensions to the left of this line. (See especially Example 4.25 and Section 6.3.1.)

Theorem 2.33. *Let $\mathcal{L}$ be a nonlattice self-similar string with scaling ratios $r_1, \ldots, r_N$. Then there exists a screen S such that $\zeta_{\mathcal{L}}$ is bounded on S and all complex dimensions to the right of S are simple with uniformly bounded residue.*

Proof. Let $\mathcal{L}'$ be a lattice string with scaling ratios $r_1', \ldots, r_N'$, approximating $\mathcal{L}$. Let D be the dimension of $\mathcal{L}$, and assume that D is also a dimension of $\mathcal{L}'$. Since the complex dimensions of $\mathcal{L}'$ lie on finitely many vertical lines, there exists $\delta > 0$ such that $\mathcal{L}'$ has no complex dimensions with $D - 2\delta < \operatorname{Re} \omega < D$. Let $\omega = \alpha + i\gamma$ be a complex dimension of $\mathcal{L}$, with $D - \delta < \alpha < D$. Then $r_1^{\alpha+i\gamma} + \cdots + r_N^{\alpha+i\gamma} = 1$. Hence, by Rouché's Theorem, $r_1^{i\gamma} r_1'^s + \cdots + r_N^{i\gamma} r_N'^s = 1$ for s close to α. But this means that the coefficients $r_1^{i\gamma}, \ldots, r_N^{i\gamma}$ are close to 1. Again by Rouché's Theorem, the order of ω is the same as the order of D, which is 1. Since the derivatives are also approximated, it follows that the residue of $\zeta_{\mathcal{L}}$ at ω is close to that of $\zeta_{\mathcal{L}'}$ at D, say $|\operatorname{res}(\zeta_{\mathcal{L}}(s); \omega)| \leq 2 \operatorname{res}(\zeta_{\mathcal{L}'}(s); \omega)$.

We now construct a screen S as follows: Initially, we choose $S(t) = D - \delta/2 + it$. But each time that $S(t)$ comes within a distance of $\delta/4$ of a complex dimension of $\mathcal{L}$, we go around this complex dimension along the shortest arc of radius $\delta/4$ to the left or to the right. Since the residue of $\zeta_{\mathcal{L}}$ at ω is bounded, it follows that $\zeta_{\mathcal{L}}$ is bounded along this screen.   □

Theorem 2.34. *Let $\mathcal{L}$ be a self-similar string. Then there exists a sequence of positive numbers $T_1, T_2, \ldots$, tending to infinity, such that $|\zeta_{\mathcal{L}}(s)|$ is uniformly bounded from above on each horizontal line $\operatorname{Im} s = T_n$ and $\operatorname{Im} s = -T_n$, for $n = 1, 2, \ldots$.*

Proof. If $\mathcal{L}$ is a lattice string, then the existence of such a sequence $\{T_n\}_{n=1}^{\infty}$ follows easily from the fact that $\zeta_{\mathcal{L}}$ is periodic (with period $\mathbf{p}$, the oscillatory period).

If $\mathcal{L}$ is a nonlattice string, we proceed as follows. Let the function m be given by Equation (2.50), as in the proof of Theorem 2.31. Choose some t_0 such that $m(t_0) \neq 0$, and let $\varepsilon < \frac{1}{3} m(t_0)$. Construct a number q and a function g approximating f, as was explained before the statement of Theorem 2.31. Then on the line $s = \sigma + it_0$, g is bounded away from 0 by at least $\frac{2}{3} m(t_0)$. By periodicity, $|g(s)| \geq \frac{2}{3} m(t_0)$ on $\lceil q^{1/2N} \rceil$ different horizontal lines of the form

$$s = \sigma + it_0 + in\mathbf{p} \quad (n = -\frac{1}{2}\left[q^{1/2N}\right],\ldots,\frac{1}{2}\left[q^{1/2N}\right]).$$

But then, $|f(\sigma + it_0 + in\mathbf{p})|$ is bounded from below by $\frac{1}{3}m(t_0)$ on these lines. By choosing ε smaller, and consequently q larger, we find infinitely many lines $s = \sigma + iT_n$ on which $|f(s)|$ is bounded from below. Since $\zeta_{\mathcal{L}}(s) = (L(1 - R))^s/f(s)$, by Equation (2.3) and definition (2.49a), this implies the corresponding statement for $\zeta_{\mathcal{L}}(s)$.   □

Remark 2.35. For future reference, we note that the conclusion of Theorem 2.34 clearly applies to the generalized Cantor strings studied in Chapter 8 and used in Chapter 9. In fact, these generalized fractal strings are lattice strings (in a generalized sense) and hence their geometric zeta function is periodic (see Equation (8.2)). In conjunction with Remark 5.9 below, this justifies the application of our explicit formulas to these generalized Cantor strings (or to the associated generalized Cantor sprays) in Chapters 8 and 9.

3
Generalized Fractal Strings Viewed as Measures

In Section 3.1 of this chapter, we introduce the notion of generalized fractal string, viewed as a measure on the half-line. We will use this notion in Chapter 4 to formulate our explicit formulas which will be applied throughout the remaining chapters. In Section 3.2, we discuss the spectrum of a generalized fractal string, and in Section 3.3, we briefly discuss the notion of generalized fractal spray, which will be used in Chapters 7 and 9.

In Section 3.4, we specialize our discussion and study the properties of the measure associated with a self-similar string (defined as in Chapter 2). Although Section 3.4 is of interest in its own right, it may be omitted on a first reading since it will not be used in the rest of this book.

3.1 Generalized Fractal Strings

In this section, we shall consider more general fractal strings than those considered in Chapter 1 and in the earlier work on this subject (see, e.g., [Lap1–3, LapPol–3, LapMal–2, HeLap1–2]).

For a measure η, we denote by $|\eta|$ the total variation measure associated with η. Recall that $|\eta|$ is a positive measure and that $|\eta| = \eta$ if η is positive. (See Remark 3.3 below and [Coh, Chapter 4] or [Ru2, Chapter 6].)

Definition 3.1. (i) A *generalized fractal string* is either a local complex or a local positive measure η on $(0, \infty)$, such that

$$|\eta|(0, x_0) = 0$$

for some positive number x_0. (See Remark 3.3 below.)

(ii) The *counting function of the 'reciprocal lengths'*, or *geometric counting function* of η, is defined as $N_\eta(x) = \int_0^x d\eta$. If the measure η has atoms, it is necessary to specify how the endpoint is counted. Throughout this book, we will adopt the convention that x is counted 'half'; i.e.,

$$N_\eta(x) = \int_0^x d\eta := \frac{\eta(0, x) + \eta(0, x]}{2}. \tag{3.1}$$

(iii) The *dimension* of η, denoted $D = D_\eta$, is the abcissa of convergence of the Dirichlet integral $\zeta_{|\eta|}(\sigma) = \int_0^\infty x^{-\sigma} |\eta|(dx)$. In other words, it is the infimum of the real numbers σ such that the improper Riemann–Lebesgue integral $\int_0^\infty x^{-\sigma} |\eta|(dx)$ converges and is finite:

$$D = D_\eta = \inf \left\{ \sigma \in \mathbb{R} : \int_0^\infty x^{-\sigma} |\eta|(dx) < \infty \right\}. \tag{3.2}$$

(iv) The *geometric zeta function* of η is the function

$$\zeta_\eta(s) = \int_0^\infty x^{-s} \eta(dx), \tag{3.3}$$

for $\operatorname{Re} s > D_\eta$.

By convention, $D_\eta = \infty$ means that $x^{-\sigma}$ is not $|\eta|$-integrable for any σ, and $D_\eta = -\infty$ means that $x^{-\sigma}$ is $|\eta|$-integrable for all σ in $\mathbb{R}$. In this last case, ζ_η is a holomorphic function, defined by its Dirichlet integral (3.3) on the whole complex plane. (See, for example, [Pos, Sections 2–4] or [Wid].)

Note that if η is a continuous measure (i.e., $\eta(\{x\}) = 0$ for all $x > 0$), then $N_\eta(x) = \eta(0, x) = \eta(0, x]$. On the other hand, if η is discrete, say

$$\eta = \sum_m w_m \delta_{\{m\}},$$

then

$$N_\eta(x) = \sum_{m < x} w_m + \frac{w_x}{2},$$

with the convention that $w_x = 0$ if x is not one of the reciprocal lengths m.

Remark 3.2. Our point of view is more general than that of [JorLan2] in the sense that we allow for 'infinitesimal multiplicities', but less general in the sense that they allow for complex 'lengths'. Of course, the interpretation in [JorLan2] is not in terms of lengths of (generalized) fractal strings.

Remark 3.3. There is a simple technical point that needs to be clarified here. Indeed, since we have to use integrals of the type $\int_0^\infty f(x)\,\eta(dx)$, for suitable functions f (see, for example, Equation (3.3) above), some caution is necessary in dealing with complex-valued measures on $(0, \infty)$.

A *local positive measure* (or a locally bounded positive measure) is just a standard positive Borel measure on $(0, \infty)$ which satisfies the following local boundedness condition:

$$\eta(J) < \infty, \text{ for all bounded subintervals } J \text{ of } (0, \infty). \tag{3.4}$$

In that case, we have $\eta = |\eta|$ on every bounded Borel subset of $(0, \infty)$.

More generally, we will say, much as in [DoFr, Definition 6.1, p. 179], that a (complex-valued or else $[0, +\infty]$-valued) set function η on the half-line $(0, \infty)$ is a *local complex measure* on $(0, \infty)$ if, for every compact subinterval $[a, b]$ of $(0, \infty)$, the following conditions are satisfied: (i) $\eta(A)$ is well defined for any Borel subset A of $[a, b]$, and (ii) the restriction of η to the Borel subsets of $[a, b]$ is a complex measure on $[a, b]$ in the traditional sense. (See, e.g., [Coh, Chapter 4] or [Ru2, Chapter 6]. It follows, by [Ru2, Theorems 6.2 and 6.4, pp. 117 and 118], that η is of bounded total variation on $[a, b]$.) We then denote by $|\eta|$ the (local) *total variation measure* of η (defined on each bounded subinterval of $(0, \infty)$); see, e.g., [Coh, p. 126] or [Ru2, p. 116]. According to the aforementioned results in [Ru2, Theorems 6.2 and 6.4], $|\eta|$ is a finite and positive measure on each bounded subinterval of $(0, \infty)$.

Note that a local positive measure is simply a local complex measure which takes its values in $[0, +\infty]$ (rather than in $\mathbb{C}$).

In addition to [DoFr, §5.5 and §1.6], we refer to the later work [JohLap, Chapters 15–19, esp. Sections 17.6, 15.2.F and 19.1] and the relevant references therein, where similar issues have to be dealt with.

A reader unconcerned with such technicalities or else unfamiliar with the notion of complex measure may assume throughout that a generalized fractal string is a positive Borel measure on $(0, \infty)$ satisfying condition (3.4) above (and which does not carry mass near 0). In Section 7.3, however, we will use generalized fractal strings associated with local complex measures.

For simplicity, we will drop from now on the adjective 'local' when referring to the measure associated with a general fractal string.

As in Chapter 1, we are interested in the meromorphic continuation of ζ_η. We define the screen S and the window W as in Section 1.2.1: Let $r: \mathbb{R} \to [-\infty, D]$ be a bounded continuous function. The *screen*, S, is the curve

$$S: t \longmapsto r(t) + it \quad (t \in \mathbb{R}) \tag{3.5}$$

(see Figure 1.4 of Section 1.2.1), and the *window*

$$W = \{s \in \mathbb{C} : \operatorname{Re} s \geq r(\operatorname{Im} s)\} \tag{3.6}$$

is the part of the plane to the right of the screen S. We assume that ζ_η has a meromorphic continuation to a neighborhood of W. We also require that ζ_η does not have any pole on S. The poles of ζ_η inside W will be called the visible complex dimensions and the associated set is denoted by $\mathcal{D}_\eta(W)$; namely, *the set of visible complex dimensions* of η is

$$\mathcal{D} = \mathcal{D}_\eta(W) = \{\omega \in W : \zeta_\eta \text{ has a pole at } \omega\}. \tag{3.7}$$

Just as in the case of ordinary fractal strings, $\mathcal{D} = \mathcal{D}_\eta(W) \subset W$ is a discrete subset of $\mathbb{C}$. Hence, its intersection with any compact subset of $\mathbb{C}$ is finite. Moreover, since, by definition, ζ_η is holomorphic for $\mathrm{Re}\, s > D$, where D is the dimension of η, it follows that $\mathcal{D}$ is contained in the closed half-plane $\mathrm{Re}\, s \leq D$. Also, if η is a positive measure, then the exact counterpart of Remark 1.10 holds.

In applying our explicit formulas, obtained in the next chapter, it will sometimes be useful to change the location of the screen S, even for a fixed choice of η. (See, for example, Section 4.5.) However, when no ambiguity may arise regarding the choice of the string η and the location of the screen S (and hence of the associated window W), we will simply write $\mathcal{D}$ or $\mathcal{D}_\eta$ instead of $\mathcal{D}_\eta(W)$.

If $W = \mathbb{C}$, then $\mathcal{D} = \mathcal{D}_\eta(\mathbb{C})$ is defined without any ambiguity and is called *the set of complex dimensions* of η. In this case, we set formally $r \equiv -\infty$ and we do not define the screen.

Remark 3.4. Assume that W is symmetric with respect to the real axis (i.e., W is equal to $\overline{W}$, the complex conjugate of W). When η is a real-valued measure, $\zeta_\eta(s)$ is real-valued for s real and hence the set of complex dimensions $\mathcal{D}_\eta(W)$ is symmetric with respect to the real axis; namely, $\omega \in \mathcal{D}_\eta(W)$ if and only if $\overline{\omega} \in \mathcal{D}_\eta(W)$. (This was the case, of course, of all standard fractal strings considered in Chapters 1 and 2, since then, $\eta = \sum_{j=1}^\infty \delta_{\{l_j^{-1}\}}$.)

3.1.1 *Examples of Generalized Fractal Strings*

Our definition of a generalized fractal string includes our previous definition of Section 1.1: With an ordinary fractal string $\mathcal{L}$, composed of the sequence of lengths $\{l_j\}_{j=1}^\infty$ (counted with multiplicity), we associate the positive measure

$$\mu_\mathcal{L} = \sum_{j=1}^\infty \delta_{\{l_j^{-1}\}}. \tag{3.8}$$

Here and in the following, we use the notation $\delta_{\{x\}}$ for the point mass (or Dirac measure) at x; i.e., for a set $A \subset (0, \infty)$, $\delta_{\{x\}}(A) = 1$ if $x \in A$ and $\delta_{\{x\}}(A) = 0$ otherwise.

In the geometric situation, as considered in Chapters 1 and 2, η is a discrete measure with integer multiplicities, because each length is repeated an integer number of times. In our theory, however, we will often need to consider discrete strings

$$\eta = \sum_l w_l \delta_{\{l^{-1}\}}$$

with noninteger multiplicities w_l. In fact, this was one of our initial motivations for introducing the notion of generalized fractal strings and for viewing them as measures.

An example of such discrete nongeometric strings is provided by the *generalized Cantor strings*: For $1 < b < a$, we define a string consisting of lengths a^{-n} with multiplicity b^n. The associated measure is

$$\sum_{n=0}^{\infty} b^n \delta_{\{a^n\}}. \tag{3.9}$$

If b is integral, this is an ordinary fractal string. For arbitrary b, we will study these generalized Cantor strings in Chapter 8 and use them in Chapter 9.

We note two additional examples: The *harmonic string* (introduced in [Lap2, Example 5.4(ii), pp. 171–172] and [Lap3, pp. 144–145]) is given by the positive measure[1]

$$h = \sum_{j=1}^{\infty} \delta_{\{j\}}. \tag{3.10}$$

This string does not have finite total length. In fact, its lengths are $1, 1/2, 1/3, \ldots, 1/n, \ldots$, each counted with multiplicity one, and hence the total length of h is $\sum_{n=1}^{\infty} 1/n = \infty$. Its dimension is 1. Since, by definition, $\zeta_h(s) = \sum_{j=1}^{\infty} j^{-s}$ (for $\operatorname{Re} s > 1$), the associated geometric zeta function is equal to the Riemann zeta function (as was noted in [Lap2, p. 171]); namely,

$$\zeta_h(s) = \zeta(s) \qquad (s \in \mathbb{C}). \tag{3.11}$$

Secondly, the *prime string* (which we now introduce for the first time in this context) is defined by the positive measure

$$\mathfrak{P} = \sum_{m \geq 1, \, p} (\log p) \, \delta_{\{p^m\}}, \tag{3.12}$$

where p runs over all prime numbers. Note that $\mathfrak{P}$ is not an ordinary fractal string because the reciprocal lengths p^m (that is, the prime powers)

---

[1]The harmonic string could also be called the *Riemann string*, as was recently suggested to the first author by Victor Kac.

have noninteger multiplicity $\log p$. Next, we use (and reinterpret) a well known identity (see, e.g., [In, Eq. (14), p. 17] or [Pat, p. 9]). By logarithmic differentiation of the Euler product representation of $\zeta(s)$,

$$\zeta(s) = \prod_p \frac{1}{1 - p^{-s}}, \qquad (3.13)$$

valid for $\operatorname{Re} s > 1$, we see that $-\zeta'(s)/\zeta(s) = \sum_{m \geq 1, p} (\log p) \, p^{-ms}$, and hence that the geometric zeta function of $\mathfrak{P}$ is given by

$$\zeta_{\mathfrak{P}}(s) = -\frac{\zeta'(s)}{\zeta(s)} \qquad (s \in \mathbb{C}). \qquad (3.14)$$

Thus this string is one-dimensional and its complex dimensions are the zeros of the Riemann zeta function, counted without multiplicity, and the simple pole at $s = 1$.

Recall that the zeros of ζ consist of the critical (or nontrivial) zeros, located in the critical strip $0 \leq \operatorname{Re} s \leq 1$, and the trivial zeros, which are simple and located at the even negative integers, $-2, -4, -6, \ldots$. It is well known that ζ does not have any zero on the vertical line $\operatorname{Re} s = 1$ (see, e.g., [In, Theorem 10, p. 28]), and hence also on $\operatorname{Re} s = 0$. The Riemann Hypothesis states that the critical zeros of ζ are all located on the critical line $\operatorname{Re} s = 1/2$. See, for example, [Ti], [Edw] or [Da].

3.2   The Frequencies of a Generalized Fractal String

In general, there is no clear interpretation for the frequencies of a generalized fractal string. Therefore, we simply adopt the following definition, motivated by the case of an ordinary fractal string (see Section 1.3 and Remark 3.6 below).

Definition 3.5. For a generalized fractal string ℓ, the *spectral measure*[2] ν of ℓ is defined by

$$\nu(A) = \ell(A) + \ell\left(\frac{A}{2}\right) + \ell\left(\frac{A}{3}\right) + \ldots, \qquad (3.15)$$

for each bounded (Borel) set $A \subset (0, \infty)$.

The *spectral zeta function of ℓ* is defined as the geometric zeta function of the measure ν.

---

[2] *Caution*: This should not be mistaken for the notion of spectral measure encountered in the spectral theory of self-adjoint operators (see, e.g., [ReSi1]), which will not be used in this book.

Note that the sum defining $\nu(A)$ is finite because A is bounded and $|\ell|$ is assumed to have no mass near 0. Indeed, choose k large enough so that $k^{-1}A \subset (0, x_0)$, where $|\ell|(0, x_0) = 0$. Then $\ell\left(k^{-1}A\right) = 0$.

For notational simplicity (see the comment following Definition 1.12), we do not explicitly indicate the dependence of ν on ℓ. In particular, ζ_ν and N_ν, the associated spectral counting function, depend on ℓ.

We will give an alternative expression for ν in Equation (3.19) below, where we should set $\eta = \ell$. By abuse of language, we will also consider ν as a generalized fractal string.[3] This way, we can conveniently formulate our explicit formulas for the geometric $(\eta = \ell)$ as well as the spectral $(\eta = \nu)$ situation.

Remark 3.6. When $\ell = \sum_{j=1}^{\infty} \delta_{\{l_j^{-1}\}}$ is the measure associated with an ordinary fractal string $\mathcal{L} = (l_j)_{j=1}^{\infty}$ as in Section 1.3, then by Equation (1.35) of Theorem 1.13, ζ_ν coincides with the spectral zeta function of $\mathcal{L}$, as defined by Equation (1.32). Also, in view of (3.1), (3.15) and footnote 2 of Chapter 1, N_ν coincides with the spectral counting function of $\mathcal{L}$, as defined by Equation (1.31).

Much as in the proof of Theorem 1.13, one shows that $\zeta_\nu(s)$, the spectral zeta function, is obtained by multiplying $\zeta_{\mathcal{L}}(s)$ by the Riemann zeta function. In other words,

$$\zeta_\nu(s) = \int_0^{\infty} x^{-s}\, \nu(dx) = \zeta_\ell(s) \cdot \zeta(s), \qquad (3.16)$$

where $\zeta(s) = \sum_{n=1}^{\infty} n^{-s}$ is the Riemann zeta function.

Definition 3.7. The *convolution*[4] of two strings η and η' is the measure $\eta * \eta'$ defined by

$$\int f\, d(\eta * \eta') = \iint f(xy)\, \eta(dx)\eta'(dy). \qquad (3.17)$$

One easily checks that

$$\zeta_{\eta * \eta'}(s) = \zeta_\eta(s)\zeta_{\eta'}(s). \qquad (3.18)$$

Thus the spectral zeta function of a string η is simply the zeta function associated with the string $\eta * h$. Indeed, in view of (3.15) and (3.17), one can easily check that the spectral measure of η is given by

$$\nu = \eta * h, \qquad (3.19)$$

---

[3]Indeed, one checks that $|\nu|(0, x_0) = 0$ if $|\ell|(0, x_0) = 0$. Thus $|\nu|$ does not have mass near 0 and so ν is a local measure on $(0, \infty)$.

[4]This is a multiplicative (rather than an additive) convolution of measures on $\mathbb{R}_+^*$. It is called the tensor product in [JorLan2].

where h is the harmonic string defined by formula (3.10) above.

In particular, the spectral measure ν of an ordinary fractal string $\mathcal{L} = \{l_j\}_{j=1}^{\infty}$, viewed as the measure $\eta = \sum_{j=1}^{\infty} \delta_{\{l_j^{-1}\}}$, is given by

$$\nu = \sum_f w_f^{(\nu)} \delta_{\{f\}}$$

$$= \eta * h \tag{3.20}$$

$$= \sum_{n,j=1}^{\infty} \delta_{\{n \cdot l_j^{-1}\}},$$

in agreement with (1.29). Here, f runs over the (distinct) frequencies of $\mathcal{L}$ and $w_f^{(\nu)}$ denotes the multiplicity of f, as in (1.30). Note that in view of (3.18), we recover Equation (1.32) (respectively, (1.35)) from the first (respectively, second) equality of (3.20).

Example 3.8 (The frequencies of the prime string). We compute in two different ways the 'frequencies' of the prime string $\mathfrak{P}$, defined by (3.12). We will first do this by evaluating the spectral zeta function, $\zeta_\nu(s) = \zeta_{\nu,\mathfrak{P}}(s)$, of $\mathfrak{P}$.

According to Equations (3.14) and (3.16), we have

$$\zeta_\nu(s) = \zeta_{\mathfrak{P}}(s) \cdot \zeta(s) = -\frac{\zeta'(s)}{\zeta(s)} \cdot \zeta(s) = -\zeta'(s),$$

where $\zeta'(s)$ is the derivative of the Riemann zeta function.

Next, since $\zeta(s) = \sum_{n=1}^{\infty} n^{-s}$ for $\operatorname{Re} s > 1$, we obtain the expression (also for $\operatorname{Re} s > 1$),

$$\zeta_\nu(s) = -\zeta'(s) = \sum_{n=1}^{\infty} (\log n) \, n^{-s}. \tag{3.21}$$

In view of Equation (3.16), we deduce from (3.21) that

$$\nu = \sum_{n=1}^{\infty} (\log n) \, \delta_{\{n\}}. \tag{3.22}$$

Hence, the prime string $\mathfrak{P}$ has for frequencies all the positive integers $1, 2, \ldots, n, \ldots$, but now, in contrast to the spectrum of an ordinary fractal string, the frequency n has noninteger multiplicity $\log n$. (Note that both the lengths and the frequencies of an ordinary fractal string have integer multiplicities. By (3.12) and (3.22), this is not so for $\mathfrak{P}$.)

Actually, it is instructive to recover this result in another way, by determining directly the spectral measure ν associated with $\mathfrak{P}$. Namely, in view of formula (3.17), we have

$$\delta_{\{x\}} * \delta_{\{y\}} = \delta_{\{xy\}}. \tag{3.23}$$

Then, in view of formulas (3.10), (3.12) and (3.19), we have, successively

$$\begin{aligned}
\nu &= \sum_{m\geq 1,\, p} (\log p)\, \delta_{\{p^m\}} * \sum_{k\geq 1} \delta_{\{k\}} \\
&= \sum_{m,k\geq 1,\, p} (\log p)\, \delta_{\{p^m \cdot k\}} \\
&= \sum_{n\geq 1} \delta_{\{n\}} \sum_{m\geq 1,\, p:\ p^m\,|\,n} \log p \\
&= \sum_{n\geq 1} (\log n)\, \delta_{\{n\}},
\end{aligned} \tag{3.24}$$

as was found in (3.22). Here, as before, p runs over all prime numbers. Note that to go from the third to the last line of (3.24), we have used the unique factorization of integers into prime powers in the form $\sum_{m\geq 1,\, p:\ p^m\,|\,n} \log p = \log n$.

In closing this example, we mention that it is the type of calculation performed in (3.24)—along with the conceptual difficulties associated with the notion of frequency with noninteger multiplicity—which led us to introduce the formalism of generalized fractal strings presented in this chapter. As was alluded to earlier, the flexibility of the language of (local) measures allows us to deal in a natural way with nonintegral multiplicities (in the case of discrete measures, such as in Equation (3.12) or (3.22)) and even (in the case of continuous measures, as we will see, for instance, in Section 7.2) to formalize the intuitive notion of infinitesimal multiplicity.

Remark 3.9 (Adelic completion of the harmonic string). It is noteworthy that the harmonic string itself is the infinite convolution over all prime numbers of what one could call the 'elementary prime strings'. Define

$$h_p = \sum_{j=0}^{\infty} \delta_{\{p^j\}}, \tag{3.25}$$

for every prime number p. Then

$$h = \underset{p}{*}\, h_p, \tag{3.26}$$

where p ranges over all prime numbers. (Here, as in (3.17), $*$ denotes the multiplicative convolution of measures.) This corresponds to the Euler product of the Riemann zeta function, recalled in formula (3.13) above. Indeed,

$$\zeta_{h_p}(s) = \frac{1}{1 - p^{-s}}, \tag{3.27}$$

the p-th Euler factor of $\zeta(s)$, and for $\operatorname{Re} s > 1$,

$$\zeta_{*_p\, h_p}(s) = \prod_p \frac{1}{1 - p^{-s}} = \zeta(s). \tag{3.28}$$

To obtain $\zeta_{\mathbb{R}}(s) = \Gamma(s/2)\pi^{-s/2}$, the Euler factor at infinity of the completed Riemann zeta function, where Γ denotes the classical gamma function (see Appendix A), one has to convolve one more time with the continuous measure

$$h_\infty(dx) = 2e^{-\pi x^{-2}} \frac{dx}{x}, \qquad (3.29)$$

which is not a string in our sense, because it has positive mass near 0.

In summary, we have the following adelic decomposition (in the sense of Tate's thesis [Ta]) of the *completed harmonic string* $h_c := h_\infty * h$:

$$h_c = h_\infty * \left(\underset{p}{*} h_p \right), \qquad (3.30)$$

where h_∞ is the continuous measure defined by (3.29) and where the infinite convolution product runs over all prime numbers p. Correspondingly, the completed Riemann zeta function (as in [Ta], but defined slightly differently from either [Da, Chapter 12] or [Edw, §1.8]) is given by

$$\xi(s) = \zeta_{h_c}(s) = \pi^{-s/2}\Gamma(s/2)\zeta(s). \qquad (3.31)$$

The function ξ is meromorphic in all of $\mathbb{C}$, with (simple) poles at $s = 0$ and at $s = 1$. We leave it to the interested reader to investigate how the functional equation for ξ, namely,

$$\xi(s) = \xi(1 - s) \quad (s \in \mathbb{C}), \qquad (3.32)$$

translates in terms of the measure h_c.

It is noteworthy that the 'partition function' (or 'theta function') of h, defined as the Laplace transform of h,

$$\theta_h(t) := \int_0^\infty e^{-tx} h(dx), \qquad (3.33)$$

for $t > 0$, is given by the function

$$\theta_h(t) = \frac{1}{e^t - 1}. \qquad (3.34)$$

We refer to Section 5.2.3 below for further discussion of the partition function of a fractal string.

3.3 Generalized Fractal Sprays

Just as in Section 1.4, we can consider basic shapes B other than the unit interval, scaled (in a formal sense) by the generalized fractal string η. Thus we define the spectral zeta function of the generalized fractal spray of η on the basic shape B as[5]

$$\zeta_\nu(s) = \zeta_\eta(s) \cdot \zeta_B(s). \tag{3.35}$$

Actually, for our purposes, we will need to extend the notion of generalized fractal spray further by going beyond the geometric situation when the basic shape B is an actual region in some geometric space. That is, we sometimes define B only virtually by its associated spectral zeta function, $\zeta_B(s)$, which can be any given generalized Dirichlet series or Dirichlet integral. Then the spectral zeta function of such a spray (of η on B) is still given by (3.35). More precisely, if

$$\zeta_B(s) = \int_0^\infty x^{-s} \rho(dx) \tag{3.36}$$

for some measure ρ, then, by definition, the spectral measure of such a virtual generalized fractal spray is given by

$$\nu = \eta * \rho, \tag{3.37}$$

from which (by (3.18)) relation (3.35) follows.

This extension will allow us to investigate the properties of any zeta function (or generalized Dirichlet series) ζ_B, as we will see, in particular, in Sections 7.3 and 9.2.1.

3.4 The Measure of a Self-Similar String

In this section, we investigate some of the properties of the measure associated (as in Section 3.1) with a self-similar string, introduced in Chapter 2. As was mentioned in the introduction to this chapter, the present section—which is of independent interest—can be omitted on a first reading.

Let $\mathcal{L}$ be a self-similar string, as in Section 2.1, constructed with the scaling ratios $r_1, \ldots, r_N$ and normalized such that the first length is 1; i.e., the total length is $L = 1/(1 - R)$, by Remark 2.5. By Equations (3.8) and (2.2), the measure $\mu_{\mathcal{L}}$ associated with $\mathcal{L}$ is

$$\mu_{\mathcal{L}} = \sum_{\nu_1 \geq 0, \ldots, \nu_N \geq 0} \binom{\Sigma\nu}{\nu} \delta_{\{\mathbf{r}^{-\nu}\}}, \tag{3.38}$$

---

[5]When η is an ordinary fractal string $\mathcal{L}$, then we recover the notion of ordinary fractal spray (of $\mathcal{L}$ on B) from [LapPo3] and Section 1.4. Further, in view of (1.41), ζ_ν coincides with the spectral zeta function of an ordinary fractal spray, as defined by (1.38).

where the sum is over all N-tuples of nonnegative integers $\nu_1, \ldots, \nu_N$. Here and in the following, we use the multi-index notation

$$\Sigma\nu = \sum_{j=1}^{N} \nu_j, \tag{3.39a}$$

$$\binom{\Sigma\nu}{\nu} = \binom{\Sigma\nu}{\nu_1 \ldots \nu_N}, \tag{3.39b}$$

$$\mathbf{r}^{-\nu} = \prod_{j=1}^{N} r_j^{-\nu_j}, \tag{3.39c}$$

and we will use $\nu \geq 0$ for $\nu_1 \geq 0, \ldots, \nu_N \geq 0$.

For a scaling factor t, we let tA be the set $\{tx \colon x \in A\}$. Further, for a predicate P, we let δ_P be 1 if P is true, and 0 otherwise.

Theorem 3.10. *The measure $\mu_{\mathcal{L}}$ defined by (3.38) satisfies the following scaling property, which we call its* self-similarity property:[6]

$$\mu_{\mathcal{L}}(A) = \delta_{1 \in A} + \sum_{j=1}^{N} \mu_{\mathcal{L}}(r_j A), \tag{3.40}$$

for every subset A of $(0, \infty)$.

Moreover, $\mu_{\mathcal{L}}$ is completely characterized by this property. In other words, every generalized fractal string satisfying this property necessarily coincides with $\mu_{\mathcal{L}}$.

Proof. The measure of A is

$$\mu_{\mathcal{L}}(A) = \sum_{\nu \geq 0} \binom{\Sigma\nu}{\nu} \delta_{\mathbf{r}^{-\nu} \in A}, \tag{3.41}$$

and that of $r_j A$ is

$$\mu_{\mathcal{L}}(r_j A) = \sum_{\nu \geq 0} \binom{\Sigma\nu}{\nu} \delta_{\mathbf{r}^{-\nu} \in r_j A}$$

$$= \sum_{\nu \geq 0} \binom{\Sigma\nu}{\nu} \delta_{\mathbf{r}^{-\nu} r_j^{-1} \in A}$$

$$= \sum_{\nu \geq 0, \, \nu_j \geq 1} \binom{\Sigma\nu - 1}{\nu_1, \ldots, \nu_j - 1, \ldots, \nu_N} \delta_{\mathbf{r}^{-\nu} \in A}.$$

---

[6] *Caution:* The measure $\mu_{\mathcal{L}}$ on $(0, \infty)$ is *not* a self-similar measure, in the sense encountered in the literature on fractal geometry; compare, e.g., [Hut; Fa1, §8.3] and [Str1–2, Lap5–6].

We sum these expressions over j and use the following generalization for multinomial coefficients of the usual property of binomial coefficients:

$$\sum_{j=1}^{N} \delta_{\nu_j \geq 1} \binom{\Sigma \nu - 1}{\nu_1, \ldots, \nu_{j-1}, \nu_j - 1, \nu_{j+1}, \ldots, \nu_N} = \binom{\Sigma \nu}{\nu},$$

where only ν_j has been decreased by one. We find

$$\sum_{j=1}^{N} \mu_{\mathcal{L}}(r_j A) = \sum_{j=1}^{N} \sum_{\nu \geq 0} \binom{\Sigma \nu - 1}{\nu_1, \ldots, \nu_j - 1, \ldots, \nu_N} \delta_{\nu_j \geq 1} \delta_{\mathbf{r}^{-\nu} \in A}$$

$$= \sum_{\nu \geq 0} \delta_{\mathbf{r}^{-\nu} \in A} \sum_{j=1}^{N} \delta_{\nu_j \geq 1} \binom{\Sigma \nu - 1}{\nu_1, \ldots, \nu_j - 1, \ldots, \nu_N}$$

$$= \sum_{\nu > 0} \binom{\Sigma \nu}{\nu} \delta_{\mathbf{r}^{-\nu} \in A}.$$

We recover formula (3.41) for $\mu_{\mathcal{L}}(A)$, up to the first term, corresponding to $\nu = (0, \ldots, 0)$. This term is $\delta_{1 \in A}$. Thus $\mu_{\mathcal{L}}$ satisfies relation (3.40), as claimed.

Next, let η be a string that satisfies this same relation. Then $\eta - \mu_{\mathcal{L}}$ is a measure η' that satisfies

$$\eta'(A) = \sum_{j=1}^{N} \eta'(r_j A). \tag{3.42}$$

We will show that this implies that $\eta' = 0$. Since every measure is determined by its values on bounded sets, we can assume that A is bounded. Since η' is a string, there exists x_0 such that $\eta' = 0$ on $(0, x_0)$. So we are done if A is contained in this interval. Suppose now that we know that $\eta' = 0$ on $(0, x_0 r_1^{-k})$. Then if A is contained in $(0, x_0 r_1^{-(k+1)})$, the sets $r_j A$ are contained in $(0, x_0 r_1^{-k})$, because $r_j \leq r_1$, for $j = 1, \ldots, N$. Hence $\eta'(A) = 0$ by relation (3.42). It follows that $\eta'(A) = 0$ for all bounded subsets A of $(0, \infty)$. Hence $\eta' = 0$ and so $\eta = \mu_{\mathcal{L}}$. This completes the proof of the theorem. □

3.4.1   Measures with a Self-Similarity Property

It follows from the above proof that a measure η on $(0, \infty)$ vanishes if it satisfies

$$\eta(A) = \sum_{j=1}^{N} \eta(r_j A) \tag{3.43}$$

and is supported away from 0. It is interesting to study how η is determined when it has mass near 0. Assume that η is absolutely continuous with respect to the measure dx/x, the Haar measure on the multiplicative group $\mathbb{R}_+^*$. Thus, there is a function f on $(0, \infty)$ such that $\eta(dx) = f(x)\, dx/x$. Then f satisfies the relation

$$f(x) = \sum_{j=1}^{N} f(r_j x), \qquad (3.44)$$

for all $x > 0$.

Choose some $r \in (0, 1)$ and write $r_j = r^{k_j}$ for positive real numbers k_j with $0 < k_1 \leq \ldots \leq k_N$. Let $g(t) = f(r^{-t})$, for $t \in \mathbb{R}$. The function g has the following periodicity property:

$$g(t) = \sum_{j=1}^{N} g(t - k_j), \qquad (3.45)$$

for all $t \in \mathbb{R}$. We cannot deal with such functions in general (see Problem 3.12 below), but we can handle them when η is a lattice string, in the sense of Definition 2.10.

In the lattice case, we choose the multiplicative generator, r, of η, and hence positive integers $k_N \geq \cdots \geq k_1$, such that $r_j = r^{k_j}$ for $j = 1, \ldots, N$. (See Definition 2.10.) Then g is determined on $\mathbb{Z}$, for example, if $g(0), g(1)$, $\ldots, g(k_N - 1)$ are chosen, and in general, g is determined on $\mathbb{R}$ when g is fixed on the interval $[0, k_N)$. We can solve the recursion by using the next proposition.

Proposition 3.11. *The solution space of the recursion relation*

$$a_n = \sum_{j=1}^{N} a_{n-k_j}$$

has dimension k_N. For each complex solution z of the polynomial equation

$$z^{k_N} = \sum_{j=1}^{N} z^{k_N - k_j}, \qquad (3.46)$$

of multiplicity $m(z)$, we obtain $m(z)$ solutions

$$n \mapsto n^q z^n \qquad (n \in \mathbb{Z}),$$

of the recursion relation, for each integer q between 0 and $m(z) - 1$.
Alternatively, for every $t \in \mathbb{R}$, we obtain $m(z)$ solutions

$$n \mapsto (n + t)^q z^{n+t} \qquad (n \in \mathbb{Z}).$$

These solutions (for fixed t, and all z and q, $0 \leq q \leq m(z) - 1$) form a basis of the solution space.

Thus, if $g(0), g(1), \ldots, g(k_N - 1)$ are chosen, there exist coefficients $c_{z,q}$ such that $g(n) = \sum_z \sum_{q=0}^{m(z)-1} c_{z,q} n^q z^n$, where z runs through the solutions of (3.46). More generally, for each $t \in [0, 1)$, if $g(t), g(t+1), \ldots, g(t+k_N-1)$ are known, then there exist coefficients $c_{z,q}(t)$ such that

$$g(n + t) = \sum_z \sum_{q=0}^{m(z)-1} c_{z,q}(t)(n + t)^q z^{n+t}.$$

We extend the definition of $c_{z,q}(t)$ by periodicity: Thus $c_{z,q}(t)$ is defined by $c_{z,q}(\{t\})$, where $\{t\} = t - [t]$ is the fractional part of t. The functions $c_{z,q}(t)$ have a Fourier series expansion: $c_{z,q}(t) = \sum_{n \in \mathbb{Z}} c_{z,q,n} e^{2\pi i n t}$. We thus find the following expansion for the general function with periodicity property (3.45):

$$g(t) = \sum_z \sum_{q=0}^{m(z)-1} \sum_{n \in \mathbb{Z}} c_{z,q,n} e^{2\pi i n t} t^q z^t. \tag{3.47}$$

This argument is justified if we impose some integrability condition on g. A convenient condition is that g is locally L^2; i.e., $|g|^2$ has a finite integral on every compact subset of the real line. Then the coefficients $c_{z,q}(t)$ are locally L^2, which is equivalent to the condition that for every z and q the sequence of Fourier coefficients $c_{z,q,n}$ is square-summable: For every z and q, $\sum_{n \in \mathbb{Z}} |c_{z,q,n}|^2 < \infty$. Since there are only finitely many z and q, this is equivalent to $\sum_z \sum_{q=0}^{m(z)-1} \sum_{n \in \mathbb{Z}} |c_{z,q,n}|^2 < \infty$.

In multiplicative terms, we obtain the following expansion for f:

$$f(x) = \sum_z \sum_{q=0}^{m(z)-1} \sum_{n \in \mathbb{Z}} c_{z,q,n} x^{in\mathbf{p}} (-\log_r x)^q z^{-\log_r x}, \quad \text{for } x > 0,$$

with $\mathbf{p} = 2\pi / \log r^{-1}$, the oscillatory period of the given lattice string. Recall from Theorem 2.13 that the complex dimensions ω of the lattice string $\mathcal{L}$ are the solutions of the equation $1 = \sum_{j=1}^{N} r^{k_j \omega}$ and that they lie periodically with period $\mathbf{p}$ on finitely many lines. Thus, to every z there corresponds an ω such that $z = r^{-\omega}$. Choose complex dimensions ω_z, one for each solution z. This means that we choose one ω_z on every line of poles of $\zeta_\mathcal{L}$. Observe that $z^{-\log_r x} = x^{\omega_z}$ and that $\omega_z + in\mathbf{p}$ runs over all complex dimensions. We find that

$$f(x) = \sum_\omega \sum_{q=0}^{m(\omega)-1} c_{\omega,q} (-\log_r x)^q x^\omega. \tag{3.48}$$

Here, ω runs over all complex dimensions of $\mathcal{L}$, and q runs from 0 to the multiplicity of ω minus one. The coefficients are determined by $c_{\omega,q} = c_{z,q,n}$ for $\omega = \omega_z + in\mathbf{p}$, and they are square-summable.

In general (that is, for a nonlattice string), there is no natural choice for r. We set $r = \exp(-1)$ and we formulate the following open problem:

Problem 3.12. Let $\mathcal{L}$ be a self-similar string, constructed with self-similarity ratios $r_1, \ldots, r_N$. Show that every function f that is locally L^2 on $(0, \infty)$ and satisfies (3.44) has an expansion of the following form:

$$f(x) = \sum_{\omega \in \mathcal{D}_{\mathcal{L}}(\mathbb{C})} \sum_{q=0}^{m(\omega)-1} c_{\omega,q} (\log x)^q x^{\omega}, \qquad \text{for } x > 0, \qquad (3.49)$$

where ω runs over the complex dimensions of the string $\mathcal{L}$ and $m(\omega)$ is the multiplicity of ω. Further, show that the coefficients $c_{\omega,q}$ in (3.49) are uniquely determined by f and that they are square-summable.

4

Explicit Formulas
for Generalized Fractal Strings

In this chapter, we present our (pointwise and distributional) explicit formulas for the lengths and frequencies of a fractal string. To unify the exposition, and with a view toward later applications, we formulate our results in the language of generalized fractal strings, introduced in Chapter 3. The explicit formulas express the counting function of the lengths or of the frequencies as a sum over the visible complex dimensions ω of the generalized fractal string η.

The plan of this chapter is as follows: After introducing some necessary notation and giving a heuristic proof of one of our formulas in Section 4.1, we discuss some technical preliminaries in Section 4.2. Our pointwise explicit formulas are proved in Section 4.3, while our distributional explicit formulas (and various useful extensions) are established in Section 4.4. Finally, in Section 4.5, we close the chapter by explaining how to apply our explicit formulas to reprove the classical Prime Number Theorem (with error term). Many additional examples illustrating our theory will be discussed in Chapter 5 and throughout the remainder of the book.

4.1 Introduction

Our explicit formulas will usually contain an error term. In most applications, this error term will be given by an integral over the vertical line $\operatorname{Re} s = \sigma_0$, for some value of σ_0. In general, it will be given by an integral over the screen S, introduced in Section 1.2.1.

We have defined $N_\eta(x)$, the counting function of the reciprocal lengths, in Definition 3.1(ii), formula (3.1). In our framework, it will be very useful to also consider the integrated versions of the counting function. We will denote by $N_\eta^{[k]}$ the k-th *primitive* (or k-th *antiderivative*) of N_η, vanishing at 0. Thus

$$N_\eta^{[k]}(x) = \int_0^x \frac{(x-y)^{k-1}}{(k-1)!} \, \eta(dy), \tag{4.1}$$

for $x > 0$ and $k = 1, 2, \dots$. In particular, $N_\eta = N_\eta^{[1]}$. Note that $N_\eta^{[k]}$ is a continuous function as soon as $k \geq 2$. In general, $N_\eta^{[k]}$ is $(k-2)$ times continuously differentiable for $k \geq 2$.

Formally, and this will be completely justified distributionally,

$$N_\eta^{[0]} = \frac{d}{dx} N_\eta = \eta. \tag{4.2}$$

The pointwise formula gives an expression for $N_\eta^{[k]}(x)$, valid for all $x > 0$ and all $k \geq 1$ sufficiently large.

The distributional formula describes η as a distribution: On a test function φ, η acts by

$$\langle \eta, \varphi \rangle = \int_0^\infty \varphi(x) \, \eta(dx). \tag{4.3}$$

The k-th *primitive* of this distribution will be denoted by $\mathcal{P}^{[k]}\eta$. More precisely, $\mathcal{P}^{[k]}\eta$ is the distribution given for all test functions φ by

$$\left\langle \mathcal{P}^{[k]}\eta, \varphi \right\rangle = (-1)^k \left\langle \eta, \mathcal{P}^{[k]}\varphi \right\rangle, \tag{4.4}$$

where $\mathcal{P}^{[k]}\varphi$ is the k-th primitive of φ that vanishes at infinity together with its derivatives. Thus, for a test function φ,

$$\left\langle \mathcal{P}^{[k]}\eta, \varphi \right\rangle = \int_0^\infty \int_y^\infty \frac{(x-y)^{k-1}}{(k-1)!} \varphi(x) \, dx \, \eta(dy), \tag{4.5}$$

and $\mathcal{P}^{[0]}\eta = \eta$.

For the general theory of distributions (or generalized functions, in the sense of Laurent Schwartz), we refer, e.g., to [Sch1–2, Hö2, ReSi1–2]. We recall from that theory that any locally integrable function f on $(0, \infty)$— that is, any measurable function on $(0, \infty)$ such that $\int_a^b |f(x)| \, dx$ is finite for every compact subinterval $[a, b]$ of $(0, \infty)$—defines a distribution in the obvious manner. Namely,

$$\langle f, \varphi \rangle = \int_0^\infty f(x)\varphi(x) \, dx, \tag{4.6}$$

for all test functions φ with compact support contained in $(0, \infty)$. This remark applies in particular, for each fixed $k \geq 1$, to the k-th integrated counting function, $f(x) = N^{[k]}(x)$, associated with an arbitrary generalized fractal string η.

Henceforth, we denote by $\mathrm{res}\,(g(s); \omega)$ the residue of a meromorphic function $g = g(s)$ at $s = \omega$.

4.1.1   Outline of the Proof

In this section, we discuss heuristically how the pointwise formula is established. We will derive a pointwise formula for η, even though, to make the argument rigorous, this formula has to be interpreted distributionally.

Our starting point is the basic formula, expressing the Dirac delta function at y as a Mellin transform,

$$\frac{1}{2\pi i} \int_{c-i\infty}^{c+i\infty} x^{s-1} y^{-s}\, ds = \delta_{\{y\}}(x),$$

for $c > 0$. This is Lemma 4.1 below, applied formally for $k = 0$. For the moment, we interpret $\delta_{\{y\}}$ as a function, which is why the present argument is not rigorous. Viewing the measure η as a superposition of shifted delta functions, we write

$$\eta(x) = \int_0^\infty \delta_{\{y\}}(x)\, \eta(dy) = \int_0^\infty \frac{1}{2\pi i} \int_{c-i\infty}^{c+i\infty} x^{s-1} y^{-s}\, ds\, \eta(dy).$$

For $c > D$, we interchange the order of integration and use the relation $\zeta_\eta(s) = \int y^{-s}\, \eta(dy)$ to deduce that

$$\eta(x) = \frac{1}{2\pi i} \int_{c-i\infty}^{c+i\infty} x^{s-1} \zeta_\eta(s)\, ds.$$

To obtain information about η, we need to push the line of integration $\mathrm{Re}\, s = c$ as far to the left as possible. When we push it to the screen S, we pick up a residue at each complex dimension ω of η. Thus, we obtain the 'density of lengths' (or 'density of geometric states') formula:

$$\eta = \sum_{\omega \in \mathcal{D}_\eta(W)} \mathrm{res}\,\left(x^{s-1} \zeta_\eta(s); \omega\right) + \frac{1}{2\pi i} \int_S x^{s-1} \zeta_\eta(s)\, ds. \tag{4.7a}$$

If the complex dimensions are simple, this becomes the formula

$$\eta = \sum_{\omega \in \mathcal{D}_\eta(W)} \mathrm{res}\,\left(\zeta_\eta(s); \omega\right) x^{\omega-1} + \frac{1}{2\pi i} \int_S \zeta_\eta(s) x^{s-1}\, ds. \tag{4.7b}$$

In order to turn this argument into a rigorous proof, we need, in particular, to assume suitable growth conditions on ζ_η, which will be stated at the beginning of Section 4.3.

4.1.2 Examples

We shall give two versions of the explicit formula. The first one is pointwise, in Section 4.3, and the second one distributional, in Section 4.4. We have in mind a number of examples to which we want to apply our explicit formulas:

1. The counting function of the lengths of a self-similar string. Then, we can choose $W = \mathbb{C}$ and therefore obtain an explicit formula involving all the complex dimensions of the string; see Sections 5.4.1 and 5.4.2.

2. The counting function of the frequencies of a self-similar string. Here, we shall obtain information up to a certain order (i.e., $W \neq \mathbb{C}$), due to the growth of the Riemann zeta function to the left of the critical strip $0 \leq \operatorname{Re} s \leq 1$; see Section 5.4.3.

3. More generally, the geometric and spectral counting functions (Sections 5.2.1, 5.2.2 and 5.3.1) and the geometric and spectral partition functions (Section 5.2.3) of an ordinary fractal string.

4. The fractal string of [Lap1, Example 5.1], also called a-string here; see Section 5.5.1.

5. The ordinary fractal string of [LapMa2], with which M. L. Lapidus and H. Maier gave a characterization of the Riemann Hypothesis. Again, we shall obtain information up to an error term since the geometric and spectral zeta functions of this string may not have an analytic continuation to all of $\mathbb{C}$. We will, however, improve significantly the error term obtained in [LapMa2]; see Chapter 7.

6. A continuous version of this string; see Section 7.2.

7. The geometric and spectral counting functions of a generalized Cantor string (Chapter 8 and Section 9.1), and of a generalized Cantor spray (Section 9.2.1).

8. The geometry and the spectrum of (generalized) fractal sprays; see Sections 5.6, 7.3 and 9.2.1.

9. The classical Prime Number Theorem and the Riemann–von Mangoldt explicit formula for the zeros of the Riemann zeta function (Section 4.5).

10. The Prime Number Theorem and the corresponding explicit formula for the primitive periodic orbits of the dynamical system naturally associated with a self-similar string (Section 5.4.4).

As was alluded to above, our explicit formulas can be applied either to the geometric zeta function of an ordinary fractal string $\mathcal{L}$, yielding explicit formulas for the counting functions of the lengths of $\mathcal{L}$, or to the spectral zeta function of $\mathcal{L}$, yielding explicit formulas for the counting functions of the frequencies of $\mathcal{L}$. See, for example, Chapters 5 and 8. The resulting explicit formulas show clearly the relationship between the counting function of the lengths and that of the frequencies. This relationship can be described as follows: The counting function of the frequencies is obtained by applying an operator, the spectral operator, to the explicit formula for the counting function of the lengths. (See Sections 5.1, 5.2, and especially 5.3.1.) This is already suggested by the results of [LapPo2] and [LapMa2], but it can be precisely formalized in our framework.

We now close this introductory section by providing some of the references on explicit formulas found in the literature and by commenting briefly on the relationships between the explicit formulas established in the present chapter and those obtained in these papers.

The first number-theoretic explicit formula is due to Riemann in his classic work [Rie1]. It was later extended and rigorously established by von Mangoldt in [vM1] and especially [vM2] (using, in particular, Hadamard's factorization for entire functions [Had1] and Euler's product representation (3.13) of the Riemann zeta function). A sample of additional references dealing with explicit formulas in number theory includes the works by Cramér [Cr], Guinand [Gui1-2], Delsarte [Del], Weil [Wei4-6], Barner [Bar], Haran [Haran], Schröter and Soulé [ScSo], Jorgenson and Lang [JorLan1, 3], Deninger [Den1-3], Deninger and Schröter [DenSc], as well as Rudnick and Sarnak [RudSar].

Further, an excellent introduction to the classical number-theoretic formulas (with varying degrees of sophistication) can be found in the books by Ingham [In, esp. Chapter IV], Edwards [Edw, esp. Chapter 3] (and the Appendix [Rie1]), Lang [Lan], Davenport [Da], Patterson [Pat, esp. Chapter 3], as well as that written by Manin and Panchishkin (and edited by Parshin and Shafarevich) [ParSh1, esp. §2.5].

In the above works, explicit formulas are given that converge pointwise or distributionally. Among the aforementioned papers, distributional-type explicit formulas (in a sense somewhat different from ours) can be found in [Wei4-6, Bar, Haran, Den1-3, DenSc]. The pointwise and distributional explicit formulas that we will give (in Sections 4.3 and 4.4 below) deviate in two ways from the usual explicit formulas found in number theory. On the one hand, we consider the density of states formula as being the more basic. This formula corresponds, for instance, to the derivative of the usual explicit formula of the prime number counting function, and exists only in a distributional sense (see Section 4.5). The integrated versions of this formula always exist as distributional formulas, and also sometimes as pointwise formulas. On the other hand, our explicit formulas will usually contain an error term (see Theorems 4.4 and 4.12). In the usual number-theoretic formulas, this error term is not present (or else is not considered as such) thanks to the use of the functional equation satisfied by the Riemann or other number-theoretic zeta functions.

Sometimes, as in the case of the counting function of the lengths of a self-similar string, this error term can be analyzed by pushing the screen away arbitrarily far to the left, and the resulting formula is an explicit formula in the classical sense. (See Theorems 4.8 and 4.17; also see Section 5.4 for the case of self-similar strings.) But already for the counting function of the frequencies of a self-similar string—and also, for example, when the geometric zeta function of a string does not have a meromorphic continuation to the whole complex plane—there is no way to avoid the presence of an error term, and our formulas are, in some sense, best possible. (See Section 5.4 and, for example, Section 5.5.) The usefulness of our explicit formulas depends very much on the possibility of giving a satisfactory analysis of this error term. We will provide such asymptotic estimates both for the pointwise error term (see Theorem 4.4, Equations (4.32) and (4.33)) and, in a suitable sense to be specified in Definition 4.22, for the distributional error term (see Theorems 4.12 and 4.23).

4.2   Preliminaries: The Heaviside Function

We refine here the basic lemma of [In, pp. 31 and 75; Da, p. 105]. This extension will be needed in the proof of Lemma 4.3, the truncated pointwise formula, which itself will be used to establish both the pointwise and the distributional formulas (Theorems 4.4, 4.8 and 4.12, 4.17).

Define the *k-th Heaviside function* for $k \geq 1$ by

$$H^{[k]}(x) = \begin{cases} \dfrac{x^{k-1}}{(k-1)!}, & \text{for } x > 0, \\ 0, & \text{for } x < 0 \text{ or } x = 0, \; k \geq 2, \\ \dfrac{1}{2}, & \text{for } x = 0, \; k = 1. \end{cases} \tag{4.8}$$

For $k \geq 2$, $H^{[k]}(x)$ is the $(k-1)$-th antiderivative (vanishing at $x = 0$) of the classical Heaviside function $H^{[1]}(x)$, equal to 1 for $x > 0$, to 0 for $x < 0$, and taking the value $\frac{1}{2}$ at $x = 0$.

Note that, in view of definition (4.1), we have

$$N_\eta^{[k]}(x) = \int_0^\infty H^{[k]}(x - t)\, \eta(dt). \tag{4.9}$$

For $k \geq 1$, we shall define the symbol $(s)_k$ by

$$(s)_k = s(s+1)\dots(s+k-1). \tag{4.10}$$

Lemma 4.1. *For $c > 0$, $T_+, -T_- \geq c + k - 1$, $x, y > 0$ and $k = 1, 2, \ldots$, the k-th Heaviside function is approximated as follows:*

$$H^{[k]}(x - y) = \frac{1}{2\pi i} \int_{c+iT_-}^{c+iT_+} x^{s+k-1} y^{-s} \frac{ds}{(s)_k} + E. \tag{4.11}$$

Putting $T_{\min} = \min\{T_+, |T_-|\}$ and $T_{\max} = \max\{T_+, |T_-|\}$, the error E of this approximation does not exceed in absolute value

$$x^{c+k-1} y^{-c} T_{\min}^{-k} \min \left\{ T_{\max}, \frac{1}{|\log x - \log y|} \right\}, \quad \text{if } x \neq y, \tag{4.12a}$$

$$x^{k-1} T_{\min}^{-k} T_{\max}, \quad \text{if } x = y, \tag{4.12b}$$

$$((c + k - 1)2^{k-1} + T_{\max} - T_{\min}) x^{k-1} T_{\min}^{-k}, \quad \text{if } x = y, \, k \text{ is odd.} \tag{4.12c}$$

Proof. Let $x < y$, so that $H^{[k]}(x - y) = 0$. We consider the integral

$$\frac{1}{2\pi i} \int x^{s+k-1} y^{-s} \frac{ds}{(s)_k}$$

over the contour $c + iT_-, c + iT_+, U + iT_+, U + iT_-, c + iT_-$, for a large positive value of U. The integral over the left side equals $-E$, and we want to show that it is small. By the Theorem of Residues [Ahl, Theorem 17, p. 150], the contour integral vanishes. Hence the integral over the left side equals the value of the integral over the contour composed of the upper, lower and right-hand sides. The integral over the right-hand side is bounded as follows:

$$\left| \frac{1}{2\pi i} \int_{U+iT_-}^{U+iT_+} x^{s+k-1} y^{-s} \frac{ds}{(s)_k} \right| \leq \frac{1}{2\pi} \int_{T_-}^{T_+} x^{U+k-1} y^{-U} \frac{dt}{|(U + it)_k|}$$

$$\leq \frac{T_+ - T_-}{2\pi} x^{U+k-1} y^{-U} U^{-k}.$$

Further, the integral over the upper side satisfies the inequality

$$\left| \frac{1}{2\pi i} \int_{c+iT_+}^{U+iT_+} x^{s+k-1} y^{-s} \frac{ds}{(s)_k} \right| \leq \frac{1}{2\pi} \frac{x^{k-1}}{T_+^k} \int_c^U (x/y)^\sigma \, d\sigma$$

$$\leq \frac{x^{c+k-1} y^{-c}}{2\pi |\log x - \log y|} T_+^{-k},$$

and, similarly, that over the lower side is bounded by

$$\frac{x^{c+k-1} y^{-c}}{2\pi |\log x - \log y|} |T_-|^{-k},$$

independent of U. On letting U go to infinity, the contribution of the right-hand side becomes arbitrarily small. The contribution of the upper and lower sides is now bounded by

$$\frac{x^{c+k-1}y^{-c}}{|\log x - \log y|}\frac{T_+^{-k}+|T_-|^{-k}}{2\pi} \leq \frac{x^{c+k-1}y^{-c}}{\pi|\log x - \log y|}T_{\min}^{-k}.$$

This proves the second inequality of (4.12a). To prove the first inequality of (4.12a), we integrate over the contour composed of the line segment from $c + iT_-$ to $c + iT_+$, a circular arc to the right with center the origin of radius $T_{\min}$, and a line segment from $c + iT_+$ to $c + iT_{\min}$, if $T_+ = T_{\max}$ (respectively, from $c - iT_{\min}$ to $c + iT_-$, if $|T_-| = T_{\max}$). Again, the integral over the line segment $c + iT_-$ to $c + iT_+$ vanishes up to the value of the integral over the circular part and the little line segment of the contour. These two integrals are bounded by

$$\frac{1}{2\pi}\cdot 2\pi T_{\min}\cdot x^{c+k-1}y^{-c}T_{\min}^{-k} = x^{c+k-1}y^{-c}T_{\min}^{1-k},$$

and

$$x^{c+k-1}y^{-c}(T_{\max}-T_{\min})T_{\min}^{-k}.$$

Note that this estimate is also valid for $x = y$, proving (4.12b).

The estimate of the error in the case $x > y$ is derived similarly, with respectively a rectangular and a circular contour going to the left. The value of the contour integral requires some consideration now. The integrand has simple poles at the points $0, -1, -2, \dots, 1 - k$. The residue at $-j$ is

$$\frac{x^{-j+k-1}y^j}{(-j)\cdot(1-j)\cdot\cdots\cdot(-1)\cdot 1\cdot 2\cdot\cdots\cdot(k-1-j)}$$

$$= \frac{1}{(k-1)!}x^{k-1-j}(-y)^j\binom{k-1}{j}.$$

Hence the sum of the residues is $\frac{(x-y)^{k-1}}{(k-1)!} = H^{[k]}(x-y)$.

It remains to derive the better order in T_+ and $|T_-|$ when k is odd and $x = y$. For simplicity of exposition, we assume that $T_+ = T_{\max}$. By a direct computation,

$$\frac{1}{2\pi i}\int_{c+iT_-}^{c+iT_+}x^{k-1}\frac{ds}{(s)_k} = \frac{x^{k-1}}{2\pi}\int_{T_-}^{T_+}\frac{dt}{(c+it)_k}$$

$$= \frac{x^{k-1}}{\pi}\int_0^{T_{\min}}\frac{\operatorname{Re}(c+it)_k}{|(c+it)_k|^2}\,dt + \frac{x^{k-1}}{2\pi}\int_{T_{\min}}^{T_+}\frac{dt}{(c+it)_k}.$$

For $k = 1$, the second term on the right is bounded by $(T_{\max}-T_{\min})T_{\min}^{-1}$. Further, the first term on the right is

$$\frac{1}{\pi}\int_0^{T_{\min}}\frac{c}{c^2+t^2}\,dt = \frac{1}{\pi}\int_0^{T_{\min}/c}\frac{du}{1+u^2}$$

$$= \frac{1}{2}-\frac{1}{\pi}\int_{T_{\min}/c}^\infty\frac{du}{1+u^2},$$

which differs from $1/2$ by at most $c/T_{\min}$.

For $k \geq 3$, the error is given by

$$\frac{x^{k-1}}{\pi} \int_{T_{\min}}^{\infty} \frac{\operatorname{Re}(c+it)_k}{|(c+it)_k|^2} \, dt + \frac{x^{k-1}}{2\pi} \int_{T_{\min}}^{T+} \frac{dt}{(c+it)_k}. \tag{4.13}$$

The last integral is bounded by $(T_{\max} - T_{\min})T_{\min}^{-k}x^{k-1}$. Next, we expand $(c+it)_k$ in powers of t:

$$(c+it)_k = (it)^k + \sum_{j=0}^{k-1} a_j (it)^j,$$

where a_j is the sum of all products of $k - j$ factors from $c, c+1, \ldots, c+k-1$. Hence for odd k, $\operatorname{Re}(c+it)_k = \sum_{j=0, \, j \text{ even}}^{k-1} a_j (it)^j$. One checks that a_j is bounded by $(c+k-1)^{k-j}\binom{k}{j} \leq (c+k-1)^{k-j}2^{k-1}$. Thus we find

$$|\operatorname{Re}(c+it)_k| \leq \sum_{j=0}^{\frac{k-1}{2}} (c+k-1)^{k-2j}2^{k-1}t^{2j}.$$

On the other hand, $|(c+it)_k|^2 \geq t^{2k}$. Thus the integrand in (4.13) is bounded by $\sum_{j=0}^{\frac{k-1}{2}} (c+k-1)^{k-2j}2^{k-1}t^{2j-2k}$. Integrating this function, we find the following upper bound for the error:

$$\sum_{j=0}^{\frac{k-1}{2}} (c+k-1)^{k-2j}2^{k-1}\frac{T_{\min}^{2j-2k+1}}{2k-2j-1} \leq (c+k-1)T_{\min}^{-k}2^{k-1}, \tag{4.14}$$

for $T > c+k-1$. This completes the proof of the lemma. $\qquad\square$

4.3 The Pointwise Explicit Formulas

In this section, we establish two different versions of our pointwise explicit formulas: one with error term (Theorem 4.4), which will be the most useful to us in this book, as well as one without error term (Theorem 4.8). The latter requires more stringent assumptions.

Let η be a generalized fractal string as in Section 3.1, with associated geometric zeta function denoted ζ_η (see Definition 3.1(iv)).

Recall from Section 3.1 and Figure 1.4 of Section 1.2 that the screen S is given as the graph of a bounded function r, with the horizontal and vertical axes interchanged:

$$S = \{r(t) + it : t \in \mathbb{R}\}.$$

We assume in addition that r is a Lipschitz continuous function; i.e., there exists a nonnegative real number, denoted by $\|r\|_{\text{Lip}}$, such that

$$|r(x) - r(y)| \le \|r\|_{\text{Lip}}|x - y| \text{ for all } x, y \in \mathbb{R}.$$

We associate with the screen the following finite quantities:

$$\sigma_l = \inf_{t \in \mathbb{R}} r(t), \tag{4.15a}$$

and

$$\sigma_u = \sup_{t \in \mathbb{R}} r(t). \tag{4.15b}$$

Further, recall from Section 3.1 that the window W is the part of the complex plane to the right of S; see formula (3.6).

Assume that ζ_η satisfies the following growth conditions:

There exist real constants $\kappa \ge 0$ and $C > 0$ and a sequence $\{T_n\}_{n \in \mathbb{Z}}$ of real numbers tending to $\pm\infty$ as $n \to \pm\infty$, with $T_{-n} < 0 < T_n$ for $n \ge 1$ and $\lim_{n \to +\infty} T_n/|T_{-n}| = 1$, such that

(H₁) For all $n \in \mathbb{Z}$ and all $\sigma \ge r(T_n)$,

$$|\zeta_\eta(\sigma + iT_n)| \le C \cdot |T_n|^\kappa, \tag{4.16}$$

(H₂) For all $t \in \mathbb{R}$, $|t| \ge 1$,

$$|\zeta_\eta(r(t) + it)| \le C \cdot |t|^\kappa. \tag{4.17}$$

Hypothesis **(H₁)** is a polynomial growth condition along horizontal lines (necessarily avoiding the poles of ζ_η), while hypothesis **(H₂)** is a polynomial growth condition along the vertical direction of the screen. We will need to assume these hypotheses to establish our (pointwise and distributional) explicit formulas with error term, Theorems 4.4 and 4.12 below.

Sometimes we can obtain an explicit formula without error term. (See Theorems 4.8 and 4.17.) In that case, in addition to **(H₁)**, we need to assume the following stronger form of hypothesis **(H₂)**:

There exists a positive number A and a sequence of screens r_m for $m \ge 1$, with $\sup_{t \in \mathbb{R}} r_m(t) \to -\infty$ as $m \to \infty$ and with a uniform Lipschitz bound $\sup_{m \ge 1} \|r_m\|_{\text{Lip}} < \infty$, such that

(H₂′) For all $t \in \mathbb{R}$ and all $m \ge 1$,

$$|\zeta_\eta(r_m(t) + it)| \le CA^{|r_m(t)|}(|t| + 1)^\kappa. \tag{4.18}$$

Definition 4.2. Given an integer $n \geq 1$, the *truncated screen* S_n is the part of the screen S restricted to the interval $[T_{-n}, T_n]$, and the *truncated window* W_n is the window W intersected with $\{s \in \mathbb{C} : T_{-n} \leq \mathrm{Im}\, s \leq T_n\}$. (See Figure 4.1 on page 84.)

The set of *truncated visible complex dimensions* is

$$\mathcal{D}(W_n) = \mathcal{D}_\eta(W_n) := \mathcal{D}_\eta(W) \cap \{s \in \mathbb{C} : T_{-n} < \mathrm{Im}\, s < T_n\}. \tag{4.19}$$

It is the set of visible complex dimensions of η with imaginary part between T_{-n} and T_n.

We begin by proving a technical lemma that summarizes the estimates that we will need in order to establish both the pointwise formulas in this section and the distributional formulas in the next section.

Note that given $\alpha, \beta \in \mathbb{R}$, $\alpha \leq \beta$, we have

$$\max\{x^\alpha, x^\beta\} = \begin{cases} x^\alpha, & \text{if } 0 < x < 1, \\ x^\beta, & \text{if } x \geq 1. \end{cases} \tag{4.20}$$

In the following, it will also be useful to keep in mind that, in view of (4.15), we have

$$\sigma_l \leq \sigma_u \leq D \tag{4.21}$$

and

$$S \subset \{s \in \mathbb{C} : \sigma_l \leq \mathrm{Re}\, s \leq \sigma_u\}. \tag{4.22}$$

Lemma 4.3 (Truncated pointwise formula). *Let $k \geq 1$ and let η be a generalized fractal string. Then, for all $x > 0$ and $n \geq 1$, the function $N_\eta^{[k]}(x)$ is approximated by*

$$\sum_{\omega \in \mathcal{D}_\eta(W_n)} \mathrm{res}\left(\frac{x^{s+k-1}\zeta_\eta(s)}{(s)_k}; \omega\right)$$

$$+ \frac{1}{(k-1)!} \sum_{\substack{j=0 \\ -j \in W \setminus \mathcal{D}_\eta}}^{k-1} \binom{k-1}{j}(-1)^j x^{k-1-j}\zeta_\eta(-j) \tag{4.23}$$

$$+ \frac{1}{2\pi i} \int_{S_n} x^{s+k-1}\zeta_\eta(s)\frac{ds}{(s)_k},$$

where $(s)_k$ is given by (4.10) and where S_n and $\mathcal{D}_\eta(W_n) = \mathcal{D}(W_n)$ are given as in Definition 4.2, while $\mathcal{D}_\eta = \mathcal{D}_\eta(W)(= \mathcal{D})$ is defined by (3.7).

More precisely, assume hypothesis $(\mathbf{H_1})$ and let $T_{\max} = \max\{T_n, |T_{-n}|\}$, $T_{\min} = \min\{T_n, |T_{-n}|\}$.[1] Let $c > D$. Then, for all $x > 0$ and all inte-

---

[1] For notational simplicity, we do not indicate explicitly the dependence of $T_{\max}$ and $T_{\min}$ on the integer n. This convention should be kept in mind when reading the proof of Theorems 4.4 and 4.8 below.

gers $n \geq 1$, the difference between $N_\eta^{[k]}(x)$ and the expression in (4.23) is bounded in absolute value by

$$d(x, n) := 2x^{k-1} T_{\min}^{-k} \left[|\eta|(\{x\}) \right.$$

$$\cdot \begin{cases} T_{\max} & \text{(for even } k\text{)} \\ (c + k - 1)2^{k-1} + T_{\max} - T_{\min} & \text{(for odd } k\text{)} \end{cases} \quad (4.24)$$

$$+ T_{\max} \cdot |\eta| \left((x(1 - T_{\min}^{-1/2}), x) \cup (x, x(1 + T_{\min}^{-1/2})) \right)$$

$$+ \left. x^c \zeta_{|\eta|}(c) T_{\min}^{1/2} + C T_{\max}^\kappa (c - \sigma_{\mathrm{l}}) \max\{x^c, x^{\sigma_{\mathrm{l}}}\} \right].$$

(See estimate (4.29) below.)

Further, for each point $s = r(t) + it$ ($|t| \geq 1$, $t \in \mathbb{R}$) and for all $x > 0$, the integrand in the integral over the truncated screen S_n occurring in (4.23) (namely, $x^{s+k-1} \zeta_\eta(s)/(s)_k$) is bounded in absolute value by

$$C x^{k-1} \max\{x^{\sigma_{\mathrm{u}}}, x^{\sigma_{\mathrm{l}}}\} |t|^{\kappa-k}, \quad (4.25)$$

when hypothesis $(\mathbf{H_2})$ holds, and by

$$C A^{-\sigma_{\mathrm{l}}} x^{k-1} \max\{x^{\sigma_{\mathrm{u}}}, x^{\sigma_{\mathrm{l}}}\} |t|^{\kappa-k}, \quad (4.25')$$

when the stronger hypothesis $(\mathbf{H_2'})$ holds. *(See Equations (4.20) and (4.21) above.)*

Proof. The proof is given in two steps. The first step consists of deriving an approximate expression for $N_\eta^{[k]}(x)$. For this, we consider the line integral

$$J(x, n) = \frac{1}{2\pi i} \int_{c+iT_{-n}}^{c+iT_n} x^{s+k-1} \zeta_\eta(s) \frac{ds}{(s)_k},$$

for some $c > D$. (See the contour $\mathcal{C}$ in Figure 4.1.) We substitute the expression of formula (3.3),

$$\zeta_\eta(s) = \int_0^\infty y^{-s} \eta(dy),$$

and interchange the order of integration. This interchange is justified since the integral is bounded by[2]

$$x^{c+k-1} \frac{1}{2\pi} \int_{T_{-n}}^{T_n} \int_0^\infty y^{-c} |\eta|(dy) \frac{dt}{c^k} \leq \frac{T_n - T_{-n}}{2\pi c^k} x^{c+k-1} \zeta_{|\eta|}(c).$$

---

[2]For simplicity, a reader unfamiliar with (local) complex-valued measures may wish to assume throughout the proofs given in this chapter that η is a positive measure on $(0, \infty)$ that is locally bounded; i.e., η is bounded for every bounded subinterval of $(0, \infty)$. Then, one may set $\eta = |\eta|$ in all the arguments presented here. See Remark 3.3.

We find that

$$J(x,n) = \int_0^\infty \frac{1}{2\pi i} \int_{c+iT_{-n}}^{c+iT_n} x^{s+k-1} y^{-s} \frac{ds}{(s)_k} \eta(dy). \tag{4.26}$$

In view of (4.9) and (4.26), the difference with $N_\eta^{[k]}(x)$ can now be bounded via an application of Lemma 4.1 (with $T_- = T_{-n}$ and $T_+ = T_n$):

$$\left| N_\eta^{[k]}(x) - J(x,n) \right|$$

$$\leq \int_0^\infty \left| H^{[k]}(x-y) - \frac{1}{2\pi i} \int_{c+iT_{-n}}^{c+iT_n} x^{s+k-1} y^{-s} \frac{ds}{(s)_k} \right| |\eta|(dy)$$

$$\leq \int_{y \neq x} x^{c+k-1} y^{-c} T_{\min}^{-k} \min \left(T_{\max}, \frac{1}{|\log x - \log y|} \right) |\eta|(dy)$$

$$+ x^{k-1} T_{\min}^{-k} |\eta|(\{x\}) \cdot \begin{cases} T_{\max} & (k \text{ even}), \\ T_{\max} - T_{\min} + 2^{k-1}(c+k-1) & (k \text{ odd}). \end{cases} \tag{4.27}$$

To obtain a good bound for the last integral on the right-hand side of (4.27), we split it as a sum of two integrals: one integral over y such that $y \leq x(1 - T_{\min}^{-1/2})$ or $y \geq x(1 + T_{\min}^{-1/2})$, where $|\log x - \log y| \gg T_{\min}^{-1/2}$, and another integral over the two open intervals in between, namely, $(x - x T_{\min}^{-1/2}, x)$ and $(x, x + x T_{\min}^{-1/2})$. The first integral is bounded by

$$x^{c+k-1} \zeta_{|\eta|}(c) T_{\min}^{-k} T_{\min}^{1/2}, \tag{4.28a}$$

and the second by

$$2x^{c+k-1} x^{-c} T_{\min}^{-k} T_{\max} |\eta| \left((x - x T_{\min}^{-1/2}, x) \cup (x, x + x T_{\min}^{-1/2}) \right). \tag{4.28b}$$

Thus, the difference $\left| N_\eta^{[k]}(x) - J(x,n) \right|$ is bounded by the sum of three terms: (4.28a), (4.28b) and the term

$$x^{k-1} T_{\min}^{-k} |\eta|(\{x\}) \cdot \begin{cases} T_{\max}, & \text{if } k \text{ is even}, \\ T_{\max} - T_{\min} + 2^{k-1}(c+k-1), & \text{if } k \text{ is odd}. \end{cases} \tag{4.28c}$$

Next we replace the right contour in Figure 4.1 (that is, the line segment from $c + iT_{-n}$ to $c + iT_n$) by $C_{\text{lower}} + S_n + C_{\text{upper}}$. Here, S_n is the truncated screen; i.e., the part of the screen for t going from T_{-n} to T_n, and the upper and lower parts of the contour are the horizontal lines $s = \sigma + iT_{\pm n}$, for $r(T_{\pm n}) \leq \sigma \leq c$. (See Figure 4.1.) By the Theorem of Residues [Ahl,

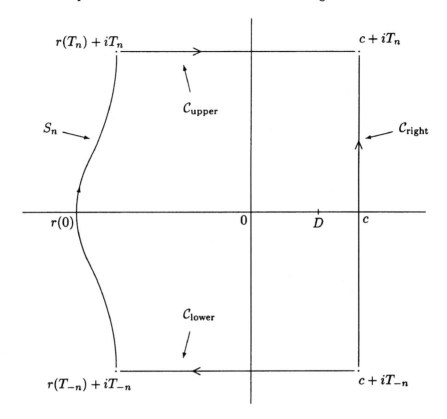

Figure 4.1: The contour $\mathcal{C}$.

Theorem 17, p. 150], we obtain

$$
\begin{aligned}
J(x,n) \\
&= \sum_{\omega \in \mathcal{D}(W_n)} \operatorname{res}\left(\frac{x^{s+k-1}\zeta_\eta(s)}{(s)_k};\omega\right) + \sum_{\substack{j=0 \\ -j\in W\setminus\mathcal{D}}}^{k-1} \operatorname{res}\left(\frac{x^{s+k-1}\zeta_\eta(s)}{(s)_k};-j\right) \\
&\quad + \frac{1}{2\pi i}\int_{S_n} x^{s+k-1}\zeta_\eta(s)\frac{ds}{(s)_k} + \frac{1}{2\pi i}\int_{\mathcal{C}_{\text{lower}}+\mathcal{C}_{\text{upper}}} x^{s+k-1}\zeta_\eta(s)\frac{ds}{(s)_k}.
\end{aligned}
$$

A computation similar to that performed in the proof of Lemma 4.1 above shows that the residue at $-j$ is equal to

$$
\frac{1}{(k-1)!}\binom{k-1}{j}(-1)^j x^{k-1-j}\zeta_\eta(-j),
$$

provided that $-j$ is not a pole of ζ_η; i.e., provided that $-j \notin \mathcal{D}$.

The integral over the upper side $\mathcal{C}_{\text{upper}}$ is bounded by

$$\left| \frac{1}{2\pi i} \int_{r(T_n)}^{c} x^{\sigma + iT_n + k - 1} \zeta_\eta(\sigma + iT_n) \frac{d\sigma}{(\sigma + iT_n)_k} \right| \le \frac{C}{2\pi} \int_{r(T_n)}^{c} x^{\sigma + k - 1} T_n^\kappa \frac{d\sigma}{T_n^k}$$

$$= T_n^{\kappa - k} x^{k - 1} \int_{r(T_n)}^{c} x^\sigma \, d\sigma$$

$$\le T_n^{\kappa - k} x^{k - 1 + c} (c - \sigma_1) \max\{x^c, x^{\sigma_1}\},$$

by hypothesis $(\mathbf{H_1})$. The integral over the lower side $\mathcal{C}_{\text{lower}}$ is bounded similarly.

In summary, we have now obtained the desired estimate:

$$\left| N_\eta^{[k]}(x) - \sum_{\omega \in \mathcal{D}(W_n)} \operatorname{res}\left(\frac{x^{s + k - 1} \zeta_\eta(s)}{(s)_k}; \omega \right) \right.$$

$$- \frac{1}{(k - 1)!} \sum_{\substack{j = 0 \\ -j \in W \setminus \mathcal{D}}}^{k - 1} \binom{k - 1}{j} (-1)^j x^{k - 1 - j} \zeta_\eta(-j) \qquad (4.29)$$

$$\left. - \frac{1}{2\pi i} \int_{S_n} x^{s + k - 1} \zeta_\eta(s) \frac{ds}{(s)_k} \right| \le d(x, n),$$

where $d(x, n)$ is given by (4.24).

Finally, at $s = r(t) + it$, we have $|(s)_k| \ge |t|^k$. Hence the integrand in (4.29) at $s = r(t) + it$ $(|t| \ge 1)$ is bounded by a constant times

$$x^{k - 1 + r(t)} |t|^{\kappa - k} \le x^{k - 1} \max\{x^{\sigma_u}, x^{\sigma_1}\} |t|^{\kappa - k}.$$

This constant is either C, when we assume $(\mathbf{H_2})$, or $C \cdot A^{-r(t)} \le C \cdot A^{-\sigma_1}$, when we assume $(\mathbf{H_2'})$. This completes the proof of Lemma 4.3. $\qquad \square$

We can now state the main result of this section, which will be used in most situations (in this and in the later chapters) to obtain pointwise explicit formulas for geometric or spectral counting functions. In the next section, we will obtain its distributional analogue, Theorem 4.12. (See also Theorem 4.20 along with Theorem 4.23.)

Theorem 4.4 (The pointwise explicit formula, with error term). *Let η be a generalized fractal string, satisfying hypotheses $(\mathbf{H_1})$ and $(\mathbf{H_2})$. (See Equations (4.16) and (4.17) above.) Let k be a positive integer such that $k > \kappa + 1$, where $\kappa \ge 0$ is the exponent occurring in the statement of $(\mathbf{H_1})$ and $(\mathbf{H_2})$. Then, for all $x > 0$, the pointwise explicit formula is given by* *the following equality:*[3]

$$N_\eta^{[k]}(x) = \sum_{\omega \in \mathcal{D}_\eta(W)} \mathrm{res}\left(\frac{x^{s+k-1}\zeta_\eta(s)}{(s)_k};\omega\right)$$

$$+ \frac{1}{(k-1)!} \sum_{\substack{j=0 \\ -j \in W \setminus \mathcal{D}_\eta}}^{k-1} \binom{k-1}{j}(-1)^j x^{k-1-j}\zeta_\eta(-j) + R_\eta^{[k]}(x).$$

$$(4.30)$$

Here, for $x > 0$, $R(x) = R_\eta^{[k]}(x)$ is the error term, given by the absolutely convergent integral

$$R(x) = R_\eta^{[k]}(x) = \frac{1}{2\pi i}\int_S x^{s+k-1}\zeta_\eta(s)\frac{ds}{(s)_k}. \qquad (4.31)$$

Further, for all $x > 0$, we have

$$R(x) = R_\eta^{[k]}(x) \le C(1 + \|r\|_{\mathrm{Lip}})\frac{x^{k-1}}{k-\kappa-1}\max\{x^{\sigma_u}, x^{\sigma_l}\} + C', \qquad (4.32)$$

where C is the positive constant occurring in $(\mathbf{H_1})$ and $(\mathbf{H_2})$ and C' is some suitable positive constant. The constants $C(1 + \|r\|_{\mathrm{Lip}})$ and C' depend only on η and the screen, but not on k. (Here, σ_l and σ_u are given by (4.15).)

In particular, we have the following pointwise error estimate:

$$R(x) = R_\eta^{[k]}(x) = O\left(x^{\sigma_u+k-1}\right), \qquad (4.33)$$

as $x \to \infty$. Moreover, if $r(t) < \sigma_u$ for all $t \in \mathbb{R}$ (i.e., if the screen lies strictly to the left of the line $\mathrm{Re}\,s = \sigma_u$), then $R(x)$ is of order less than x^{σ_u+k-1} as $x \to \infty$:

$$R(x) = R_\eta^{[k]}(x) = o\left(x^{\sigma_u+k-1}\right), \qquad (4.34)$$

as $x \to \infty$.

Remark 4.5. The proof of Theorem 4.4 does not give information about the nature of the convergence of the sum over the complex dimensions ω in (4.30). Therefore, we need to specify the meaning of $\sum_{\omega \in \mathcal{D}_\eta(W)}$ as $\lim_{n \to +\infty} \sum_{\omega \in \mathcal{D}_\eta(W_n)}$, where W_n is the truncated window, given by Definition 4.2. (The fact that this limit exists follows from the proof of the theorem.) The same remark applies to the sum occurring in formula (4.36) of Theorem 4.8 below.

---

[3]Recall that $\mathcal{D}_\eta = \mathcal{D}_\eta(W)(= \mathcal{D})$ denotes the set of visible complex dimensions of η (for the window W), and that $(s)_k$ is defined by (4.10). The complement of $\mathcal{D}_\eta(W)$ in W is denoted by $W \setminus \mathcal{D}_\eta$.

Further, the precise form of the term corresponding to ω in this sum will change depending on the multiplicity of the pole of ζ_η at $s = \omega$. In particular, a multiple pole will give rise to logarithmic terms. On the other hand, if $\omega \in \mathcal{D}_\eta(W)$ is a simple pole, then the corresponding term becomes

$$\operatorname{res}\left(\frac{x^{s+k-1}\zeta_\eta(s)}{(s)_k}; \omega\right) = \operatorname{res}\left(\zeta_\eta(s); \omega\right)\frac{x^{\omega+k-1}}{(\omega)_k}. \tag{4.35}$$

See Section 5.1 for a more elaborate discussion of the local term associated with ω.

Remark 4.6. In most applications, we have a lot of freedom for choosing the sequence $\{T_n\}_{n \in \mathbb{Z}}$ satisfying hypothesis $(\mathbf{H_1})$. If η is a real-valued (rather than a complex-valued) measure, we can always choose W to be symmetric and set $T_{-n} = -T_n$ for all $n \geq 1$. Then, in the sum over the complex dimensions ω in (4.30), we can collect the terms in complex conjugate pairs $\omega, \overline{\omega}$ (as is done in classical number-theoretic explicit formulas involving the Riemann zeta function [Da, Edw, In, Pat]).

Finally, we note that, in practice, we can often choose the screen S to be a vertical line (with equation $\operatorname{Re} s = \sigma_0$, say). Then $r(t) \equiv \sigma_0$ is clearly Lipschitz continuous and bounded, and we have $\sigma_u = \sigma_l = \sigma_0$.

Similar remarks apply to all explicit formulas in this chapter.

Remark 4.7. Note that our hypotheses $(\mathbf{H_1})$ and $(\mathbf{H_2})$ imply, in particular, that ζ_η has an analytic continuation to a neighborhood of $[\operatorname{Re} s \geq D]$. It is in general hard to determine whether this condition is satisfied. Indeed, this condition is essentially equivalent to the existence of a suitable explicit formula for the counting function N_η. Establishing such a result would entail developing a theory of almost periodic functions with amplitude of polynomial growth (rather than with constant amplitude as in [Bohr]). This theory would be of independent interest. It would be naturally motivated by our explicit formulas, in which the real parts of the underlying complex dimensions give rise to generalized Fourier series with variable amplitudes.

Proof of Theorem 4.4. For a given $n \geq 1$, we apply Lemma 4.3, the truncated pointwise formula, with $T_+ = T_n$, $T_- = T_{-n}$. For $k \geq 2$, the first term (4.28c) of $d(x, n)$ above tends to zero as $n \to \infty$. Also the second term (4.28a) and the middle term (4.28b) go to 0 as $n \to \infty$. (Note that $k \geq 2$ because k is an integer such that $k > \kappa + 1$, with $\kappa \geq 0$.) Finally, for $k > \kappa$, the last term tends to zero. All these terms tend to zero at the rate of some negative power of $T_{\min}$.[4]

Finally, the error term is absolutely convergent if $k > \kappa + 1$. Note that r is differentiable almost everywhere since it is assumed to be Lipschitz. Furthermore, since for almost every $t \in \mathbb{R}$, the derivative of $r(t) + it$ is bounded

---

[4] Recall from Lemma 4.1 that $T_{\max} = \max\{T_n, |T_{-n}|\}$ and $T_{\min} = \min\{T_n, |T_{-n}|\}$ both depend on the integer n.

by $1 + \|r\|_{\mathrm{Lip}}$, where $\|r\|_{\mathrm{Lip}}$ denotes the Lipschitz norm of r, and since

$$\int_1^\infty t^{\kappa-k}\, dt = \frac{1}{k-\kappa-1}$$

(because $\kappa - k < -1$, by assumption), the estimate (4.32) of the error term follows from (4.25). The constant C' on the right-hand side of (4.32) comes from a bound on the integral over the part of the screen for $-1 \le t \le 1$.

To obtain the better estimate (4.34) when S approximates the line $\mathrm{Re}\, s = \sigma_{\mathrm{u}}$, but stays to the left of it, we use a well known method to estimate (4.31); see, e.g., [In, pp. 33–34]. Given $\varepsilon > 0$, we want to show that (4.31) is bounded by $\varepsilon \cdot x^{\sigma_{\mathrm{u}}+k-1}$. Write this integral as the sum of the integral over the part of S contained in $-T \le \mathrm{Im}\, s \le T$, and the part of S contained in $|\mathrm{Im}\, s| > T$. Since the second integral is absolutely convergent, it is bounded by $\frac{1}{2}\varepsilon \cdot x^{\sigma_{\mathrm{u}}+k-1}$, provided we choose T sufficiently large. Then, since r has a maximum strictly less than σ_{u} on the compact interval $[-T, T]$, we can find $\delta > 0$ such that $r(t) \le \sigma_{\mathrm{u}} - \delta$ for $|t| \le T$. It follows that the integral over this part of the screen is of order $O\left(x^{\sigma_{\mathrm{u}}-\delta+k-1}\right)$ as $x \to \infty$. Hence for large x, it is less than $\frac{1}{2}\varepsilon \cdot x^{\sigma_{\mathrm{u}}+k-1}$. This proves that $R(x) = o\left(x^{\sigma_{\mathrm{u}}+k-1}\right)$ as $x \to \infty$. $\square$

As was noted above, the assumptions made in Theorem 4.4 imply that $k \ge 2$. However, in certain situations (namely, when ζ_η satisfies the stronger hypothesis $(\mathbf{H}_2')$ rather than $(\mathbf{H}_2)$), we can 'push the screen S to $-\infty$' and thereby obtain a pointwise explicit formula without error term, valid for a lower level $k > \kappa$ (rather than $k > \kappa + 1$, as in Theorem 4.4).

We will apply this result, in particular, to the geometric counting function (at level 1) of a self-similar string, as defined in Section 2.1. (See Section 5.4; we will see that we can set $\kappa = 0$, so that $k = 1$ will indeed be allowed in applying formula (4.36) below.)

Theorem 4.8 (The pointwise formula, without error term).   *Let η be a generalized fractal string satisfying hypotheses $(\mathbf{H}_1)$ and $(\mathbf{H}_2')$. (See Equations (4.16) and (4.18) above.) Let k be a positive integer such that $k > \kappa$. Then, for all $x > A$, the pointwise explicit formula is given by the following equality:*

$$N_\eta^{[k]}(x) = \sum_{\omega \in \mathcal{D}_\eta(W)} \mathrm{res}\left(\frac{x^{s+k-1}\zeta_\eta(s)}{(s)_k}; \omega\right)$$

$$+ \frac{1}{(k-1)!} \sum_{\substack{j=0 \\ -j \in W \setminus \mathcal{D}_\eta}}^{k-1} \binom{k-1}{j}(-1)^j x^{k-1-j}\zeta_\eta(-j). \tag{4.36}$$

Here, A is the positive number given by hypothesis $(\mathbf{H}_2')$ and $\kappa \ge 0$ is the exponent occurring in the statement of hypotheses $(\mathbf{H}_1)$ and $(\mathbf{H}_2')$.

Proof. For a fixed integer $n \geq 1$, we apply Lemma 4.3 with the screen r_m given by hypothesis $(\mathbf{H_2'})$. We now assume that $k > \kappa$ (instead of $k > \kappa + 1$ as in the proof of Theorem 4.4). We first let m tend to ∞, while keeping n fixed. Since the functions r_m have a uniform Lipschitz bound, the sequence of integrals over the screens tends to 0 as $m \to \infty$, provided $x > A$.

Then we let $n \to \infty$. For $k \geq 2$, in Equation (4.24), the first three terms of $d(x, n)$ tend to zero as $n \to \infty$, as we have seen in the proof of Theorem 4.4, and for $k > \kappa$, the last term also tends to zero.

Further, for $k = 1$, the first term (4.28c) of $d(x, n)$ tends to 0, by our assumption that $T_{\max}/T_{\min} \to 1$ as $n \to \infty$. (Note that this assumption implies that $(T_{\max} - T_{\min})/T_{\min} \to 0$ as $n \to \infty$.) The situation is more complicated for the middle term (4.28b), in case $k = 1$, since η can have a large portion of its mass close to x. However, by writing the interval $(x, 2x)$ as a disjoint union

$$(x, 2x) = \bigcup_{j=1}^{\infty} \left[x + \frac{x}{j+1}, x + \frac{x}{j} \right),$$

and noting that $|\eta|(x, 2x) = \sum_{j=1}^{\infty} |\eta| \left[x + \frac{x}{j+1}, x + \frac{x}{j} \right)$ is finite, we see that

$$|\eta| \left(x, x + \frac{x}{m} \right) = \sum_{j=m}^{\infty} |\eta| \left[x + \frac{x}{j+1}, x + \frac{x}{j} \right)$$

goes to 0 when $m \to \infty$, at some rate depending on the positive measure $|\eta|$ (on the given interval $(x, 2x)$). Thus also $|\eta|(x, x + x T_{\min}^{-1/2}) \to 0$ as $n \to \infty$. Similarly, we check that $|\eta|(x - x T_{\min}^{-1/2}, x) \to 0$ as $n \to \infty$. This completes the proof of Theorem 4.8. $\qquad\square$

Remark 4.9. As can be seen from the above proof, the full strength of the condition $\lim_{n \to +\infty} T_n/|T_{-n}| = 1$ is needed only when $k = 1$, which is only allowed in Theorem 4.8.

Remark 4.10. There is another situation in which we can push the screen S to $-\infty$ and obtain an explicit formula without error term. Indeed, sometimes the geometric zeta function satisfies a functional equation (in the sense of [JorLan3]). Then the theory of [JorLan3] (see also [JorLan1–2]) applies to yield an explicit formula without error term, but with a Weil term [Wei4, 6], coming from the 'fudge factor' involved in the statement of the functional equation in [JorLan3].

4.3.1 The Order of the Sum over the Complex Dimensions

In the applications given in Chapters 6, 7 and 9 below, we often want to single out the sum over the complex dimensions on the line $\operatorname{Re} s = D$, and estimate the sum over those to the left of $\operatorname{Re} s = D$. We also want the resulting error term to be of lower order. If we can choose a screen that makes all complex dimensions to the left of this line invisible, then we apply the estimate (4.34). But as Examples 4.25 and 4.26 show, we cannot always choose such a screen. In that case, we use the following argument:

Theorem 4.11. *Let*

$$\sum_{\operatorname{Re}\omega<D} a_\omega x^\omega (\log x)^{m_\omega} \tag{4.37}$$

be an absolutely convergent sum over the visible complex dimensions with real part less than D, arising from Theorem 4.4 or 4.8 (hence $a_\omega \in \mathbb{C}$ and $m_\omega \in \mathbb{N}$ for each ω). Then this sum is of order $o\left(x^D\right)$, as $x \to \infty$, and the corresponding sum

$$\sum_{\operatorname{Re}\omega=D} a_\omega x^\omega (\log x)^{m_\omega}$$

converges.

Proof. The (conditional) convergence of the sum over the complex dimensions with real part D follows from Theorem 4.4 or 4.8, and the assumption that (4.37) converges absolutely.

It remains to estimate the sum in (4.37). This is done by adapting the method of [In, pp. 82 and 33] as follows. Let $\varepsilon > 0$. Since (4.37) is absolutely convergent, the sum over ω with $|\operatorname{Im}\omega| > T$ is less than εx^D for T large enough. Next, since there are only finitely many visible complex dimensions ω with $|\operatorname{Im}\omega| < T$, their real parts attain a maximum $M < D$. Then (4.37) is bounded by $\varepsilon x^D + Cx^M < 2\varepsilon x^D$ for large enough x. This shows that the expression in (4.37) is $o\left(x^D\right)$, as $x \to \infty$. $\square$

4.4   The Distributional Explicit Formulas

In this section, we will give a distributional counterpart of the pointwise explicit formulas obtained in the previous section. We will also obtain several refinements of the distributional formula that will be needed in the rest of this book. (See Sections 4.4.2 and 4.4.3.)

We view η as a distribution, acting on a test function φ by $\int_0^\infty \varphi\,d\eta$. More generally, for $k \geq 1$, the *k-th primitive* $\mathcal{P}^{[k]}\eta$ of η is the distribution given by Equation (4.5) or, equivalently, (4.4). For $k \leq 0$, we extend this definition by differentiating $|k| + 1$ times the distribution $\mathcal{P}^{[1]}\eta$. Thus, in particular, $\mathcal{P}^{[0]}\eta = \eta$.

In the following,

$$(s)_k = \frac{\Gamma(s+k)}{\Gamma(s)}, \tag{4.38}$$

for $k \in \mathbb{Z}$. This extends the previous definition (4.10) to $k \leq 0$. Thus, $(s)_0 = 1$ and, for $k \geq 1$, $(s)_k = s(s+1)\ldots(s+k-1)$.

We shall denote by $\tilde{\varphi}$ the *Mellin transform* of a (suitable) function φ on $(0, \infty)$; it is defined by

$$\tilde{\varphi}(s) = \int_0^\infty \varphi(x)x^{s-1}\, dx. \tag{4.39}$$

As before, we denote by $\mathrm{res}(g(s); \omega)$ the residue of a meromorphic function $g = g(s)$ at ω. It vanishes unless ω is a pole of g. Since $\mathrm{res}(g(s); \omega)$ is linear in g, we have that

$$\int_0^\infty \varphi(x)\, \mathrm{res}\left(x^{s+k-1}g(s); \omega\right)\, dx = \mathrm{res}\left(\tilde{\varphi}(s+k)g(s); \omega\right), \tag{4.40}$$

for φ in any of the classes of test functions considered in this chapter.

We first formulate our distributional explicit formulas, Theorems 4.12 and 4.17, with error term and without error term respectively, for (complex-valued) test functions in the class $C^\infty(0, \infty)$ (or $C^\infty(A, \infty)$, respectively),[5] with $\varphi^{(q)}(t)t^m \to 0$ for all $m \in \mathbb{Z}$ and $q \in \mathbb{N}$, as $t \to 0^+$ (respectively, as $t \to A^+$, for Theorem 4.17) or as $t \to \infty$. Then we will formulate the most general conditions under which our explicit formula applies; see Theorems 4.20 and 4.21. In particular, we will want to weaken the decay condition on φ at 0 and ∞ and the differentiability assumption on φ. (See especially Section 4.4.2 below, the results of which will be applied, in particular, in Section 5.2.3 and in Chapter 6.)

The following result provides a distributional analogue of the pointwise formula obtained in Theorem 4.4. This result will be complemented below by Theorem 4.23, which will provide an estimate for the distributional error term.

Theorem 4.12 (The distributional formula, with error term). *Let η be a generalized fractal string satisfying hypotheses* $(\mathbf{H_1})$ *and* $(\mathbf{H_2})$. *(See Equations (4.16) and (4.17) above.) Then, for every $k \in \mathbb{Z}$. the distribution $\mathcal{P}^{[k]}\eta$ is given by*

$$\mathcal{P}^{[k]}\eta(x) = \sum_{\omega \in \mathcal{D}_\eta(W)} \mathrm{res}\left(\frac{x^{s+k-1}\zeta_\eta(s)}{(s)_k}; \omega\right)$$

$$+ \frac{1}{(k-1)!} \sum_{\substack{j=0 \\ -j \in W \backslash \mathcal{D}_\eta}}^{k-1} \binom{k-1}{j}(-1)^j x^{k-1-j}\zeta_\eta(-j) + \mathcal{R}_\eta^{[k]}(x). \tag{4.41a}$$

---

[5]Given an open interval J, $C^\infty(J)$ denotes the space of infinitely differentiable functions on J.

That is, the action of $\mathcal{P}^{[k]}\eta$ on a test function φ is given by

$$\left\langle \mathcal{P}^{[k]}\eta, \varphi \right\rangle = \sum_{\omega \in \mathcal{D}_\eta(W)} \operatorname{res}\left(\frac{\zeta_\eta(s)\widetilde{\varphi}(s+k)}{(s)_k}; \omega \right)$$

$$+ \frac{1}{(k-1)!} \sum_{\substack{j=0 \\ -j \in W \setminus \mathcal{D}_\eta}}^{k-1} \binom{k-1}{j}(-1)^j \zeta_\eta(-j)\widetilde{\varphi}(k-j) \quad (4.41\text{b})$$

$$+ \left\langle \mathcal{R}_\eta^{[k]}, \varphi \right\rangle.$$

Here, the distribution $\mathcal{R} = \mathcal{R}_\eta^{[k]}$ is the error term, given by

$$\langle \mathcal{R}, \varphi \rangle = \left\langle \mathcal{R}_\eta^{[k]}, \varphi \right\rangle = \frac{1}{2\pi i} \int_S \zeta_\eta(s)\widetilde{\varphi}(s+k)\frac{ds}{(s)_k}. \quad (4.42)$$

Remark 4.13. In (4.41), the sum over j is interpreted as being equal to 0 if $k \leq 0$. The same comment applies to the corresponding sum in Equation (4.43) of Theorem 4.17 below.

Proof of Theorem 4.12. First, given an integer $n \geq 1$, we apply Lemma 4.3 for $k > \kappa + 1$. Let φ be a test function. Then

$$\left\langle \mathcal{P}^{[k]}\eta, \varphi \right\rangle = \int_0^\infty N_\eta^{[k]}(x)\varphi(x)\, dx,$$

which is approximated by

$$\sum_{\omega \in \mathcal{D}_\eta(W_n)} \operatorname{res}\left(\frac{\zeta_\eta(s)\widetilde{\varphi}(s+k)}{(s)_k}; \omega \right)$$

$$+ \frac{1}{(k-1)!} \sum_{\substack{j=0 \\ -j \in W \setminus \mathcal{D}_\eta}}^{k-1} \binom{k-1}{j}(-1)^j \zeta_\eta(-j)\widetilde{\varphi}(k-j)$$

$$+ \frac{1}{2\pi i} \int_{S_n} \zeta_\eta(s)\widetilde{\varphi}(s+k)\frac{ds}{(s)_k}.$$

(Here, we use the notation $\mathcal{D}_\eta(W_n)$ and S_n introduced in Definition 4.2.) Since $k > \kappa + 1$, the integral converges and the error of the approximation vanishes as $n \to \infty$, by the same argument as in the proof of Theorem 4.4. Then we derive formula (4.41a) for every $k \in \mathbb{Z}$ by differentiating, as a distribution, the above formula sufficiently many times. $\qquad\square$

Remark 4.14. The method used above (which may be coined the 'descent method') shows that from a pointwise formula applied at a high enough level (so as to avoid any problem of convergence in the sum or in the integrals involved), we can deduce a corresponding distributional formula on every level. This is reminiscent of the method used to establish the convergence of the Fourier series associated to a periodic distribution. (See, for example, [Sch1, §VII, I, esp. p. 226] or [Sch2, Chapter IV].)

Remark 4.15. The (infinite) sum over the visible complex dimensions of η appearing on the right-hand side of (4.41a) defines a distribution. Hence, (4.41a) is not only an equality between distributions, but each term on the right-hand side of (4.41a) also defines a distribution. Indeed, this is a simple consequence of the following well known fact about the convergence of distributions. (See, for example, [Sch1–2] or [Hö2, Theorem 2.18, p. 39]; this property was also used in a related context in [DenSc, p. 50].) For definiteness, we will work with the space of test functions $\mathbf{D} = \mathbf{D}(0, \infty)$, consisting of all infinitely differentiable functions with compact support contained in $(0, \infty)$, and the associated classical space of distributions $\mathbf{D}' = \mathbf{D}'(0, \infty)$, the topological dual of $\mathbf{D}$; see [Sch1, Hö2]. (Note that $\mathbf{D}$ is contained in our space of admissible test functions.) Let $\{T_n\}_{n=1}^\infty$ be a sequence of distributions in $\mathbf{D}'$ such that $\langle T, \varphi \rangle := \lim_{n \to \infty} \langle T_n, \varphi \rangle$ exists for every test function $\varphi \in \mathbf{D}$. Then T is a distribution in $\mathbf{D}'$. (This follows from the Uniform Boundedness Principle [Ru2, Theorem 5.8, p. 58], also known as the Banach–Steinhaus Theorem.)

In our present setting, we can apply this result to an appropriate sequence of partial sums (the sequence of partial sums mentioned in Remark 4.5, applied to φ) to deduce that the sum over $\mathcal{D}_\eta(W)$ on the right-hand side of (4.41a) is really a distribution in $\mathbf{D}'$, as stated above, and hence that each term taken separately on the right-hand side of (4.41a) defines a distribution in $\mathbf{D}'$.[6]

An entirely analogous comment applies to Theorem 4.17 below, the distributional explicit formula without error term, provided that we work instead with $\mathbf{D} = \mathbf{D}(A, \infty)$ and $\mathbf{D}' = \mathbf{D}'(A, \infty)$.

Remark 4.16. The class of test functions considered in Theorem 4.12 is the counterpart for $(0, +\infty)$ of the Schwartz space $\mathcal{S}(\mathbb{R})$ [Sch1; ReSi2, §IX.1], where the additive real line $\mathbb{R} = (-\infty, +\infty)$ has been replaced by the multiplicative half-line $(0, +\infty)$; namely, the (smooth) test functions $\varphi = \varphi(x)$ must decay faster than any polynomial at $x = 0$ and $x = +\infty$. By making the change of variable $y = \log x$ ($y \in \mathbb{R}$, $x > 0$), we see that this corresponds to (smooth) test functions $\psi = \psi(y)$ on $\mathbb{R}$ decaying faster than any polynomial in $\log y$ at $y = -\infty$ and at $y = +\infty$. (Note, however, that both classes of test functions contain all C^∞ functions with compact support on $(0, +\infty)$ or on $\mathbb{R}$, respectively.)

---

[6]Note that, for every $k \geq 1$, $N_\eta^{[k]}(x)$ defines $\mathcal{P}^{[k]}\eta$ as a distribution in $\mathbf{D}'$ because $N_\eta^{[k]}(x)$ is a locally integrable function. Also, since η is a (local) measure, $\mathcal{P}^{[0]}\eta = \eta$ is a distribution in $\mathbf{D}'$ having $\mathcal{P}^{[k]}\eta$ as its k-th primitive.

Next, we obtain the distributional analogue of the pointwise formula without error term obtained in Theorem 4.8. This is an 'asymptotic' distributional formula; i.e., it applies to test functions that are supported on the right of $x = A$, where $A > 0$ is as in hypothesis $(\mathbf{H}_2')$.

Theorem 4.17 (The distributional formula, without error term). *Let η be a generalized fractal string satisfying hypotheses $(\mathbf{H}_1)$ and $(\mathbf{H}_2')$. (See Equations (4.16) and (4.18) above.) Then, for every $k \in \mathbb{Z}$ and on test functions supported on (A, ∞), the distribution $\mathcal{P}^{[k]}\eta$ is given by*

$$\mathcal{P}^{[k]}\eta(x) = \sum_{\omega \in \mathcal{D}_\eta(W)} \operatorname{res}\left(\frac{x^{s+k-1}\zeta_\eta(s)}{(s)_k} ; \omega \right)$$

$$+ \frac{1}{(k-1)!} \sum_{\substack{j=0 \\ -j \in W \setminus \mathcal{D}_\eta}}^{k-1} \binom{k-1}{j}(-1)^j x^{k-1-j}\zeta_\eta(-j). \tag{4.43a}$$

That is, the action of $\mathcal{P}^{[k]}\eta$ on a test function φ supported on (A, ∞) is given by

$$\left\langle \mathcal{P}^{[k]}\eta, \varphi \right\rangle = \sum_{\omega \in \mathcal{D}_\eta(W)} \operatorname{res}\left(\frac{\zeta_\eta(s)\tilde{\varphi}(s+k)}{(s)_k} ; \omega \right)$$

$$+ \frac{1}{(k-1)!} \sum_{\substack{j=0 \\ -j \in W \setminus \mathcal{D}_\eta}}^{k-1} \binom{k-1}{j}(-1)^j \zeta_\eta(-j)\tilde{\varphi}(k-j). \tag{4.43b}$$

Proof. Again, given $n \geq 1$, we apply Lemma 4.3, but now for $k > \kappa$ (instead of $k > \kappa + 1$ as in the proof of Theorem 4.12). Let φ be a test function whose support is contained in $[A + \delta, \infty)$, for some $\delta > 0$. Then $\tilde{\varphi}(s)$ is bounded by

$$(A+\delta)^\sigma \int_0^\infty |\varphi(x)| \frac{dx}{x}.$$

It follows that the integral over the screen $r_m(t) + it$ tends to 0 as $m \to \infty$. The rest of the argument is exactly the same as in the proof of Theorem 4.12 above. $\square$

By applying Theorem 4.12 (respectively, 4.17) at level $k = 0$, we obtain the following result, which is central to our theory and will be used repeatedly in the rest of this book. (See Sections 5.3.1, 5.3.2 and 4.1.1 for further discussion and interpretation of (4.44) and, for example, Section 4.5 as well as Chapters 7 and 9 for applications of this result.)

Corollary 4.18 (The density of states formula). *Under the same hypotheses as in Theorem 4.12 (respectively, Theorem 4.17), we have the following* *distributional explicit formula for* $\mathcal{P}^{[0]}\eta = \eta$:

$$\eta = \sum_{\omega \in \mathcal{D}_\eta(W)} \text{res}\left(\frac{x^{s+k-1}\zeta_\eta(s)}{(s)_k}; \omega\right) + \mathcal{R}_\eta^{[0]}, \tag{4.44}$$

where $\mathcal{R}_\eta^{[0]}$ *is the distribution given by* (4.42) *with* $k = 0$ (*respectively,* $\mathcal{R}_\eta^{[0]} \equiv 0$).

4.4.1 Alternate Proof of Theorem 4.12

We now provide an alternate proof of Theorem 4.12, the distributional explicit formula with error term. It is a direct proof of this theorem, in the sense that by contrast to that given above in Section 4.4, it does not involve the descent method from Remark 4.14.

Alternate proof of Theorem 4.12. Note that

$$\begin{aligned}
\left\langle \mathcal{P}^{[k]}\eta, \varphi \right\rangle &= (-1)^q \left\langle \mathcal{P}^{[k+q]}\eta, \varphi^{(q)} \right\rangle \\
&= (-1)^q \int_0^\infty \varphi^{(q)}(x) N_\eta^{[k+q]}(x)\, dx,
\end{aligned} \tag{4.45}$$

when $k + q > \kappa + 1$. We then apply the pointwise explicit formula, Theorem 4.4. Using the fact that $\widetilde{\varphi}(s) = \frac{(-1)^q}{(s)_q}\widetilde{\varphi^{(q)}}(s + q)$ (see Equation (4.52) below), we obtain the explicit formula of Theorem 4.12. $\square$

4.4.2 Extension to More General Test Functions

We extend our distributional explicit formulas (Theorems 4.12 and 4.17) to a broader class of test functions φ, not necessarily C^∞ and decaying less rapidly near 0 and ∞ than considered previously. In particular, the allowed test functions will be assumed to have a suitable asymptotic expansion at $x = 0$. (See Theorem 4.20.) In the corresponding distributional explicit formula without error term, Theorem 4.21, we only allow test functions having a very special expansion on $(0, A + \delta)$, for some $\delta > 0$.

We will use this extension to obtain an explicit formula for the volume of the tubular neighborhoods of the boundary of a fractal string. (See especially the proof of Theorem 6.1 in Section 6.1.) Moreover, although this aspect will not be stressed as much in this work, we mention that we can also use this extension to obtain explicit formulas for other geometric or spectral functions, such as the partition function. (See Section 5.2.3.)

Given $v \in \mathbb{R}$, we will say that a function φ on $(0, \infty)$ has an asymptotic expansion of order v at 0 if there are finitely many complex exponents α, with $\operatorname{Re}\alpha < v$, and complex coefficients a_α such that

$$\varphi(x) = \sum_\alpha a_\alpha x^\alpha + O(x^v), \qquad \text{as } x \to 0^+. \tag{4.46}$$

Note that if $\varphi(x) = O(x^a)$ as $x \to 0^+$ and $\varphi(x) = O(x^b)$ as $x \to \infty$, with $a > b$, then $\tilde{\varphi}(s)$ is defined and holomorphic in the strip $-a < \operatorname{Re} s < -b$. Moreover, for any given integer $k \geq 1$,

$$x^\alpha = x^\alpha e^{-x} \left(1 + x + \frac{x^2}{2!} + \cdots + \frac{x^k}{k!} \right) + O\left(x^{\alpha+k+1}\right), \qquad \text{as } x \to 0^+. \tag{4.47}$$

The Mellin transform of $x^\alpha e^{-x}$ is $\Gamma(s + \alpha)$, with simple poles at $s = -\alpha, -\alpha - 1, \ldots$, and with residue $(-1)^k/k!$ at $s = -\alpha - k$. Consequently, the Mellin transform of the first term of the right-hand side of (4.47) is meromorphic in $\operatorname{Re} s > -\alpha - k - 1$, with a simple pole with residue 1 at $s = -\alpha$, the residues at the other possible poles adding up to 0. It follows that the Mellin transform of the function φ in (4.46) is meromorphic in $-v < \operatorname{Re} s < -b$, with simple poles at $s = -\alpha$ with residue a_α, provided that $\varphi(x) = O(x^b)$ at ∞.

We will need the following lemma.

Lemma 4.19. *Let $f(s)$ and $g(s)$ be meromorphic in a disc around 0. Then the function $z \mapsto \operatorname{res}\left(f(s - z)g(s); 0\right) + \operatorname{res}\left(f(s - z)g(s); z\right)$ is holomorphic in the same disc. Its value at $z = 0$ is $\operatorname{res}\left(f(s)g(s); 0\right)$.*

Proof. This follows since the function can be written as the integral over a small circle around $s = 0$ and $s = z$ of $f(s - z)g(s)$, and this function is analytic in z. $\qquad\square$

Theorem 4.20 (Extended distributional formula, with error term). *Let η be a generalized fractal string satisfying* $(\mathbf{H}_1)$ *and* $(\mathbf{H}_2)$. *Let $k \in \mathbb{Z}$ and let $q \in \mathbb{N}$ be such that $k + q \geq \kappa + 1$. Further, let φ be a test function that is q times continuously differentiable on $(0, \infty)$, and assume that its j-th derivative satisfies, for each $0 \leq j \leq q$ and some $\delta > 0$,*

$$\varphi^{(j)}(x) = O(x^{-k-j-D-\delta}), \qquad \text{as } x \to \infty, \tag{4.48a}$$

and

$$\varphi^{(j)}(x) = \sum_\alpha a_\alpha^{(j)} x^{\alpha-j} + O(x^{-k-j-\sigma_1+\delta}), \qquad \text{as } x \to 0^+. \tag{4.48b}$$

(So that φ has an asymptotic expansion of order $-k - j - \sigma_1 + \delta$ at 0, where σ_1 is given by (4.15a). Note also that $a_\alpha^{(j)} = \alpha \cdots (\alpha - j + 1)a_\alpha^{(0)}$.)

Then we have the following distributional explicit formula with error term for $\mathcal{P}^{[k]}\eta$ (applied to φ):

$$\left\langle \mathcal{P}^{[k]}\eta, \varphi \right\rangle = \sum_{\omega \in \mathcal{D}_\eta(W)} \operatorname{res}\left(\frac{\zeta_\eta(s)\widetilde{\varphi}(s+k)}{(s)_k}; \omega \right)$$

$$+ \frac{1}{(k-1)!} \sum_{\substack{j=0 \\ -j \in W \setminus \mathcal{D}_\eta}}^{k-1} \binom{k-1}{j}(-1)^j \zeta_\eta(-j)\widetilde{\varphi}(k-j) \qquad (4.49)$$

$$+ \sum_{\substack{-\alpha \in W \setminus \mathcal{D}_\eta \\ \alpha \notin \{0,\ldots,k-1\}}} a_\alpha^{(0)} \frac{\zeta_\eta(-\alpha)}{(-\alpha)_k} + \left\langle \mathcal{R}_\eta^{[k]}, \varphi \right\rangle,$$

where $\mathcal{R}_\eta^{[k]}$ is the distribution given by (4.42).

For the extended distributional formula without error term, we require that the test function is a finite sum of terms $x^\alpha e^{-c_\alpha x}$ in a neighborhood of the entire interval $(0, A]$. Here, the constants c_α are complex numbers with positive real part.

Theorem 4.21 (Extended distributional formula, without error term). *Let η be a generalized fractal string satisfying $(\mathbf{H_1})$ and $(\mathbf{H_2'})$. Let $k \in \mathbb{Z}$ and let $q \in \mathbb{N}$ be such that $k + q > \max\{1, \kappa\}$. Further, let φ be a test function that is q times continuously differentiable on $(0, \infty)$. Assume that as $x \to \infty$, the j-th derivative $\varphi^{(j)}(x)$ satisfies (4.48a), and that for some $\delta > 0$,*

$$\varphi^{(j)}(x) = \sum_\alpha a_\alpha^{(j)} x^\alpha e^{-c_\alpha x}, \quad \textit{for } x \in (0, A+\delta) \textit{ and } 0 \le j \le q. \qquad (4.50)$$

Then formula (4.49), with $\mathcal{R}_\eta^{[k]} \equiv 0$, gives the distributional explicit formula without error term on level k for φ.

Proof of Theorems 4.20 and 4.21. First of all, the condition at infinity on φ implies that $\left\langle \mathcal{P}^{[k+j]}\eta, \varphi^{(j)} \right\rangle$ is well defined. Furthermore, since

$$\left\langle \mathcal{P}^{[k]}\eta, \varphi \right\rangle = (-1)^q \left\langle \mathcal{P}^{[k+q]}\eta, \varphi^{(q)} \right\rangle, \qquad (4.51)$$

and

$$\widetilde{\varphi}(s) = \frac{(-1)^q}{(s)_q} \widetilde{\varphi^{(q)}}(s+q) \qquad (4.52)$$

is meromorphic in $k + \sigma_1 \le \operatorname{Re} s \le k + D$, it suffices to establish the theorem when $q = 0$ and $k > \kappa + 1$ (respectively, $k > \max\{1, \kappa\}$ for Theorem 4.21).

We first assume that $\varphi(x) = O\left(x^{-k-\sigma_1+\delta}\right)$ as $x \to 0^+$ (or $\varphi(x) = 0$ for all $x \in (0, A + \delta)$, in the case of Theorem 4.21). Using Lemma 4.3, with $c = D + \delta/2$, we obtain a truncated explicit formula,

$$
\left\langle \mathcal{P}^{[k]}\eta, \varphi \right\rangle \approx \sum_{\omega \in \mathcal{D}_\eta(W_n)} \operatorname{res}\left(\frac{\zeta_\eta(s)\widetilde{\varphi}(s+k)}{(s)_k}; \omega\right)
$$

$$
+ \frac{1}{(k-1)!} \sum_{\substack{j=0 \\ -j \in W_n \setminus \mathcal{D}_n}}^{k-1} \binom{k-1}{j}(-1)^j \zeta_\eta(-j)\widetilde{\varphi}(k-j)
$$

$$
+ \sum_{\substack{-\alpha \in W_n \setminus \mathcal{D}_n \\ \alpha \notin \{0, \ldots, k-1\}}} a_\alpha^{(0)} \frac{\zeta_\eta(-\alpha)}{(-\alpha)_k} + \frac{1}{2\pi i} \int_{S_n} \zeta_\eta(s)\widetilde{\varphi}(s+k) \frac{ds}{(s)_k},
$$

up to an error not exceeding a constant times

$$
\int_0^\infty |\varphi(x)| x^{k-1} T^{-k} \left(Tx^{D+\delta/2} + T^\kappa \max\{x^{D+\delta/2}, x^{\sigma_1}\}\right) dx. \tag{4.53}
$$

Here, we have used the fact that for fixed $0 < \beta_1 < \beta_2$,

$$
|\eta|\left(\beta_1 x, \beta_2 x\right) = O\left(x^{D+\delta/2}\right), \qquad \text{as } x \to \infty,
$$

since D is the abcissa of convergence of $\zeta_{|\eta|}$. Firstly, one checks that the integral (4.53) converges, due to the conditions we imposed on φ at 0 and ∞. Then we see that it vanishes as $T \to \infty$, provided $k > 1$ and $k > \kappa$. The integral over the truncated screen converges as $T \to \infty$ if $k > \kappa + 1$. If $(\mathbf{H_2'})$ is satisfied, we first let $m \to \infty$; i.e., we derive a truncated formula without an integral over the truncated screen. Then we do not need the assumption that $k > \kappa + 1$ to ensure convergence. This establishes the formula when $\varphi(x) = O\left(x^{-k-\sigma_1+\delta}\right)$ as $x \to 0^+$.

We now prove the theorem for the special test function $\varphi(x) = x^\alpha e^{-cx}$, with $\alpha \in \mathbb{C}$. Then, in light of Equation (4.47), we obtain the formula for general test functions by subtracting sufficiently many functions of this type, depending on the asymptotic expansion at 0 of the test function.

Let $\varphi(x) = x^\alpha e^{-cx}$. For the formula with error term, varying c does not give any extra generality, and we simply put $c = 1$, but when $(\mathbf{H_2'})$ is satisfied, we let c be any complex number with positive real part. Then

$$
\widetilde{\varphi}(s) = c^{-s-\alpha}\Gamma(\alpha + s).
$$

This function has poles at $-\alpha, -\alpha - 1, \ldots$, and it has the right decay as $\operatorname{Im} s \to \pm\infty$, so that the integral defining the error term converges. Note in addition that by Stirling's formula [In, p. 57], we have

$$
|\Gamma(\alpha + s)| \ll_a \exp(a \operatorname{Re} s)
$$

for every $a > 0$, as $\operatorname{Re} s \to -\infty$ away from the poles $-\alpha, -\alpha - 1, \dots$. Thus we can first let $m \to \infty$ to obtain a formula without error term. In this case, as $m \to \infty$, we pick up a residue $\zeta_\eta(-\alpha - l - k)c^l(-1)^l/(l!(-\alpha - l - k)_k)$ at each point $-\alpha - l - k$ where $\widetilde{\varphi}(s + k)$ has a pole. Since the integral over the screen converges and the expression for the error (4.53) vanishes as $T \to \infty$, the sum of these residues converges.

For the formula with error term, we first apply the explicit formula for $\operatorname{Re} \alpha$ small enough, so that none of the points $-\alpha, -\alpha - 1, \dots$ lies inside W. The left-hand side of the explicit formula is clearly an analytic function in α, taking into account the fact that η is supported away from 0. The proof will be complete when we have shown that the right-hand side is also analytic in α. For this, we need to show that the right-hand side changes analytically when one of the points $-\alpha, -\alpha - 1, \dots$ crosses the screen S or coincides with a complex dimension or with one of the points $-j$, for $j = 0, \dots, k - 1$. Indeed, when one of the points $-\alpha, -\alpha - 1, \dots$ crosses S, the integral over the screen changes by minus the residue of the integrand at this point, and this cancels the corresponding term in the last sum. Secondly, when one of the points $-\alpha, -\alpha - 1, \dots$ coincides with a complex dimension or with one of the points $-j$, $j = 0, \dots, k - 1$, the analyticity follows from Lemma 4.19. This completes the proof of Theorems 4.20 and 4.21. $\qquad\qquad\qquad\qquad\qquad\qquad\qquad\qquad\qquad\qquad\qquad\qquad\square$

4.4.3 The Order of the Distributional Error Term

We now provide a more quantitative version of our distributional explicit formulas with error term (Theorem 4.12, or more generally, Theorem 4.20), which will play a key role in the remainder of this book. (See, for example, Chapters 6, 7 and 9.)

Given $a > 0$ and a test function φ, we set

$$\varphi_a(x) = \frac{1}{a}\varphi\left(\frac{x}{a}\right). \tag{4.54}$$

In light of Equation (4.39), the Mellin transform of $\varphi_a(x)$ is given by

$$\widetilde{\varphi_a}(s) = a^{s-1}\widetilde{\varphi}(s). \tag{4.55}$$

Definition 4.22. We will say that a distribution $\mathcal{R}$ is of *asymptotic order* at most x^α (respectively, less than x^α)—and we will write $\mathcal{R}(x) = O(x^\alpha)$ (respectively, $\mathcal{R}(x) = o(x^\alpha)$), as $x \to \infty$—if applied to a test function φ, we have that

$$\langle \mathcal{R}, \varphi_a \rangle = O\left(a^\alpha\right) \quad (\text{respectively, } \langle \mathcal{R}, \varphi_a \rangle = o\left(a^\alpha\right)), \quad \text{as } a \to \infty. \tag{4.56}$$

When we apply Definition 4.22, we use an arbitrary test function φ of the type considered in Theorem 4.12 (respectively, Theorem 4.20).

The following theorem completes Theorems 4.12 and 4.20 by specifying the asymptotic order of the distributional error term $\mathcal{R}_\eta^{[k]}$ obtained in our distributional explicit formula (4.41). It shows that, in a suitable sense, Theorem 4.12 and its extension, Theorem 4.20, are as flexible as and more widely applicable than their pointwise counterpart, Theorem 4.4. As was mentioned above, we will take advantage of this fact on many occasions in the rest of this book.

Recall that the screen S is the curve $S(t) = r(t) + it$, for some bounded continuous function r, and that the least upper bound of r is denoted σ_u. (See Section 3.1 and formula (4.15b).)

Theorem 4.23 (Order of the distributional error term). *Fix $k \in \mathbb{Z}$. Assume that the hypotheses of Theorem 4.12 (or more generally, of Theorem 4.20, with $k+q > \kappa+1$) are satisfied, and let the distribution $\mathcal{R} = \mathcal{R}_\eta^{[k]}$ be given by (4.42). Then $\mathcal{R}$ is of asymptotic order at most x^{σ_u+k-1} as $x \to \infty$:*

$$\mathcal{R}_\eta^{[k]}(x) = O\left(x^{\sigma_u+k-1}\right), \quad \text{as } x \to \infty, \tag{4.57}$$

in the sense of Definition 4.22.

Moreover, if $r(t) < \sigma_u$ for all $t \in \mathbb{R}$ (i.e., if the screen lies strictly to the left of the line $\operatorname{Re} s = \sigma_u$), then $\mathcal{R}$ is of asymptotic order less than x^{σ_u+k-1} as $x \to \infty$:

$$\mathcal{R}_\eta^{[k]}(x) = o\left(x^{\sigma_u+k-1}\right), \quad \text{as } x \to \infty. \tag{4.58}$$

Proof. The integral (4.42) for $\langle \mathcal{R}, \varphi \rangle$ converges absolutely. Let φ be a localized test function. When we replace φ by φ_a in (4.42), we see, by (4.55), that the absolute value of the integrand is multiplied by $a^{\operatorname{Re} s+k-1} \leq a^{\sigma_u+k-1}$ for $s = r(t) + it$. Hence $|\langle \mathcal{R}, \varphi_a \rangle|$ is bounded by a constant times a^{σ_u+k-1}.

To obtain the better estimate (4.58) when S approximates the line $\operatorname{Re} s = \sigma_u$, but stays strictly to the left of it, we use an argument similar to the one used to derive estimate (4.34) of Theorem 4.4.   □

The analysis of the error term given in Theorem 4.23 also allows us to estimate the sum over the complex dimensions occurring in our distributional explicit formulas.

Theorem 4.24. *Let $v \leq D$. Assume that the hypotheses of Theorem 4.12 (or of Theorem 4.20, with $k+q > \kappa+1$) are satisfied, with a screen contained in the open half-plane $\operatorname{Re} s < v$. Assume, in addition, that there exists a screen S_0 contained in $\operatorname{Re} s < v$, satisfying $(\mathbf{H}_2)$ and such that every complex dimension to the right of S_0 has real part $\geq v$. Then*

$$\sum_{\omega \in \mathcal{D}_\eta(W),\, \operatorname{Re}\omega < v} \operatorname{res}\left(\frac{x^{s+k-1}\zeta_\eta(s)}{(s)_k}; \omega\right) = o\left(x^{v+k-1}\right), \quad \text{as } x \to \infty,$$

$$\tag{4.59}$$

in the sense of Definition 4.22.

Proof. We write this distribution as $\mathcal{R}_{0,\eta}^{[k]}(x) - \mathcal{R}_{\eta}^{[k]}(x)$, where $\mathcal{R}_{0,\eta}^{[k]}$ is the error term associated with the screen S_0. The result then follows from Theorem 4.23 applied to $\mathcal{R}_{0,\eta}^{[k]}$ and $\mathcal{R}_{\eta}^{[k]}$. $\square$

We will apply Theorem 4.24 in Chapters 6, 7 and 9 with $v = D$. However, the hypotheses of this theorem are not always satisfied: It is not always possible to choose a screen S_0 passing between $\mathrm{Re}\, s = D$ and the complex dimensions to the left of this line. Nevertheless, in the case of self-similar strings, we can still obtain information from the complex dimensions; see Remark 5.14. In the following example, we construct a nonlattice string that does not satisfy $(\mathbf{H_2})$ for any such screen S_0.

Example 4.25. Let $\alpha > 1$ be an irrational number with continued fraction expansion

$$\alpha = [a_0, a_1, a_2, \dots] := a_0 + \cfrac{1}{a_1 + \cfrac{1}{a_2 + \cdots}}.$$

Here, the integers a_j (also called 'partial quotients' in [HardW]) are determined recursively by

$$\alpha_0 := \alpha, \quad a_j = [\alpha_j], \quad \alpha_{j+1} = \frac{1}{\alpha_j - a_j}, \qquad \text{for } j = 1, 2, \dots.$$

The convergents p_n/q_n of α are defined by

$$\frac{p_n}{q_n} = [a_0, a_1, \dots, a_n].$$

Recall that these rational numbers are given by the recursion relations

$$p_{n+1} = a_{n+1}p_n + p_{n-1},$$
$$q_{n+1} = a_{n+1}q_n + q_{n-1}, \qquad \text{for } n = -1, 0, 1, \dots,$$

with $p_{-2} = q_{-1} = 0$, $p_{-1} = q_{-2} = 1$. Note that p_n and q_n are integers. Moreover, p_n and q_n satisfy the identity

$$q_n\alpha - p_n = \frac{(-1)^n}{\alpha_{n+1}q_n + q_{n-1}}.$$

We will also write

$$q'_{n+1} = \alpha_{n+1}q_n + q_{n-1}.$$

We refer the interested reader to [HardW, Chapter X] for an introduction to the theory of continued fractions.

We construct the nonlattice self-similar string $\mathcal{L}$ with two scaling ratios $r_1 = e^{-1}$, $r_2 = e^{-\alpha}$, where α will be specified below.[7] Consider the function $f(s) = 1 - e^{-s} - e^{-\alpha s}$. Let D be the real zero of f; i.e., the dimension of $\mathcal{L}$. Let p_n/q_n be a convergent of α. Then $f(D + 2\pi i q_n)$ is very close to 0. Indeed, we have

$$
\begin{aligned}
f(D + 2\pi i q_n) &= 1 - e^{-(D+2\pi i q_n)} - e^{-(D+2\pi i q_n)\alpha} \\
&= 1 - e^{-D} - e^{-D\alpha} e^{-2\pi i q_n \alpha + 2\pi i p_n} \\
&= 1 - e^{-D} - e^{-D\alpha} e^{2\pi i (-1)^n / q'_{n+1}} \\
&= 1 - e^{-D} - e^{-D\alpha} \left(1 + O\left(\frac{2\pi}{q'_{n+1}} \right) \right) \\
&= O\left(\frac{2\pi}{q'_{n+1}} \right), \quad \text{as } n \to \infty.
\end{aligned}
$$

The reason for this near vanishing is that there is a zero of f close to $s = D + 2\pi i q_n$, but slightly to the left. It follows that

$$
|\zeta_{\mathcal{L}}(D + 2\pi i q_n)| \gg q'_{n+1}.
$$

Note that

$$
q'_{n+1} = \alpha_{n+1} q_n + q_{n-1} > a_{n+1} q_n.
$$

We next choose the integers a_j as follows: $a_0 = 1$, and if $a_0, a_1, \ldots, a_n$ are already constructed, then we compute q_n and set $a_{n+1} = q_n^n$. (Thus $a_1 = 1$, $a_2 = 1$, $a_3 = 4$, $a_4 = 9^3$, $a_5 \approx 1.8 \cdot 10^{15}$, ...) Then

$$
|\zeta_{\mathcal{L}}(D + 2\pi i q_n)| \gg q_n^{n+1},
$$

so that on the vertical line $\operatorname{Re} s = D$, condition ($\mathbf{H_2}$) is violated for any value of $\kappa \geq 0$. (See Equation (4.17).)

Naturally, the value of $f(s)$ will be even smaller between $s = D + 2\pi i q_n$ and the nearby zero of this function. Thus ($\mathbf{H_2}$) will not be satisfied on any screen passing between $\operatorname{Re} s = D$ and the poles of $\zeta_{\mathcal{L}}$.

The reader sees that the integers a_j in this example, and hence the numbers q_n, grow extremely fast. Thus the set of points where $\zeta_{\mathcal{L}}$ is large is very sparse.

We note that our theory extends to the situation when hypothesis ($\mathbf{H_2}$) in (4.17) is replaced by the following integrability condition: The function

$$
t \longmapsto |\zeta_{\mathcal{L}}(D + it)|(|t| + 1)^{-\kappa - 1} \tag{4.60}
$$

is integrable on $\mathbb{R}$.

---

[7]Here, $e = \exp(1)$ is the base of the natural logarithm.

However, we can even choose α such that $|\zeta_{\mathcal{L}}(D + it)|/g(|t|)$ is not integrable as a function of $t \in \mathbb{R}$ no matter how fast the function g grows. Namely, we choose α such that $\log q_{n+1}/g(q_n)$ grows unboundedly. This is the case, for example, if we choose the coefficients of α so as to satisfy

$$a_{n+1} \geq e^{ng(q_n)},$$

for $n = 0, 1, 2, \dots$.

We again give an example, now of a non self-similar string, showing that condition (4.60) is not satisfied for all generalized fractal strings.

Example 4.26. Consider the measure μ on $[1, \infty)$ defined by

$$\mu = \sum_{n \in \mathbb{Z} \setminus \{0\}} \frac{x^{D-1-d_n+in}}{n^2} \, dx,$$

where the real numbers $d_n = d_{-n}$ are small, and will be specified below. The geometric zeta function of this generalized fractal string has poles at $D - d_n \pm in$, with residue $1/n^2$, for $n \in \mathbb{Z} - \{0\}$. Hence the value of the integral of $|\zeta_\mu(s)|$ over an interval from $D + i\left(n - \frac{1}{2}\right)$ to $D + i\left(n + \frac{1}{2}\right)$ is close to

$$\frac{1}{n^2} \int_{-1/2}^{1/2} \frac{dt}{|t + id_n|} \approx -\frac{\log d_n}{n^2},$$

as $|n| \to \infty$. Therefore, we have

$$\int_{D+i(n-1/2)}^{D+i(n+1/2)} |\zeta_\mu(D + it)| \, t^{-\kappa} \, dt \approx -\frac{\log d_n}{n^{2+\kappa}},$$

as $n \to \infty$.

We conclude that if $d_n = \exp(-n^n)$, then for every value of $\kappa \geq 0$, the function $|\zeta_\mu(D + it)|t^{-\kappa}$ is not integrable along the vertical line $\operatorname{Re} s = D$, and hence along any screen passing between $\operatorname{Re} s = D$ and the poles of ζ_μ. It follows that condition (4.60) (or its obvious analogue along a suitable screen S) is never satisfied for this string μ.

The following example explains rather clearly why, in general, our explicit formulas must have an error term. Indeed, in this extreme situation, there is only an error term because the set $\mathcal{D}_\eta$ of visible complex dimensions is empty and therefore the corresponding sum over $\mathcal{D}_\eta$ vanishes.

Example 4.27. The generalized fractal string

$$\eta = \sum_{n=1}^{\infty} (-1)^{n-1} \delta_{\{n\}} \tag{4.61}$$

has no complex dimensions. Indeed, by [Ti, Eq. (2.2.1), p. 16], the associated geometric zeta function is

$$\zeta_\eta(s) = \sum_{n=1}^{\infty} (-1)^{n-1} n^{-s} = \left(1 - 2^{1-s}\right) \zeta(s). \tag{4.62}$$

Since the pole of $\zeta(s)$ at $s = 1$ is canceled by the corresponding zero of $\left(1 - 2^{1-s}\right)$ in Equation (4.62), $\zeta_\eta(s)$ is holomorphic in all of $\mathbb{C}$ and hence $\mathcal{D}_\eta$ is empty, as claimed above. Thus the explicit formula for $N_\eta(x)$ (and for $N_\eta^{[k]}(x)$ at level k) has no sum over ω, only an error term and a constant term.

We now explain this in more detail in the case when $k = 1$. Choose a screen to the left of $\mathrm{Re}\, s = 0$. Since $\zeta(\sigma + it)$ grows like $|t|^{1/2-\sigma}$ (see Equation (5.18) in Chapter 5 below), we have to interpret $N_\eta(x)$ as a distribution. Theorem 4.12 yields[8]

$$N_\eta(x) = \frac{1}{2} + \mathcal{R}^{[1]}(x). \tag{4.63}$$

By construction, we have

$$N_\eta(x) = \begin{cases} 0, & \text{for } 2n \le x < 2n+1, \ n = 0, 1, \ldots, \\ 1, & \text{for } 2n+1 \le x < 2n+2, \ n = 0, 1, \ldots. \end{cases} \tag{4.64}$$

Observe that $N_\eta(x)$ is an additively periodic function, with period 2. Its Fourier series is

$$N_\eta(x) = \frac{1}{2} + \frac{1}{\pi i} \sum_{n=-\infty}^{\infty} \frac{e^{\pi i(2n+1)x}}{2n+1}. \tag{4.65}$$

We point out that the reason why our explicit formula gives only the constant term $1/2$ and an error term in (4.63) is that this function is not multiplicatively periodic. (In general, our explicit formula yields the Fourier series of a multiplicatively periodic counting function.)

Remark 4.28. By (4.64), the average value of $N_\eta(x)$ in the previous example is equal to $1/2$, as expressed by the first term of (4.63). Further, $N_\eta(x)$ jumps by $+1$ at every odd integer, and by -1 at every even integer. Hence, as $x \to \infty$, $N_\eta(x) = 1/2 + O(1)$ as a function, and no better pointwise estimate holds as $x \to \infty$. On the other hand, as a distribution, $\mathcal{R}^{[1]}(x) = O(x^\sigma)$ for every $\sigma < 0$, since we can choose screens arbitrarily far to the left. Thus $N_\eta(x)$ does not oscillate. This seems to be a contradiction.

The resolution of this apparent paradox is important since in Chapters 7 and 9, we will make use of the fact that a certain type of oscillations,

---

[8]The more stringent assumptions of Theorem 4.17, needed to obtain a distributional explicit formula without error term, are not satisfied here.

multiplicative oscillations, are reflected in our explicit formulas. We discuss this paradox here in the case of the harmonic string h, which will serve as a paradigm for this phenomenon.

Since by (3.10), $h = \sum_{n=1}^{\infty} \delta_{\{n\}}$, we have

$$N_h(x) = \#\{n \geq 1 : n \leq x\} = [x]. \tag{4.66}$$

Hence $N_h(x)$ jumps at every positive integer. On the other hand, according to Theorem 4.12 applied at level $k = 1$ and $k = 0$ respectively, we have

$$N_h(x) = x - \frac{1}{2} + \mathcal{R}_h^{[1]}(x), \tag{4.67a}$$

for the counting function of the lengths, and

$$h = 1 + \mathcal{R}_h^{[0]}(x), \tag{4.67b}$$

for the density of states. In the sense of Definition 4.22, the error terms are estimated by

$$\mathcal{R}_h^{[1]}(x) = O(x^{\sigma}), \quad \text{as } x \to \infty, \tag{4.68a}$$

and

$$\mathcal{R}_h^{[0]}(x) = O(x^{\sigma-1}), \quad \text{as } x \to \infty, \tag{4.68b}$$

for every $\sigma < 0$. As in Example 4.27, this seems to be in contradiction with the jumps of $N_h(x)$ and the point masses of h at the positive integers.

The meaning of the distributional equalities (4.67) and (4.68) becomes clear when we do not apply them to a localized test function, but to a test function having an asymptotic expansion at 0,

$$\varphi(x) = \sum_{\text{Re}\,\alpha > -1} a_\alpha x^\alpha, \quad \text{as } x \to 0^+,$$

as in (4.48b), with $a_\alpha = 0$ for $\text{Re}\,\alpha \leq -1$. Write

$$\varphi_t(x) = \frac{1}{t}\varphi(x/t)$$

and $J = \int_0^\infty \varphi(x)\,dx$. Then, by Theorem 4.20, applied at level $k = 0$,

$$\sum_{n=1}^{\infty} \varphi_t(n) = J + \sum_{\text{Re}\,\alpha \leq \sigma} a_\alpha \zeta(-\alpha)t^{-\alpha-1} + \left\langle \mathcal{R}_h^{[0]}, \varphi_t \right\rangle,$$

and the error term is of order $O(t^{\sigma-1})$ as $t \to \infty$, for every $\sigma < 0$.[9]

On a logarithmic scale, the points $1, 2, 3, \ldots$ become dense on the real line, which is why h is distributionally approximated by the density 1 up

Figure 4.2: The integers viewed additively.

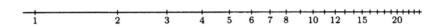

Figure 4.3: The integers viewed multiplicatively.

to every order. (See Figures 4.2 and 4.3.) On the other hand, multiplicative oscillations of η, like, for example, in the case of the Cantor string, do give rise to oscillatory terms in the explicit formula for η, as we will see in Section 6.3.1 and in Chapter 8.

4.5   Example: The Prime Number Theorem

In the next chapter, we will discuss a number of examples illustrating our pointwise and distributional explicit formulas. In the present section, we will concentrate on one particular application of these formulas; namely, to the Prime Number Theorem and issues surrounding it, particularly the Riemann–von Mangoldt formula. Our main purpose is to indicate by means of this example the flexibility provided by our distributional formula and by the introduction of a screen in the formulation of our explicit formulas.

This example is also included in order to clarify the relationships between our explicit formulas and the traditional ones from number theory. Our proof of the Prime Number Theorem with Error Term obtained in this manner is essentially the same as the usual one, except for the analysis of the error term. We note, however, that in the present context, the systematic use of the screen S in our theory makes particularly transparent the connection between a given zero-free region for $\zeta(s)$ and the corresponding error term $\mathcal{R}_\pi(x)$ in the asymptotic formula (4.71).

The reader familiar with the classical explicit formulas from number theory—the first one of which was discovered by Riemann in his famous 1858 paper [Rie1]—may wonder why the theorems obtained in this chapter are called explicit formulas. Indeed, Riemann's explicit formula—as inter-

---

[9]This is a very useful classical formula, which does not appear in the literature. It was explained heuristically to the second author by Don Zagier [Zag].

preted and proved thirty-six years later by von Mangoldt [vM1–2]—relates a suitable counting function associated with the prime numbers to an infinite sum involving the zeros of the Riemann zeta function (that is, the poles of its logarithmic derivative ζ'/ζ); see, for example, [Pat, esp. Chapters 1 and 3], [Da, Chapter 17], [In, Chapters II–IV], [Edw, Chapters 1 and 3]. (A similar statement holds for the various types of number-theoretic explicit formulas found in the literature; see, for instance, the references given at the end of Section 4.1 above.)

However, we will show in this section that we can easily recover the Riemann–von Mangoldt formula from our own explicit formulas. Actually, we can obtain different versions of it, either pointwise or distributional, at various levels k, the most basic of which is the distributional formula obtained when $k = 0$, corresponding to the density of prime powers viewed as the generalized fractal string $\mathfrak{P} = \sum_{m\geq 1,\,p}(\log p)\,\delta_{\{p^m\}}$, the prime string, introduced in (3.12); see Equation (4.76) in Section 4.5.1 below.

In turn, now standard arguments (see, e.g., [Da, Chapter 18], [Edw, Chapter 4], [Pat, Chapter 3] or [In, p. 5 and Chapter II])—first used in slightly different forms by Hadamard [Had2] and de la Vallée Poussin [dV1–2][10]—enable us to deduce from this explicit formula the Prime Number Theorem with Error Term; namely, if

$$\pi(x) = \#\left\{p \leq x : p \text{ is a prime number}\right\} \tag{4.70}$$

denotes the prime number counting function, then

$$\pi(x) = \text{Li}(x) + \mathcal{R}_\pi(x), \quad \text{as } x \to \infty, \tag{4.71}$$

where Li denotes the logarithmic integral

$$\text{Li}(x) = \lim_{\varepsilon \to 0^+}\left(\int_0^{1-\varepsilon} + \int_{1+\varepsilon}^x\right)\frac{1}{\log t}\,dt, \quad \text{for } x > 1, \tag{4.72}$$

and estimates for the error term $\mathcal{R}_\pi(x)$ depend on our current knowledge of the zero-free region for the Riemann zeta function. For example, in (4.71),

---

[10]Their results, obtained independently and published almost simultaneously in 1896, relied on von Mangoldt's work [vM1–2], as well as on Hadamard's product formula for entire functions [Had1]. In their proof, both Hadamard [Had2] and de la Vallée Poussin [dV1] showed (also independently) that $\zeta(s) \neq 0$ for all s on the vertical line $\text{Re}\,s = 1$. Hadamard's classical proof of the Prime Number Theorem is the simplest of the two, but in his second paper [dV2], de la Vallée Poussin established the existence of a zero-free region to the left of $\text{Re}\,s = 1$ and thus went further in his investigation of the approximation of $\pi(x)$, in the form of the error estimate (4.71). (See, e.g., [In, p. 5], [Edw, Chapter 5], [Ti, Chapter III] and [Pat, Chapter 3].) We note that the original form of the Prime Number Theorem obtained in [Had2, dV1] was as follows:

$$\pi(x) = \text{Li}(x)\,(1 + o(1)) = \frac{x}{\log x}\,(1 + o(1)), \text{ as } x \to \infty. \tag{4.69}$$

we have

$$\mathcal{R}_\pi(x) = O\left(xe^{-c\sqrt{\log x}}\right), \quad \text{as } x \to \infty, \tag{4.73}$$

for some explicit positive constant c. Better error estimates—following from a refined analysis of the zero-free regions of ζ—can be found, for instance, in Ivić's book [Ivi] (see also, e.g., [In, pp. xi–xii] and the references therein) and can also be deduced from our explicit formula, simply by choosing a different screen S and thus a different window W suitably adapted to the zero-free region.

4.5.1   The Riemann–von Mangoldt Formula

We now briefly explain how to interpret and derive from our explicit formula the Riemann–von Mangoldt formula. (Compare, for example, with the discussion in [Edw, Chapter 3, esp. §3.5] or else in [In, Pat].) As was seen in Section 3.1.1, the geometric zeta function of the prime string

$$\mathfrak{P} = \sum_{m \geq 1,\, p} (\log p)\delta_{\{p^m\}} \tag{4.74}$$

is

$$\zeta_\mathfrak{P}(s) = -\frac{\zeta'(s)}{\zeta(s)}. \tag{4.75}$$

(In (4.74) and in (4.77) below, p runs over the prime numbers.) This function satisfies $(\mathbf{H_1})$ and $(\mathbf{H_2'})$ for every $\kappa > 0$ and with $W = \mathbb{C}$. By Theorem 4.17, applied at level $k = 0$, and with $A = 1$, the density of lengths formula without error term for the prime string reads (see also Corollary 4.18 above and Section 5.3.1 below):

$$\mathfrak{P} = 1 - \sum_\rho x^{\rho-1} - \sum_{n=1}^{\infty} x^{-1-2n}, \quad \text{for } x > 1, \tag{4.76}$$

where ρ runs through the sequence of critical zeros of ζ. (See the last comment at the end of Section 3.1.1.) This formula should be interpreted distributionally on the open interval $(1, \infty)$, as in Theorem 4.17. Further, note that the last sum on the right-hand side of (4.76) coincides with $\frac{1}{x(x^2-1)}$ and corresponds to the trivial zeros of ζ, located at $-2, -4, \ldots$.

The corresponding formula for the measure

$$\eta = \sum_{m \geq 1,\, p} \frac{1}{m} \delta_{\{p^m\}}, \tag{4.77}$$

with geometric zeta function $\log \zeta(s)$, can be derived by means of the following artefact: One first establishes that the string $\frac{dx}{\log x}$ has a geometric zeta function with a logarithmic singularity at 1 that cancels that of $\log \zeta(s)$. Hence, our explicit formula applies to the string $\eta - \frac{dx}{\log x}$, with a screen to the left of $\operatorname{Re} s = 1$, but to the right of all zeros of ζ. We thus find (with $\sigma_u \leq 1$ given as in formula (4.15b)):

$$\eta = \left(\frac{1}{\log x} + o\left(x^{\sigma_u - 1}\right)\right) dx = \frac{1}{\log x}\left(1 + o\left(1\right)\right) dx, \quad \text{as } x \to \infty. \quad (4.78)$$

This provides a distributional interpretation of (part of) Riemann's original explicit formula [Rie1].[11]

Remark 4.29. Using the explicit formula (4.76), together with a deep analysis of the location of the critical zeros ρ of the Riemann zeta function, one can prove the Prime Number Theorem with Error Term, stating that

$$\pi(x) = \operatorname{Li}(x) + O\left(x e^{-c\sqrt{\log x}}\right), \quad \text{as } x \to \infty, \quad (4.79)$$

for some constant $c > 0$. This result does not seem to be attainable by a Tauberian argument. See [Pos, Section 27, pp. 109–112; Sh, Theorem 14.1, p. 115] for the Wiener–Ikehara Tauberian Theorem. Also see [Va] for a concise exposition of the logic of the proof of the Prime Number Theorem by means of the Wiener–Ikehara Tauberian Theorem.

---

[11]The latter is difficult to establish pointwise (see, e.g., [Edw, §1.18, p. 36]) and was not properly justified in Riemann's original paper [Rie1].

5

The Geometry and the Spectrum of Fractal Strings

In this chapter, we give various examples of explicit formulas for the counting function of the lengths and the frequencies of (generalized) fractal strings and sprays.

In Section 5.1, we give a detailed discussion of the oscillatory term associated with a single complex dimension of the (generalized) fractal string η. Then, in Section 5.2, we derive the explicit formulas for the geometric and spectral counting functions of η, and in Section 5.2.3, for the geometric and spectral partition functions, which are sometimes a useful substitute for the corresponding counting functions. In Section 5.3.1, we obtain the explicit formulas for the geometric and spectral density of states of η, and use them in Section 5.3.2 to define the spectral operator, which formalizes the so-called direct spectral problem.

In Section 5.4, we explore the important special case of self-similar strings studied in Chapter 2. We discuss, in particular, the geometry and the spectrum of both lattice and nonlattice strings. As an application, in Section 5.4.4, we prove a Prime Orbit Theorem for suspended flows. We study two examples of non self-similar strings in Section 5.5. Finally, in Section 5.6, we close the chapter by explaining how our explicit formulas apply to the study of the geometry and the spectrum of fractal sprays, a higher-dimensional analogue of fractal strings.

5.1   The Local Terms in the Explicit Formulas

Our explicit formulas[1] give expansions of various functions associated with
a fractal string as a sum over the complex dimensions of this string. The
term corresponding to the complex dimension ω of multiplicity one is of the
form Cx^ω, where C is a constant depending on ω. If ω is real, the function
x^ω simply has a certain asymptotic behavior as $x \to \infty$. If, on the other
hand, $\omega = \beta + i\gamma$ has a nonzero imaginary part γ, then $x^\omega = x^\beta \cdot x^{i\gamma}$ is
of order $O(x^\beta)$ as $x \to \infty$, with a multiplicatively periodic behavior: The
function $x^{i\gamma} = \exp(i\gamma \log x)$ takes the same value at the points $e^{2\pi n/\gamma}x$
($n \in \mathbb{Z}$). Thus, the term corresponding to ω will be called an oscillatory
term. If there are complex dimensions with higher multiplicity, there will
also be terms of the form $Cx^\omega (\log x)^m$, $m \in \mathbb{N}^*$, which have a similar
oscillatory behavior.

5.1.1   The Geometric Local Terms

Let η be a generalized fractal string and let ω be a complex dimension of η;
i.e., $s = \omega$ is a pole of $\zeta_\eta(s)$. The term corresponding to ω in the explicit
formula of Theorems 4.4, 4.8, 4.12 and 4.17 is

$$\text{res}\left(\frac{x^{s+k-1}\zeta_\eta(s)}{(s)_k}; \omega\right). \tag{5.1}$$

The nature of this term depends on the multiplicity of ζ_η at ω and on the
expansion of ζ_η around ω. Let $m = m(\omega)$ be the multiplicity of ω, and let

$$\frac{a_m}{(s-\omega)^m} + \frac{a_{m-1}}{(s-\omega)^{m-1}} + \frac{a_{m-2}}{(s-\omega)^{m-2}} + \cdots + \frac{a_1}{s-\omega} \tag{5.2}$$

be the principal part of the Laurent series of $\zeta_\eta(s)$ at $s = \omega$. The expansion
of x^{s-1} around $s = \omega$ is given by

$$x^{s-1} = x^{\omega-1}\left(1 + (s-\omega)\log x + \frac{(s-\omega)^2}{2!}(\log x)^2 + \cdots\right). \tag{5.3}$$

In general, for $k \geq 1$, the expansion of $x^{s+k-1}/(s)_k$ around $s = \omega$ is given
by

$$\frac{x^{s+k-1}}{(s)_k} = x^{\omega+k-1}\sum_{n=0}^{\infty}(s-\omega)^n\sum_{\mu=0}^{n}\frac{(\log x)^\mu}{\mu!}$$
$$\sum_{\nu=0}^{k-1}\frac{(-1)^{n+\nu-\mu}}{\nu!(k-1-\nu)!}\frac{1}{(\omega+\nu)^{n+1-\mu}}. \tag{5.4}$$

---

[1]Throughout this discussion, we assume implicitly that η satisfies the appropriate
growth conditions; namely, hypotheses $(\mathbf{H_1})$ and $(\mathbf{H_2})$ (respectively, $(\mathbf{H_1})$ and $(\mathbf{H_2'})$) if
the explicit formula in Theorem 4.4 or 4.12 (respectively, 4.8 or 4.17) is applied.

To derive this formula, we first compute the partial fraction expansion of $1/(s)_k$. The residue at $s = -\nu$ is $(-1)^\nu/(\nu!(k - \nu - 1)!)$. Hence, for $k \geq 1$,

$$\frac{1}{(s)_k} = \sum_{\nu=0}^{k-1} \frac{(-1)^\nu}{\nu!(k-1-\nu)!(s+\nu)}. \tag{5.5}$$

Next, we substitute the power series

$$\frac{1}{s+\nu} = \sum_{n=0}^{\infty}(-1)^n \frac{(s-\omega)^n}{(\omega+\nu)^{n+1}}$$

and multiply by the power series of x^{s+k-1} to obtain formula (5.4).

Theorem 5.1. *Let ω be a complex dimension of multiplicity m and let the principal part of $\zeta_\eta(s)$ be given by formula (5.2). Then the local term (5.1) in the explicit formula at ω is given by*

$$x^{\omega-1} \sum_{j=1}^{m} a_j \frac{(\log x)^{j-1}}{(j-1)!}, \tag{5.6}$$

for $k = 0$, and by

$$x^{\omega+k-1} \sum_{j=1}^{m} a_j \sum_{\nu=0}^{k-1} \sum_{\mu=0}^{j-1} \frac{(\log x)^\mu}{\mu!} \frac{(-1)^{j-1+\nu-\mu}}{\nu!(k-1-\nu)!} \frac{1}{(\omega+\nu)^{j-\mu}}, \tag{5.7}$$

for $k \geq 1$.

In particular, if ω is a simple pole, then the local term is given by these formulas for $m = 1$:

$$a_1 x^{\omega-1}, \tag{5.8}$$

if $k = 0$, and by

$$x^{\omega+k-1} a_1 \sum_{\nu=0}^{k-1} \frac{(-1)^\nu}{\nu!(k-1-\nu)!} \frac{1}{\omega+\nu}, \tag{5.9}$$

if $k \geq 1$.

5.1.2   The Spectral Local Terms

To study the spectral local terms, we need to compute the principal part of the Laurent series of $\zeta_\eta(s)\zeta(s)$ at $s = \omega$. Let this principal part be

$$\frac{b_m}{(s-\omega)^m} + \frac{b_{m-1}}{(s-\omega)^{m-1}} + \frac{b_{m-2}}{(s-\omega)^{m-2}} + \cdots + \frac{b_1}{s-\omega}. \tag{5.10}$$

The coefficients b_j can be expressed in terms of the a_j's of formula (5.2):

$$b_j = \sum_{q=1}^{j} a_q \frac{\zeta^{(j-q)}(\omega)}{(j-q)!}. \tag{5.11}$$

Note that $b_m = a_m \zeta(\omega)$. Thus, if $\zeta(s)$ vanishes at $s = \omega$, the multiplicity of the complex dimension ω diminishes in the spectrum. The complex dimension ω could even disappear altogether, which will be the subject of Chapter 7.

5.1.3 The Weyl Term

There is one special complex dimension associated with the spectrum, namely at $s = 1$. We call it the spectral complex dimension of the Bernoulli string.[2] It is never a pole of ζ_η, thus we can easily compute the corresponding term in the explicit formulas.

Definition 5.2. The *Weyl term* associated with the Bernoulli string is the term at $s = 1$ in the explicit formula for the spectrum. It is given by

$$W^{[k]}(x) = \zeta_\eta(1) \frac{x^k}{k!}, \tag{5.12}$$

for $k \geq 0$.

5.1.4 The Distribution $x^\omega \log^m x$

The local terms are finite sums of terms proportional to $x^{\omega+k-1} (\log x)^\mu$. In Theorems 4.12 and 4.17, these terms have to be interpreted as distributions, which act on a test function φ in the following way:

$$\langle x^{\omega+k-1} (\log x)^\mu, \varphi \rangle = \int_0^\infty x^{\omega+k-1} (\log x)^\mu \, \varphi(x) \, dx. \tag{5.13}$$

Now, $\tilde{\varphi}(s) = \int_0^\infty x^{s-1} \varphi(x) \, dx$ is an analytic function (it is meromorphic in a strip with finitely many poles under the more general assumptions of Theorems 4.20 and 4.21). Since this integral is absolutely convergent, we can differentiate under the integral sign. The μ-th derivative with respect to s is

$$\tilde{\varphi}^{(\mu)}(s) = \int_0^\infty x^{s-1} (\log x)^\mu \, \varphi(x) \, dx. \tag{5.14}$$

---

[2]The Bernoulli string $\mathcal{B}$ (called the 'Sturm–Liouville string' in [Lap2, Remark 2.5, p. 144]) is defined by $\Omega = (0, 1)$; hence it consists of a single open interval of length 1, and its spectral zeta function $\zeta_{\nu,\mathcal{B}}$ is given by $\zeta_{\nu,\mathcal{B}}(s) = \zeta(s)$, the Riemann zeta function.

Combining (5.13) and (5.14), putting $s = \omega + k$, we find that

$$\langle x^{\omega+k-1} (\log x)^{\mu}, \varphi \rangle = \widetilde{\varphi}^{(\mu)}(\omega + k), \tag{5.15}$$

where $\widetilde{\varphi}^{(\mu)}$ is the μ-th derivative of the holomorphic function $\widetilde{\varphi}$.

5.2   Explicit Formulas for Lengths and Frequencies

In this section, we give the explicit formulas for the geometric and spectral counting functions $N_\eta(x)$ and $N_\nu(x)$. We also give these formulas on the k-th level, for $k \geq 2$.

5.2.1   The Geometric Counting Function of a Fractal String

Choose a screen S such that $(\mathbf{H_1})$ and $(\mathbf{H_2})$ are satisfied for some $\kappa \geq 0$. When $k > \kappa + 1$, the pointwise explicit formula (Theorem 4.4) for the counting function is valid. Otherwise (that is, if $k \leq \kappa + 1$, $k \in \mathbb{Z}$), we have to interpret $N_\eta^{[k]}(x)$ as a distribution. Note that for the usual counting function $N_\eta(x) = N_\eta^{[1]}(x)$, we do not have a pointwise formula with error term. But as a distribution, according to Theorem 4.12, it is given by

$$N_\eta(x) = \sum_{\omega \in \mathcal{D}_\eta} \operatorname{res}\left(\frac{x^s \zeta_\eta(s)}{s}; \omega\right) + \{\zeta_\eta(0)\} + R_\eta^{[1]}(x), \tag{5.16a}$$

and, if the poles are simple, by

$$N_\eta(x) = \sum_{\omega \in \mathcal{D}_\eta} \frac{x^\omega}{\omega} \operatorname{res}\left(\zeta_\eta(s); \omega\right) + \{\zeta_\eta(0)\} + R_\eta^{[1]}(x), \tag{5.16b}$$

where the term in braces is included only if $0 \in W \backslash \mathcal{D}$.

These are special cases of the following more general formulas, for $k \geq 1$:

$$N_\eta^{[k]}(x) = \sum_{\omega \in \mathcal{D}_\eta} \operatorname{res}\left(\frac{x^{s+k-1}\zeta_\eta(s)}{(s)_k}; \omega\right)$$

$$+ \frac{1}{(k-1)!} \sum_{\substack{j=0 \\ -j \in W \backslash \mathcal{D}_\eta}}^{k-1} \binom{k-1}{j}(-1)^j x^{k-1-j}\zeta_\eta(-j) + R_\eta^{[k]}(x),$$

$$\tag{5.17a}$$

and, if the complex dimensions are simple,

$$N_\eta^{[k]}(x) = \sum_{\omega \in \mathcal{D}_\eta} \frac{x^{\omega+k-1}}{(\omega)_k} \operatorname{res}\left(\zeta_\eta(s); \omega\right)$$

$$+ \frac{1}{(k-1)!} \sum_{\substack{j=0 \\ -j \in W \setminus \mathcal{D}_\eta}}^{k-1} \binom{k-1}{j} (-1)^j x^{k-1-j} \zeta_\eta(-j) + R_\eta^{[k]}(x).$$

$$(5.17\text{b})$$

We note that the distributional error term occurring in formulas (5.16a)–(5.17b) can be estimated by means of Theorem 4.23.

Remark 5.3. From now on, for notational simplicity, we will no longer distinguish between the distributional error terms at level k (previously denoted by $\mathcal{R}_\eta^{[k]}(x)$) and their pointwise counterparts, $R_\eta^{[k]}(x)$. In other words, a pointwise or distributional error term at level k will indifferently be denoted by $R_\eta^{[k]}(x)$, as we have done in formulas (5.16a)–(5.17b) above. (As usual, the case $k \leq 0$ is allowed in the distributional case.)

5.2.2 The Spectral Counting Function of a Fractal String

For the spectral zeta function $\zeta_\nu(s)$, hypothesis $(\mathbf{H_2'})$ is never satisfied. Hence only Theorems 4.4 and 4.12 apply; i.e., our explicit formulas for the spectrum will always contain an error term. Indeed, by [Edw, §9.2, p. 185], the Riemann zeta function satisfies

$$|\zeta(\sigma + it)| \leq C_\sigma \left(|t| + 1\right)^{\frac{1}{2}-\sigma} \qquad (5.18)$$

for $\sigma < 0$, and it does not satisfy a better estimate in this half-plane. (The constant C_σ decreases like $(2\pi e)^{\sigma-1/2}$ as $\sigma \to -\infty$. Moreover, for $0 \leq \sigma \leq 1$, one has the estimate $|\zeta(\sigma + it)| \leq K \cdot t^{(1-\sigma)/2} \log t$ for some constant K, throughout the half-strip $0 \leq \sigma \leq 1, t \geq 1$.)

It follows from this discussion that if $\mathcal{L}$ is a fractal string satisfying hypotheses $(\mathbf{H_1})$ and $(\mathbf{H_2})$ for some value of $\kappa \geq 0$, then, by Equation (1.35), its spectral zeta function $\zeta_\nu(s) = \zeta_\mathcal{L}(s)\zeta(s)$ satisfies the same assumptions for the same screen, except with κ replaced by $\kappa + \frac{1}{2} - \sigma_1$ if $\sigma_1 \leq 0$ (or by $\kappa + \frac{1}{2}$ if $\sigma_1 > 0$).

Given $\sigma_0 \leq 0$, we apply Theorem 4.4 to find the following pointwise explicit formula, valid for all positive integers $k > \frac{3}{2} - \sigma_0 + \kappa$:

$$N_\nu^{[k]}(x) = W_\eta^{[k]}(x) + \sum_{\omega \in \mathcal{D}_\eta} \text{res} \left(\frac{x^{s+k-1}\zeta_\nu(s)}{(s)_k}; \omega \right)$$

$$+ \frac{1}{(k-1)!} \sum_{\substack{j=0 \\ -j \in W\backslash\mathcal{D}_\eta}}^{k-1} \binom{k-1}{j}(-1)^j x^{k-1-j}\zeta_\nu(-j) + R_\nu^{[k]}(x),$$

$$(5.19a)$$

where $W_\eta^{[k]}(x) = \zeta_\eta(1)x^k/k!$ is the Weyl term (at level k), introduced in formula (5.12). If the poles are simple, this becomes

$$N_\nu^{[k]}(x) = W_\eta^{[k]}(x) + \sum_{\omega \in \mathcal{D}_\eta} \frac{x^{\omega+k-1}}{(\omega)_k}\zeta(\omega)\, \text{res}\,(\zeta_\eta(s); \omega)$$

$$+ \frac{1}{(k-1)!} \sum_{\substack{j=0 \\ -j \in W\backslash\mathcal{D}_\eta}}^{k-1} \binom{k-1}{j}(-1)^j x^{k-1-j}\zeta_\nu(-j) + R_\nu^{[k]}(x).$$

$$(5.19b)$$

Alternatively, if we apply Theorem 4.12, we obtain a distributional explicit formula (with error term) valid for all $k \in \mathbb{Z}$ (and in particular for $k = 1$), which is still given by Equation (5.19), but which is now interpreted distributionally. (See Remark 5.3 above.) This distributional formula will be very useful later on in the book; see, in particular, Chapter 7 and especially Section 9.2.1 below.

Remark 5.4. Let us set $k = 1$ in (5.19a) and (5.19b), which is only possible when these formulas are interpreted distributionally. Then Equation (5.19a) becomes

$$N_\nu(x) = \zeta_\eta(1)x + \sum_{\omega \in \mathcal{D}_\eta} \text{res}\left(\frac{x^s \zeta_\nu(s)}{s}; \omega \right) - \{\zeta_\eta(0)/2\} + R_\nu(x), \quad (5.20a)$$

while (if all the complex dimensions of $\mathcal{L}$ are simple), Equation (5.19b) becomes

$$N_\nu(x) = \zeta_\eta(1)x + \sum_{\omega \in \mathcal{D}_\eta} \frac{x^\omega}{\omega}\zeta(\omega)\, \text{res}\,(\zeta_\eta(s); \omega) - \{\zeta_\eta(0)/2\} + R_\nu(x).$$

$$(5.20b)$$

(In both formulas, the term in braces is included only if $0 \in W\backslash\mathcal{D}_\eta$.)

5.2.3   The Geometric and Spectral Partition Functions

The *geometric partition function* of an ordinary fractal string $\mathcal{L} = (l_j)_{j=1}^{\infty}$ is

$$\theta_{\mathcal{L}}(t) = \sum_{j=1}^{\infty} e^{-tl_j^{-1}}, \qquad \text{for } t > 0. \tag{5.21}$$

For a generalized fractal string η, it is given by

$$\begin{aligned}
\theta_{\eta}(t) &= \int_{0}^{\infty} e^{-xt}\eta(dx) \\
&= \left\langle \mathcal{P}^{[0]}\eta, \varphi_t \right\rangle,
\end{aligned} \tag{5.22}$$

where $\varphi_t(x) = e^{-xt}$ for $t > 0$. Hence, if η satisfies $(\mathbf{H_1})$ and $(\mathbf{H_2})$, the explicit formula of Theorem 4.20 applies, and we obtain the following result (note that $\widetilde{\varphi_t}(s) = \Gamma(s)t^{-s}$, where Γ is the gamma function):

$$\begin{aligned}
\theta_{\eta}(t) = &\sum_{\omega \in \mathcal{D}_{\eta}(W)} \operatorname{res}\left(\zeta_{\eta}(s)\Gamma(s)t^{-s}; \omega\right) \\
&+ \sum_{\substack{k=0 \\ -k \in W \setminus \mathcal{D}_{\eta}}}^{\infty} \frac{(-1)^k}{k!} t^k \zeta_{\eta}(-k) + \left\langle R_{\eta}^{[0]}, \varphi_t \right\rangle.
\end{aligned} \tag{5.23a}$$

If the complex dimensions of η are simple, this yields

$$\begin{aligned}
\theta_{\eta}(t) = &\sum_{\omega \in \mathcal{D}_{\eta}(W)} \operatorname{res}\left(\zeta_{\eta}(s); \omega\right) \Gamma(\omega)t^{-\omega} \\
&+ \sum_{\substack{k=0 \\ -k \in W \setminus \mathcal{D}_{\eta}}}^{\infty} \frac{(-1)^k}{k!} t^k \zeta_{\eta}(-k) + \left\langle R_{\eta}^{[0]}, \varphi_t \right\rangle.
\end{aligned} \tag{5.23b}$$

Note that by Theorem 4.21, if η satisfies the stronger hypothesis $(\mathbf{H_2'})$ instead of $(\mathbf{H_2})$, we can choose $W = \mathbb{C}$ and set $\left\langle R_{\eta}^{[0]}, \varphi_t \right\rangle = 0$ in both (5.23a) and (5.23b).

Recall from (3.20) that the spectral measure of the ordinary fractal string $\mathcal{L}$ is given by

$$\nu = \sum_{f} w_f^{(\nu)} \delta_{\{f\}} = \sum_{k,j=1}^{\infty} \delta_{\{kl_j^{-1}\}}, \tag{5.24}$$

where f runs over the sequence of (distinct) frequencies of $\mathcal{L}$ and $w_f^{(\nu)}$ denotes the multiplicity of f, as in (1.30). Therefore, according to definition (5.22), the *spectral partition function* of $\mathcal{L}$ is given by

$$\theta_{\nu}(t) = \sum_{f} w_f^{(\nu)} e^{-ft} = \sum_{k,j=1}^{\infty} e^{-kl_j^{-1}t}, \qquad \text{for } t > 0. \tag{5.25}$$

In general, the spectral measure ν associated with a generalized fractal string η (that is, the measure ν representing the frequency spectrum of η) is defined by (3.15) or (3.19). Then, the spectral partition function of η is given by (5.22), with η replaced by ν. When we apply formula (5.23a) to ν, we obtain the following formula for the spectral partition function:

$$
\theta_\nu(t) = \zeta_\eta(1)\frac{1}{t} + \sum_{\omega \in \mathcal{D}_\eta(W)} \mathrm{res}\left(\zeta_\eta(s)\zeta(s)\Gamma(s)t^{-s}; \omega\right)
$$

$$
+ \sum_{\substack{k=0 \\ -k \in W \setminus \mathcal{D}_\eta}}^{\infty} \frac{(-1)^k}{k!} t^k \zeta_\eta(-k)\zeta(-k) + \left\langle R_\nu^{[0]}, \varphi_t \right\rangle. \tag{5.26a}
$$

If the complex dimensions of η are simple, this yields

$$
\theta_\nu(t) = \zeta_\eta(1)\frac{1}{t} + \sum_{\omega \in \mathcal{D}_\eta(W)} \mathrm{res}\left(\zeta_\eta(s); \omega\right)\zeta(\omega)\Gamma(\omega)t^{-\omega}
$$

$$
+ \sum_{\substack{k=0 \\ -k \in W \setminus \mathcal{D}_\eta}}^{\infty} \frac{(-1)^k}{k!} t^k \zeta_\eta(-k)\zeta(-k) + \left\langle R_\nu^{[0]}, \varphi_t \right\rangle. \tag{5.26b}
$$

Remark 5.5. In the literature on spectral geometry or on mathematical physics (see, for example, [Gi] or [BaltHi, Si]), the spectral partition function $\theta_\nu(t)$ is frequently used as a substitute for the spectral counting function $N_\nu(x)$. The interested reader can find in Appendix B below (and the relevant references therein) some information about the short time asymptotics of $\theta_\nu(t)$ in the classical case of a smooth manifold, along with their relationship with the spectral zeta function $\zeta_\nu(s)$. We note that for the spectral problems, one usually needs information about the asymptotic behavior of $\theta_\nu(t)$ as $t \to 0^+$, which corresponds to the asymptotic behavior of $N_\nu(x)$ as $x \to \infty$.

5.3 The Direct Spectral Problem for Fractal Strings

5.3.1 The Density of Geometric and Spectral States

Of special interest are the explicit formulas at level $k = 0$, the so-called density of states formulas. They only have a distributional interpretation and they express the measure of the generalized fractal string, or the spectral measure, as a sum over the complex dimensions.

We apply our distributional explicit formula with error term (Theorem 4.12) to $\eta = \mathcal{P}^{[0]}\eta$, where η is a generalized fractal string satisfying hypotheses $(\mathbf{H_1})$ and $(\mathbf{H_2})$. We then obtain that, as a distribution, the measure η is given by the following *density of geometric states* (or *density of lengths*) formula:

$$\eta = \sum_{\omega \in \mathcal{D}_\eta(W)} \mathrm{res}\left(\zeta_\eta(s)x^{s-1}; \omega\right) + R_\eta^{[0]}(x). \tag{5.27a}$$

If the complex dimensions of η are simple, this becomes

$$\eta = \sum_{\omega \in \mathcal{D}_\eta(W)} \mathrm{res}\left(\zeta_\eta(s); \omega\right) x^{\omega-1} + R_\eta^{[0]}(x). \tag{5.27b}$$

The error term is given by the integral over the screen,

$$R_\eta^{[0]}(x) = \frac{1}{2\pi i} \int_S \zeta_\eta(s)x^{s-1}\, ds. \tag{5.28}$$

Further, by Theorem 4.17, $R_\eta^{[0]}(x)$ vanishes identically if $(\mathbf{H_2'})$, rather than $(\mathbf{H_2})$, is satisfied and $W = \mathbb{C}$.

Next, let ν be the generalized fractal string representing the frequencies of η. Note that we can use the same window W for ν, since ζ grows polynomially in vertical strips (see Equation (5.18) above or [In]). Hence, by Theorem 4.12 applied to $\nu = \mathcal{P}^{[0]}\nu$, we have the following *density of spectral states* (or *density of frequencies*) formula:

$$\nu = \zeta_\eta(1) + \sum_{\omega \in \mathcal{D}_\eta(W)} \mathrm{res}\left(\zeta_\eta(s)\zeta(s)x^{s-1}; \omega\right) + R_\nu^{[0]}(x). \tag{5.29a}$$

By evaluating these residues, we find, if the complex dimensions of η are simple,

$$\nu = \zeta_\eta(1) + \sum_{\omega \in \mathcal{D}_\eta(W)} \mathrm{res}\left(\zeta_\eta(s); \omega\right) \zeta(\omega)x^{\omega-1} + R_\nu^{[0]}(x). \tag{5.29b}$$

The error term is given by the integral over the screen:

$$R_\nu^{[0]}(x) = \frac{1}{2\pi i} \int_S \zeta_\eta(s)\zeta(s)x^{s-1}\, ds. \tag{5.30}$$

Note that in this case, $(\mathbf{H_2'})$ is never satisfied. (See also Remark 5.4 in Section 5.2.2 for a closely related comment.)

Heuristically, this distributional formula describes what physicists call the 'density of states' of the string η (or, quantum-mechanically, the 'density of energy levels' of η). (See, e.g., [Berr3, BogKe, Ke] and the relevant references therein.)[3]

---

[3]The authors of these physical references study the spectrum of suitable Hamiltonians and not that of (generalized) fractal strings. Moreover, from the mathematical point of view, it is clear that their formal treatment of the density of spectral states should be interpreted distributionally (as could be done using, for example, our explicit formulas). We do not claim, however, to be able to recover from our explicit formulas all the results contained in those papers. (See Section 10.4.3 for a related discussion.)

5.3.2   The Spectral Operator

Our explicit formulas for the density of geometric and spectral states obtained in the previous section—along with previous work on related subjects (especially [LapPo1–2], [LapMa1–2] and Part II of [Lap3])—suggest introducing the following definition, which fits naturally within our present framework of generalized fractal strings. In turn, it helps conceptualize aspects of the aforementioned work and will become useful in our own investigations. In particular, it formalizes the direct spectral problem which consists of deducing spectral information from the geometry.

Definition 5.6. The *spectral operator* maps η onto ν. Specifically, it adds the Weyl term $\zeta_\eta(1)$, and locally, if the complex dimensions are simple, it multiplies each $x^{\omega-1}$-term by $\zeta(\omega)$. Further, the integrand of the error term is multiplied by $\zeta(s)$.

Remark 5.7. We could use the geometric and spectral partition function of Section 5.2.3, instead of the geometric and spectral measure, to formulate the definition of the spectral operator.

Remark 5.8. The *inverse spectral operator* maps ν back onto η. Since it involves dividing by $\zeta(\omega)$ at the complex dimensions and by $\zeta(s)$ in the error term, the extent to which it exists depends on the location of the critical zeros of ζ. This sheds new light on the work of the first author with H. Maier [LapMa1–2], to be revisited and extended in Chapter 7 below.

It follows from our results in Section 8.2 below that, in some sense, the inverse spectral operator is well defined on the class of generalized Cantor strings studied in Chapter 8. (See Remark 8.10.) This is the guiding principle underlying our proof in Chapter 9 of the fact that many number-theoretic zeta functions and other Dirichlet series do not have an infinite vertical sequence of critical zeros forming an arithmetic progression.

5.4   Self-Similar Strings

We now consider the case of self-similar strings studied in Chapter 2. The reader may first wish to review briefly some of the main results and definitions of Sections 2.1–2.4, particularly Theorems 2.3 and 2.13, as well as Definition 2.10.

According to Theorem 2.3, the geometric zeta function of a (normalized) self-similar string $\mathcal{L}$ with scaling ratios $r_1, r_2, \ldots, r_N$ is given by

$$\zeta_{\mathcal{L}}(s) = \frac{1}{1 - r_1^s - \cdots - r_N^s}. \tag{5.31}$$

(Here, for notational simplicity, we assume that $\mathcal{L}$ is normalized as in Remark 2.5, so that its first length is equal to 1.)

We deduce from (5.31) that

$$|\zeta_{\mathcal{L}}(s)| \ll r_N^{-\sigma} \quad \text{as } \sigma = \operatorname{Re} s \to -\infty. \tag{5.32}$$

It follows that we can let $W = \mathbb{C}$ and that $\zeta_{\mathcal{L}}(s)$ satisfies $(\mathbf{H_1})$ and $(\mathbf{H_2'})$ with $\kappa = 0$ and $A = r_N$. (See Remark 5.9 below.) Thus Theorems 4.8 and 4.17 apply to obtain asymptotic pointwise and distributional formulas without error term, valid for $x > r_N$.

Remark 5.9. It may be helpful to explain in more detail why assumptions $(\mathbf{H_1})$ and $(\mathbf{H_2'})$ are satisfied with $W = \mathbb{C}$ and the above choices of κ and A. First, Theorem 2.34 (from the very end of Chapter 2) enables us to find a suitable sequence $\{T_n\}_{n\in\mathbb{Z}}$, with $T_n \to \pm\infty$ as $n \to \pm\infty$, such that $\zeta_{\mathcal{L}}(s)$ is uniformly bounded on the vertical lines $\operatorname{Im} s = T_n$ ($n \in \mathbb{Z}$), and hence such that hypothesis $(\mathbf{H_1})$ given by (4.16) is satisfied with $\kappa = 0$. Secondly, in view of estimate (5.32) above, we can simply set $r_m(t) = -m$ for $m \geq 1$ and every $t \in \mathbb{R}$ to verify that hypothesis $(\mathbf{H_2'})$ given by (4.18) holds with $\kappa = 0$ and $A = r_N$. (Of course, in light of (5.32), many other choices of the screens r_m are possible.) See also Theorem 2.33, along with its application given in Section 6.3.1 below.

Remark 5.10. In the following, it will be useful to recall from Theorem 2.13 that for a self-similar string, D is always simple, where D denotes the dimension of $\mathcal{L}$.[4] Moreover, all the other complex dimensions of $\mathcal{L}$ are located to the left of the vertical line $\operatorname{Re} s = D$. According to the basic dichotomy of Theorem 2.13, for a nonlattice string, they all lie strictly to the left of this line, and a subsequence of complex dimensions lies arbitrarily close to it, whereas if $\mathcal{L}$ is a lattice string, there is an infinite sequence of complex dimensions ω on the line $\operatorname{Re} s = D$ (with constant residue), namely $\omega = D + in\mathbf{p}$ ($n \in \mathbb{Z}$), where $\mathbf{p}$ denotes the oscillatory period of $\mathcal{L}$.

5.4.1 Lattice Strings

The explicit formulas for self-similar lattice strings are particularly simple, since by Theorem 2.13, the complex dimensions of $\mathcal{L}$ are located on finitely many vertical lines, and the residues of $\zeta_{\mathcal{L}}$ at the complex dimensions on a given line are all equal.

If the complex dimensions $\omega + in\mathbf{p}$ ($n \in \mathbb{Z}$) on the vertical line $\operatorname{Re} s = \operatorname{Re} \omega$ are simple (i.e., if ω is a simple pole of $\zeta_{\mathcal{L}}$), then the corresponding sum in the pointwise formula for $N_{\mathcal{L}}(x)$ gives the multiplicatively periodic function

$$\operatorname{res}\left(\zeta_{\mathcal{L}}; \omega\right) \sum_{n=-\infty}^{\infty} \frac{x^{\omega+in\mathbf{p}}}{\omega + in\mathbf{p}}. \tag{5.33}$$

---

[4] D is the Minkowski dimension (or the only real dimension) of $\mathcal{L}$, also called the similarity dimension of $\mathcal{L}$. See Remark 2.17.

Using the Fourier series (1.13), with $\log b = 2\pi\omega/\mathbf{p}$ and $u = \mathbf{p}\log x/2\pi$, we see that this sum is equal to

$$\text{res}\,(\zeta_{\mathcal{L}};\omega) \sum_{n=-\infty}^{\infty} \frac{x^{\omega+in\mathbf{p}}}{\omega + in\mathbf{p}} = \text{res}\,(\zeta_{\mathcal{L}};\omega) \frac{b}{b-1} b^{-\{u\}} x^{\omega} \frac{2\pi}{\mathbf{p}}. \tag{5.34}$$

In particular, since D is simple (see Remark 5.10 above), the local term corresponding to D in the explicit formula for $N_{\mathcal{L}}(x)$ is given by (5.33) (with the infinite sum expressed as in (5.34)) with $\omega = D$.

Likewise, these complex dimensions contribute the quantity

$$\text{res}\,(\zeta_{\mathcal{L}};\omega) \sum_{n=-\infty}^{\infty} \Gamma\,(\omega + in\mathbf{p})\,t^{-\omega-in\mathbf{p}} \tag{5.35}$$

to the asymptotic expansion of the geometric partition function $\theta_{\mathcal{L}}(t)$, as defined by Equation (5.21).

For example, for the Cantor string defined in Section 2.2.1, we have $\omega = D = \log_3 2$ and $\mathbf{p} = 2\pi/\log 3$. Hence we can set $b = 2$ in the Fourier expansion (1.13), and we recover formula (1.28) stated at the end of Section 1.2.2:[5]

$$N_{\text{CS}}(x) = 2^{1-\{\log_3 x\}} x^D - 1$$

$$= \frac{1}{\log 3} \sum_{n=-\infty}^{\infty} \frac{x^{D+in\mathbf{p}}}{D + in\mathbf{p}} - 1. \tag{5.36}$$

In the same way, we find the geometric partition function of the Cantor string, in view of formula (5.23b) and the comment following it:

$$\theta_{\text{CS}}(t) = \sum_{n=0}^{\infty} 2^n e^{-3^n t}$$

$$= \frac{1}{\log 3} \sum_{n=-\infty}^{\infty} \Gamma\,(D + in\mathbf{p})\,t^{-D-in\mathbf{p}} + \sum_{k=0}^{\infty} \frac{(-1)^k}{k!} t^k \frac{1}{1 - 2\cdot 3^k}$$

$$= t^{-D} g_{\text{CS}}\left(\log_3 t^{-1}\right) + \sum_{k=0}^{\infty} \frac{(-1)^k}{k!} t^k \frac{1}{1 - 2\cdot 3^k}, \tag{5.37}$$

where g_{CS} is the nonconstant periodic function (of period 1) given by

$$g_{\text{CS}}(u) = \frac{1}{\log 3} \sum_{n=-\infty}^{\infty} \Gamma\,(D + in\mathbf{p})\,e^{2\pi inu}. \tag{5.38}$$

---

[5]Recall that our definition of the Cantor string in Chapter 1 was slightly different from the present one in that the first length was equal to $1/3$ instead of 1.

Remark 5.11. The periodic function g_{CS} occurring in the explicit formula (5.37) for $\theta_{CS}(t)$ is smooth on $(-\infty, +\infty)$. (In light of a well known theorem about Fourier series, this follows from the fast decay of $\Gamma(D + in\mathbf{p})$ as $|n| \to \infty$; see, e.g., [Sch1, §VII.1] along with Stirling's formula [In, p. 57].) In contrast, the corresponding periodic function occurring in the explicit formula (5.36) (or (1.28)) for the geometric counting function $N_{CS}(x)$ has infinitely many discontinuities.

An analogous comment applies to the Fibonacci string studied below—and, more generally, to every lattice string.

Remark 5.12. It is instructive to see to what extent formula (5.37) could have been derived by a direct computation (even formally). Indeed, writing

$$\frac{1}{1 - 2 \cdot 3^k} = -\sum_{l=1}^{\infty} 2^{-l} 3^{-kl},$$

we find that the sum over k in (5.37) is equal to $-\sum_{n=-\infty}^{-1} 2^n e^{-3^n t}$. Hence, it remains to establish the formula

$$\sum_{n=-\infty}^{-\infty} 2^n e^{-3^n t} = t^{-D} g_{CS}\left(\log_3 t^{-1}\right).$$

Now, since the left-hand side changes by a factor of 2 when t is divided by 3, it is immediate that g_{CS} is periodic with period 1. But to compute the Fourier series of g_{CS}, as is done in (5.37) and (5.38), is not an easy task.

For the Fibonacci string, introduced in Section 2.2.2, there are two lines of complex dimensions. (See Figure 2.3 on page 31.) Also, we have $\mathbf{p} = 2\pi/\log 2$. For the sum corresponding to the line of complex dimensions above $D = \log_2 \phi$, we have $b = \phi$, and for the other line, above $-D + \frac{1}{2} i\mathbf{p}$, we have $b = -\phi^{-1}$ (in formula (1.13)). Thus we find that

$$\begin{aligned}
N_{\text{Fib}}(x) &= \frac{3 + 4\phi}{5} \phi^{-\{\log_2 x\}} x^D - 1 + \frac{7 - 4\phi}{5} (-\phi)^{\{\log_2 x\}} x^{-D + \frac{1}{2} i\mathbf{p}} \\
&= \frac{3 + 4\phi}{5} \phi^{-\{\log_2 x\}} x^D - 1 + \frac{7 - 4\phi}{5} \phi^{\{\log_2 x\}} x^{-D} (-1)^{[\log_2 x]}.
\end{aligned}$$

(5.39)

In view of formula (5.23b) and the comment following it, the geometric partition function of the Fibonacci string has an asymptotic expansion given by

$$\theta_{\mathrm{Fib}}(t) = \sum_{n=0}^{\infty} F_{n+1} e^{-2^n t}$$

$$= \frac{\phi + 2}{5 \log 2} \sum_{n=-\infty}^{\infty} \Gamma(D + in\mathbf{p}) t^{-D-in\mathbf{p}} + \sum_{k=0}^{\infty} \frac{(-1)^k}{k!} t^k \frac{1}{1 - 2^k - 4^k}$$

$$+ \frac{3 - \phi}{5 \log 2} \sum_{n=-\infty}^{\infty} \Gamma(-D + in\mathbf{p}) t^{D-in\mathbf{p}}.$$

$$(5.40)$$

(Recall that the Fibonacci numbers F_{n+1} occurring in (5.40) are defined by the recursive equation (2.11).)

If the complex dimensions of the lattice string $\mathcal{L}$ on a line are not simple, then formula (5.33) has to be changed. Moreover, the resulting function is no longer multiplicatively periodic in that case. For example, if the complex dimensions on the (discrete) line $\omega + in\mathbf{p}$ are double poles of $\zeta_{\mathcal{L}}$, and the Laurent series of $\zeta_{\mathcal{L}}$ around the pole at ω is given by

$$\zeta_{\mathcal{L}}(s) = \alpha_{-2}(s - \omega)^{-2} + \alpha_{-1}(s - \omega)^{-1} + \dots,$$

then the sum over these complex dimensions in the explicit formula for $N_{\mathcal{L}}(x)$ is the function

$$\alpha_{-2} \frac{2\pi}{\mathbf{p}} \frac{b}{b-1} b^{-\{u\}} x^{\omega} \log x + \alpha_{-1} \frac{2\pi}{\mathbf{p}} \frac{b}{b-1} b^{-\{u\}} x^{\omega}, \qquad (5.41)$$

where, as above, we write $u = \mathbf{p} \log x / 2\pi$ and $b = \exp(2\pi\omega/\mathbf{p})$.

In particular, for the lattice string $\mathcal{L}$ with multiple poles introduced in Section 2.2.3, we thus find

$$N_{\mathcal{L}}(x) = \frac{4}{9} 2^{1-\{u\}} x^D + (-1)^{[\log_3 x]} \frac{3 \log_3 x + 5}{18} - \frac{1}{4}. \qquad (5.42)$$

The asymptotic expansion of the geometric partition function is given by

$$\theta_{\mathcal{L}}(t) = \frac{4}{9 \log 3} \sum_{n=-\infty}^{\infty} \Gamma(D + in\mathbf{p}) t^{-D-in\mathbf{p}}$$

$$+ \sum_{k=0}^{\infty} \frac{(-1)^k}{k!} t^k \frac{1}{1 - 3 \cdot 9^k - 2 \cdot 27^k} + \frac{1}{9 \log 3} \sum_{n=-\infty}^{\infty} t^{-\frac{1}{2} i \mathbf{p} - in\mathbf{p}}$$

$$\cdot \left((5 - 3 \log_3 t) \Gamma\left(\tfrac{1}{2} i\mathbf{p} + in\mathbf{p}\right) + \frac{3}{\log 3} \Gamma'\left(\tfrac{1}{2} i\mathbf{p} + in\mathbf{p}\right) \right),$$

$$(5.43)$$

where $\Gamma'(s)$ denotes the derivative of the gamma function.

5.4.2 Nonlattice Strings

As was recalled in Remark 5.10, according to the nonlattice case of Theorem 2.13, there is only one complex dimension with real part $\geq D$, namely D itself. Further, D is always simple. Consequently, by estimate (4.58) of Theorem 4.23, if there exists a screen passing between $\operatorname{Re} s = D$ and the complex dimensions to the left of this line (for the general case, see Remark 5.14 below), we have for $k \geq 1$,

$$N_{\mathcal{L}}^{[k]}(x) = \operatorname{res}\left(\zeta_{\mathcal{L}}; D\right) \frac{x^{D+k-1}}{(D)_k} + o\left(x^{D+k-1}\right), \quad \text{as } x \to \infty. \tag{5.44}$$

In particular, for $k = 1$, we have, again provided a suitable screen exists,

$$N_{\mathcal{L}}(x) = \operatorname{res}\left(\zeta_{\mathcal{L}}; D\right) \frac{x^D}{D} + o\left(x^D\right), \quad \text{as } x \to \infty. \tag{5.45}$$

With the same restriction, we find for the partition function

$$\theta_{\mathcal{L}}(t) = \operatorname{res}\left(\zeta_{\mathcal{L}}; D\right) \Gamma\left(D\right) t^{-D} + o\left(t^{-D}\right), \quad \text{as } t \to 0^+. \tag{5.46}$$

Similarly, nonlattice strings are Minkowski measurable, since (as will follow from Theorem 6.20 below),

$$V(\varepsilon) = \operatorname{res}\left(\zeta_{\mathcal{L}}; D\right) \frac{(2\varepsilon)^{1-D}}{D(1-D)} + o\left(\varepsilon^{1-D}\right), \quad \text{as } \varepsilon \to 0^+. \tag{5.47}$$

Remark 5.13. We point out that the estimates in (5.44)–(5.47) are the best possible, since by the nonlattice case of Theorem 2.13, there always exist complex dimensions of $\mathcal{L}$ arbitrarily close to, but strictly to the left of, the vertical line $\operatorname{Re} s = D$.

Remark 5.14. Recall that we cannot always choose a screen passing between the line $\operatorname{Re} s = D$ and all complex dimensions strictly to the left of this line; see Example 4.25. Hence the above analysis is valid only for nonlattice strings that allow the choice of such a screen. However, if $\mathcal{L}$ does not allow such a choice, we apply Theorem 2.33 to write, for all small positive numbers δ,

$$N_{\mathcal{L}}^{[k]}(x) = \operatorname{res}\left(\zeta_{\mathcal{L}}; D\right) \frac{x^{D+k-1}}{(D)_k} + \sum_{D-\delta/2 < \operatorname{Re}\omega < D} \operatorname{res}\left(\zeta_{\mathcal{L}}; \omega\right) \frac{x^{\omega+k-1}}{(\omega)_k} \tag{5.48}$$
$$+ o\left(x^{D-\delta/4+k-1}\right), \quad \text{as } x \to \infty,$$

and

$$V(\varepsilon) = \operatorname{res}\left(\zeta_{\mathcal{L}}; D\right) \frac{(2\varepsilon)^{1-D}}{D(1-D)} + \sum_{D-\delta/2 < \operatorname{Re}\omega < D} \operatorname{res}\left(\zeta_{\mathcal{L}}; \omega\right) \frac{(2\varepsilon)^{1-\omega}}{\omega(1-\omega)} \tag{5.49}$$
$$+ o\left(\varepsilon^{1-D+\delta/4}\right), \quad \text{as } \varepsilon \to 0^+.$$

The sum in (5.49) converges absolutely, since the residues are bounded, by Theorem 2.33. For the same reason, the sum in (5.48) converges absolutely provided that $k \geq 2$. The same argument applies to the geometric partition function.

Remark 5.15. Finally, we point out that the above formulas and estimates for $V(\varepsilon)$ are to be understood distributionally. On the other hand, the estimates for $N_{\mathcal{L}}^{[k]}(x)$ can be understood pointwise for $k \geq 1$, except when we are in the situation of the previous remark, in which case one needs $k \geq 2$. In the latter situation, the estimate for $N_{\mathcal{L}}^{[1]}(x)$ can still be interpreted distributionally.

5.4.3   The Spectrum of a Self-Similar String

Lattice Case

Let $\mathcal{L}$ be a lattice self-similar string with oscillatory period $\mathbf{p}$. Suppose that ω is a simple complex dimension. In the explicit formula for the spectral counting function, the sum over the complex dimensions on the line $\omega + in\mathbf{p}$ gives the periodic distribution

$$\mathrm{res}\,(\zeta_{\mathcal{L}}; \omega) \sum_{n=-\infty}^{\infty} \frac{x^{\omega+in\mathbf{p}}}{\omega + in\mathbf{p}} \zeta(\omega + in\mathbf{p}). \qquad (5.50)$$

In contrast to the geometric formula, there is no nice closed formula for this expression, because $0 < \mathrm{Re}\,\omega < 1$ and hence $\sum_{j=1}^{\infty} j^{-\omega}$ does not converge.

Again, in view of Remark 5.10 above, we can apply (5.50) to $\omega = D$ itself. We find that the spectrum of a lattice self-similar string is, as a distribution, given by its counting function,

$$N_{\nu}(x) = |\mathcal{L}| \cdot x + \mathrm{res}\,(\zeta_{\mathcal{L}}; D) \sum_{n=-\infty}^{\infty} \zeta(D + in\mathbf{p}) \frac{x^{D+in\mathbf{p}}}{D + in\mathbf{p}}$$
$$+ O\big(x^{\Theta}(\log x)^{m-1}\big) + O(1), \qquad (5.51)$$

as $x \to \infty$, where $\mathrm{Re}\,s = \Theta$ is the first line of complex dimensions to the left of D, and m is the maximal multiplicity of the complex dimensions of $\mathcal{L}$ on this line.[6] Note that $|\mathcal{L}| = \zeta_{\mathcal{L}}(1)$ is the one-dimensional total length of $\mathcal{L}$, and $\mathrm{res}\,(\zeta_{\mathcal{L}}; D)$ is related to the D-dimensional volume of the boundary of the string; see Chapter 6, Theorem 6.12. Further note that the error term $O(1)$ in (5.51) (or in (5.52) below) is needed if $\Theta < 0$, as is the case, for example, for the Fibonacci string.

---

[6]Observe that in the notation of Theorem 2.13, different ω_u's can have the same real part.

Similarly, the spectral partition function has an asymptotic expansion

$$\theta_\nu(t) = |\mathcal{L}| \cdot t^{-1} + \operatorname{res}(\zeta_{\mathcal{L}}; D) \sum_{n=-\infty}^{\infty} \Gamma(D+in\mathbf{p})\zeta(D+in\mathbf{p})t^{-D-in\mathbf{p}}$$
$$+ O\left(t^{-\Theta}(\log t^{-1})^{m-1}\right) + O(1),$$

$$(5.52)$$

as $t \to 0^+$, where Θ and m are as above.

Remark 5.16. We note that $\zeta(D+in\mathbf{p})$ does not vanish for infinitely many values of n, by Theorem 9.1 of Chapter 9. Hence the sum over the line of complex dimensions $D + in\mathbf{p}$ contains infinitely many oscillatory terms. In other words, the spectral counting function and the spectral partition function of every lattice string have oscillations of order D. This additional information cannot be obtained by means of renewal theory or by means of a Tauberian-type argument.

For the Cantor string of Section 2.2.1, we obtain (with $D = \log_3 2$ and $\mathbf{p} = 2\pi/\log 3$):

$$N_\nu(x) = 3x + \frac{1}{\log 3} \sum_{n=-\infty}^{\infty} \zeta(D+in\mathbf{p}) \frac{x^{D+in\mathbf{p}}}{D+in\mathbf{p}} + O(1) \qquad (5.53)$$

as $x \to \infty$, and

$$\theta_\nu(t) = 3t^{-1} + \frac{1}{\log 3} \sum_{n=-\infty}^{\infty} \Gamma(D+in\mathbf{p})\zeta(D+in\mathbf{p})t^{-D-in\mathbf{p}} + O(1),$$

$$(5.54)$$

as $t \to 0^+$. In light of Remark 5.16, the (multiplicatively) periodic functions occurring in (5.53) and (5.54) are both nonconstant.

We mention that a detailed analysis of the spectral counting function $N_\nu(x)$ of the Cantor string (and of other integral Cantor strings) is provided in Section 8.2.1 below. (Set $a = 3$ and $b = 2$ in Theorem 8.6 and Corollary 8.7; also see [LapPo2, Theorem 4.6, p. 65] for an earlier, although less precise, study of this particular case.)

As was pointed out above, each line of complex dimensions of a lattice string yields a corresponding sum in our explicit formulas. For example, for the Fibonacci string of Section 2.2.2, we obtain the following expression (with $D = \log_2 \phi$ and $\mathbf{p} = 2\pi/\log 2$, where $\phi = (1+\sqrt{5})/2$ is the golden ratio):

$$N_\nu(x) = 4x + \frac{\phi + 2}{5\log 2} \sum_{n=-\infty}^{\infty} \zeta(D + in\mathbf{p}) \frac{x^{D+in\mathbf{p}}}{D + in\mathbf{p}} + \frac{1}{2}$$

$$+ \frac{3 - \phi}{5\log 2} \sum_{n=-\infty}^{\infty} \zeta(-D + i(n + 1/2)\mathbf{p}) \frac{x^{-D+i(n+1/2)\mathbf{p}}}{-D + i(n + 1/2)\mathbf{p}} \quad (5.55)$$

$$+ O(x^\rho)$$

as $x \to \infty$, and

$$\theta_\nu(t) = 4t^{-1} + \frac{\phi + 2}{5\log 2} \sum_{n=-\infty}^{\infty} \Gamma(D + in\mathbf{p}) \zeta(D + in\mathbf{p}) t^{-D-in\mathbf{p}} + \frac{1}{2}$$

$$+ \frac{3 - \phi}{5\log 2} \sum_{n=-\infty}^{\infty} \Gamma(-D + i(n + 1/2)\mathbf{p}) \zeta(-D + i(n + 1/2)\mathbf{p})$$

$$\cdot t^{D-i(n+1/2)\mathbf{p}}$$

$$+ O(t^{-\rho}),$$

$$(5.56)$$

as $t \to 0^+$, for every $\rho < -D$.

Remark 5.17. We leave it as an exercise for the interested reader to write down, possibly with the help of a symbolic computation package, the explicit formulas for the spectral counting and partition functions of the lattice string with multiple poles introduced in Section 2.2.3.

Nonlattice Case

Next, let $\mathcal{L}$ be a nonlattice self-similar string. After separating the term corresponding to the dimension of the string, we find

$$N_\nu(x) = |\mathcal{L}| \cdot x - \operatorname{res}(\zeta_\mathcal{L}; D)(-\zeta(D)) \frac{x^D}{D} + o(x^D), \quad \text{as } x \to \infty, \quad (5.57)$$

interpreted distributionally. Again, when there is no screen passing between $\operatorname{Re} s = D$ and the complex dimensions to the left of this line, one has to apply the technique of Remark 5.14 to derive this formula, still interpreted distributionally.

For the spectral partition function, one obtains

$$\theta_\nu(t) = |\mathcal{L}| \cdot t^{-1} - \operatorname{res}(\zeta_\mathcal{L}; D)(-\zeta(D)) \Gamma(D) t^{-D} + o(t^{-D}), \quad \text{as } t \to 0^+.$$

$$(5.58)$$

We leave it as an exercise for the interested reader to apply the formulas obtained for the geometric and spectral density of states in Section 5.3.1.

Remark 5.18. We point out that in (5.57) and (5.58), the error term is the best possible. Indeed, by [Ti, Theorem 9.19(C), p. 204], the density of the zeros off the critical line $\operatorname{Re} s = 1/2$ of the Riemann zeta function is less than linear. On the other hand, by Theorem 2.13, every nonlattice string has complex dimensions with real part arbitrary close to D (from the left) with linear density. Hence these complex dimensions cannot all be cancelled by zeros of $\zeta(s)$. (A moment's reflection shows that this argument is valid for any D in $(0,1)$, including $D = 1/2$.)

5.4.4 The Prime Number Theorem for Suspended Flows

Let $\mathcal{L}$ be a self-similar string, normalized as in Remark 2.5. Our explicit formulas allow us to derive an estimate for the number of weighted primitive periodic orbits of the dynamical system associated with $\mathcal{L}$ and introduced in Section 2.1.1.

Let the weight of a periodic orbit $\mathfrak{x}$ be

$$w(\mathfrak{x}) = \left(r(\mathfrak{x}) r(\sigma\mathfrak{x}) \dots r(\sigma^{l(\mathfrak{x})-1}\mathfrak{x}) \right)^{-1},$$

where $l(\mathfrak{x})$ denotes the length of $\mathfrak{x}$, as in Section 2.1.1. It is most natural to count the periodic orbits by their weight, as follows:[7]

$$\psi_{\mathcal{L}}(x) = \sum_{w(\mathfrak{x}) \leq x} \frac{1}{l(\mathfrak{x})} \log w(\mathfrak{x}). \tag{5.59}$$

Recall that we have shown in Theorem 2.6 that the geometric zeta function $\zeta_{\mathcal{L}}(s)$ could be interpreted as a dynamical (or Ruelle) zeta function; see formula (2.6). Further, in Theorem 2.7, we have obtained an Euler product for $\zeta_{\mathcal{L}}(s)$, expressed in terms of the primitive periodic orbits; see formula (2.7).

It follows from (2.7) that

$$-\frac{\zeta_{\mathcal{L}}'(s)}{\zeta_{\mathcal{L}}(s)} = \sum_{\mathfrak{x}} \frac{\log w(\mathfrak{x})}{l(\mathfrak{x})} w(\mathfrak{x})^{-s}. \tag{5.60}$$

We see that $\psi_{\mathcal{L}}(x)$ is the geometric counting function of this zeta function, and we thus obtain the following explicit formula for $\psi_{\mathcal{L}}(x)$:

$$\psi_{\mathcal{L}}(x) = \sum_{\omega \in \mathcal{D}_{\mathcal{L}}} -\operatorname{ord}\left(\zeta_{\mathcal{L}}; \omega\right) \frac{x^{\omega}}{\omega} + R(x), \tag{5.61}$$

---

[7] The function $\psi_{\mathcal{L}}(x)$ is the counterpart of $\psi(x) = \sum_{p^k \leq x} \log p$, which counts prime powers p^k with a weight $\log p$.

where ord $(\zeta_{\mathcal{L}}; \omega)$ denotes the order of $\zeta_{\mathcal{L}}$ at ω, and

$$R(x) = -\int_S \frac{\zeta'_{\mathcal{L}}(s)}{\zeta_{\mathcal{L}}(s)} x^s \frac{ds}{s} = O\left(x^{\sigma_u}\right), \tag{5.62}$$

where σ_u is associated by formula (4.15b) with the screen of $\zeta_{\mathcal{L}}$.

In the nonlattice case, we use Theorem 2.33 according to which there exists a screen S lying to the left of the line $\operatorname{Re} s = D$, such that $\zeta_{\mathcal{L}}$ is bounded on S and all the complex dimensions ω to the right of S are simple: $\operatorname{ord}(\zeta_{\mathcal{L}}; \omega) = -1$. We then obtain the Prime Number Theorem for nonlattice suspended flows:

$$\psi_{\mathcal{L}}(x) = \frac{x^D}{D} + o\left(x^D\right), \tag{5.63}$$

as $x \to \infty$. We note that this estimate is best possible, since by the nonlattice case of Theorem 2.13, there always exist complex dimensions of $\mathcal{L}$ arbitrarily close to the vertical line $\operatorname{Re} s = D$.

In the lattice case, we obtain the Prime Number Theorem for lattice suspended flows:

$$\psi_{\mathcal{L}}(x) = x^D g_1(\log x) + O\left(x^{D-\alpha}\right), \tag{5.64}$$

as $x \to \infty$. Here, $D - \alpha$ is the abcissa of the first vertical line of complex dimensions next to D, and the periodic function g_1, of period $2\pi/\mathbf{p}$, is given by

$$g_1(u) = \sum_{n=-\infty}^{\infty} \frac{e^{inpu}}{D + in\mathbf{p}} = \frac{b}{b-1} b^{-\{up/2\pi\}} \frac{2\pi}{\mathbf{p}}, \tag{5.65}$$

where $\log b = 2\pi D/\mathbf{p}$, by (5.34). By choosing a screen located to the left of all the complex dimensions of $\mathcal{L}$, we can even obtain more precise information about $\psi_{\mathcal{L}}$. In the notation of Theorem 2.13, we obtain

$$\psi_{\mathcal{L}}(x) = \sum_{j=1}^{q} -\operatorname{ord}\left(\zeta_{\mathcal{L}}; \omega_j\right) \sum_{n \in \mathbb{Z}} \frac{x^{\omega_j + in\mathbf{p}}}{\omega_j + in\mathbf{p}}$$
$$= \sum_{j=1}^{q} g_j(\log x) x^{\operatorname{Re} \omega_j}, \tag{5.66}$$

where for each $j = 1, \ldots, q$, the function g_j is periodic of period $2\pi/\mathbf{p}$, given by

$$g_j(u) = -\operatorname{ord}\left(\zeta_{\mathcal{L}}; \omega_j\right) \sum_{n \in \mathbb{Z}} \frac{e^{i(n\mathbf{p} + \operatorname{Im} \omega_j)u}}{\omega_j + in\mathbf{p}}. \tag{5.67}$$

Here, $\omega_1 (= D), \omega_2, \ldots, \omega_q$ are given as in the lattice case of Theorem 2.13 and $\operatorname{ord}(\zeta_{\mathcal{L}}; \omega_1) = -1$.

For example, for the Cantor string, we obtain (with $D = \log_3 2$ and $\mathbf{p} = 2\pi/\log 3$)

$$\psi_{\mathrm{CS}}(x) = g_1(\log x) x^D, \tag{5.68}$$

while for the Fibonacci string of Section 2.2.2, we have (with $D = \log_2 \phi$ and $\mathbf{p} = 2\pi/\log 2$)

$$\psi_{\mathrm{Fib}}(x) = g_1(\log x) x^D + g_2(\log x) x^{-D}, \tag{5.69}$$

where g_1 is given by (5.65) and

$$g_2(u) = \sum_{n \in \mathbb{Z}} \frac{e^{i(n+1/2)\mathbf{p}u}}{-D + i(n + 1/2)\mathbf{p}}.$$

Finally, for the lattice string with multiple poles, introduced in Section 2.2.3, the explicit formula becomes (with $D = \log_3 2$ and $\mathbf{p} = 2\pi/\log 3$):

$$\psi_{\mathcal{L}}(x) = g_1(\log x) x^D + g_2(\log x), \tag{5.70}$$

where

$$g_2(u) = 2 \sum_{n \in \mathbb{Z}} \frac{e^{i(n+1/2)\mathbf{p}u}}{i(n + 1/2)\mathbf{p}}.$$

Remark 5.19. We can apply our results to obtain explicit formulas for the more general dynamical systems considered, for example, in [PaPol] or [Lal2–3] and the relevant references therein. We intend to do this in a later work. Even in the present situation of suspended flows associated with self-similar strings, our results are significantly more precise than those available in the literature because the explicit formula (5.61) not only yields a Prime Number Theorem with Error Term, but also a full expansion of the prime power counting function in terms of the complex dimensions of $\mathcal{L}$ (that is, by Theorem 2.6, the poles of the associated dynamical zeta function).

5.5   Examples of Non-Self-Similar Strings

Let η be a generalized fractal string with geometric zeta function ζ_η. In general, $\zeta_\eta(s)$ does not need to have a continuation as a meromorphic function beyond the line $\operatorname{Re} s = D$. Or the meromorphic continuation does not need to satisfy $(\mathbf{H}_1)$ and $(\mathbf{H}_2)$ beyond a certain screen. Thus, the analysis of ζ_η and the choice of a good screen become the key to understanding the geometry and the spectrum of a fractal string. We give here an example of such an analysis.

5.5.1   The a-String

Given $a > 0$, an arbitrary positive real number, we consider the ordinary fractal string $\mathcal{L}$ with lengths

$$l_j = j^{-a} - (j+1)^{-a}, \quad j = 1, 2, \ldots . \tag{5.71}$$

This string (which we call the a-string) has already been discussed in a related context in [Lap1, Example 5.1, pp. 512–513] and was later on revisited in [LapPo2, pp. 64–65]. However, thanks to Theorem 5.20 below and our explicit formulas, we will be able to obtain much more precise results than in these earlier papers.

Recall from [Lap1, Example 5.1] that $\mathcal{L}$ can be realized as the open set $\Omega \subset \mathbb{R}$ obtained by removing the points j^{-a} ($j = 1, 2, \ldots$) from the interval $(0, 1)$; namely,

$$\Omega = \bigcup_{j=1}^{\infty} \left((j+1)^{-a}, j^{-a} \right). \tag{5.72}$$

Hence, its boundary is the countable compact subset of $\mathbb{R}$ given by

$$\partial\Omega = \left\{ j^{-a} : j = 1, 2, \ldots \right\} \cup \{0\}. \tag{5.73}$$

We begin by determining the (potential) complex dimensions of $\mathcal{L}$, that is, the poles of $\zeta_{\mathcal{L}}$.[8]

Theorem 5.20. *Let $a > 0$ and let $\mathcal{L}$ be the ordinary fractal string with lengths l_j given by (5.71). Then $\zeta_{\mathcal{L}}(s) = \sum_{j=1}^{\infty} l_j^s$ has a meromorphic continuation to all of $\mathbb{C}$. The poles of $\zeta_{\mathcal{L}}$ are located at $\frac{1}{a+1}$ and at (a subset of) the points $-\frac{1}{a+1}, -\frac{2}{a+1}, -\frac{3}{a+1}, \ldots$, and they are all simple. In particular, the dimension of $\mathcal{L}$ is $D = \frac{1}{a+1}$, and this is the only pole of $\zeta_{\mathcal{L}}$ with positive real part. The residue of $\zeta_{\mathcal{L}}$ at this pole is equal to a^D.*

Further, for any screen not passing through a pole, $\zeta_{\mathcal{L}}$ satisfies (H$_1$) and (H$_2$) with $\kappa = \frac{1}{2} - \sigma_1(a+1)$ or $\kappa = \frac{1}{2}$ if $\sigma_1 \le 0$ or $\sigma_1 \ge 0$, respectively. (Here, σ_1 is defined as in (4.15a).)

Proof. We compute the first term of an asymptotic expansion of l_j:

$$l_j = j^{-a} - (j+1)^{-a} = a \int_j^{j+1} x^{-a-1}\, dx = aj^{-a-1} + H(j),$$

where $H(j) = a \int_j^{j+1} \left(x^{-a-1} - j^{-a-1} \right) dx$. It follows that

$$h_j := a^{-1} j^{a+1} H(j) = j \int_0^{1/j} \left((1+t)^{-a-1} - 1 \right) dt. \tag{5.74}$$

---

[8]We are grateful to Driss Essouabri [Es1–2] for providing us with the main idea of the proof of Theorem 5.20.

Note that $h_j = O(1/j)$, as $j \to \infty$. Choose an integer $M \geq 0$. Then

$$l_j^s = \left(aj^{-a-1}\left(1 + h_j\right)\right)^s$$

$$= a^s j^{-s(a+1)} \left(\sum_{n=0}^{M} \binom{s}{n} h_j^n + O\left(\frac{(|s|+1)^{M+1}}{j^{M+1}}\right)\right),$$

where we have set

$$\binom{s}{n} = \frac{s(s-1)\ldots(s-n+1)}{n!}, \qquad \text{for } s \in \mathbb{C}. \tag{5.75}$$

We thus obtain

$$\zeta_{\mathcal{L}}(s) = \sum_{n=0}^{M} a^s \binom{s}{n} \sum_{j=1}^{\infty} h_j^n j^{-s(a+1)} + f(s), \tag{5.76}$$

where $f(s)$ is defined and holomorphic for $\operatorname{Re} s > -\frac{M}{a+1}$. The first term of this sum, for $n = 0$, is $a^s \zeta((a+1)s)$. Thus we find the first pole at $s = \frac{1}{a+1}$. Note that the first term grows on vertical lines $\operatorname{Re} s = \sigma < 0$ as $(|t|+1)^{\frac{1}{2}-\sigma(a+1)}$.

It remains to analyze the functions

$$\sum_{j=1}^{\infty} h_j^n j^{-s(a+1)}, \tag{5.77}$$

for $n \geq 1$. We will show that these functions are meromorphic with simple poles at the points $0, -\frac{1}{a+1}, -\frac{2}{a+1}, \ldots$.
Using the asymptotic expansion $(1+t)^{-a-1} = \sum_{k=0}^{M} \binom{-a-1}{k} t^k + O\left(t^{M+1}\right)$, as $t \to 0$, we obtain, in view of (5.74),

$$h_j = j \int_0^{1/j} \sum_{k=1}^{M} \binom{-a-1}{k} t^k \, dt + O\left(j^{-M-1}\right)$$

$$= -\frac{1}{a} \sum_{k=1}^{M} \binom{-a}{k+1} j^{-k} + O\left(j^{-M-1}\right), \qquad \text{as } j \to \infty.$$

By taking the n-th power of this expansion, we find an asymptotic expansion for h_j^n. Substituting this expansion, we write each of the functions in (5.77) as a sum of constant multiples of $\zeta(k+(a+1)s)$ (which has a pole at $s = \frac{1-k}{a+1}$), for $n \leq k \leq M$. In view of Equation (5.76), we thus deduce that $\zeta_{\mathcal{L}}$ has a meromorphic continuation to $\operatorname{Re} s > -\frac{M}{a+1}$, with simple poles at $s = \frac{1-k}{a+1}$, $k = 0, 2, 3, 4, \ldots, M$. (Note that 0 is not a pole of $\zeta_{\mathcal{L}}$, due to the factor $\binom{s}{1} = s$ on the right-hand side of (5.76).) Since M is arbitrary, it follows that $\zeta_{\mathcal{L}}$ has a meromorphic continuation to all of $\mathbb{C}$. A direct computation shows that the residue of $\zeta_{\mathcal{L}}$ at $D = 1/(a+1)$ is equal to a^D.

Finally, for $k \geq 1$, the growth of $\zeta(k + (a+1)s)$ is superseded by the growth of the first term $a^s \zeta((a+1)s)$. Thus we can choose $\kappa = \frac{1}{2} - \sigma_1(a+1)$ for any screen $\text{Re } s = \sigma_1 < 0$. $\qquad \square$

It follows from Theorem 5.20 (as well as from Theorems 4.12 and 4.23) that the geometric counting function of $\mathcal{L}$ satisfies

$$N_{\mathcal{L}}(x) = a^D x^D + \zeta_{\mathcal{L}}(0) + O\left(x^{-D}\right), \qquad \text{as } x \to \infty, \qquad (5.78)$$

while its spectral counting function satisfies

$$N_\nu(x) = x + \zeta(D)a^D x^D - \frac{1}{2}\zeta_{\mathcal{L}}(0) + O\left(x^{-D}\right), \qquad \text{as } x \to \infty. \qquad (5.79)$$

(Recall from [Ti] that $\zeta(D) < 0$ since $D = \frac{1}{a+1} \in (0,1)$. Also note that $\zeta_{\mathcal{L}}(0) = -1/2$ because the first term in (5.76) is $a^s \zeta((a+1)s)$, $\zeta(0) = -1/2$, and all other terms vanish at $s = 0$ since $\binom{0}{n} = 0$ for $n \geq 1$.)

We note that in [LapPo2, Example 4.3], the error term is only of the form $o(x^D)$ in the counterpart of (5.79) (or of (5.78)). (See, in particular, [LapPo2, Equation (4.25), p. 65].)

Actually, since the poles of $\zeta_{\mathcal{L}}$ are all simple, it follows from Theorem 5.20 and our explicit formulas that Equations (5.78) and (5.79) can be replaced respectively by the much more precise expressions

$$N_{\mathcal{L}}(x) = a^D x^D + \zeta_{\mathcal{L}}(0) + \sum_{k=1}^{M} \text{res}\left(\zeta_{\mathcal{L}}; -kD\right) x^{-kD}$$
$$+ O\left(x^{-(M+1)D}\right), \qquad \text{as } x \to \infty, \qquad (5.78')$$

and

$$N_\nu(x) = x + \zeta(D)a^D x^D - \frac{1}{2}\zeta_{\mathcal{L}}(0) + \sum_{k=1}^{M} \text{res}\left(\zeta_{\mathcal{L}}; -kD\right)\zeta(-kD)x^{-kD}$$
$$+ O\left(x^{-(M+1)D}\right), \qquad \text{as } x \to \infty, \qquad (5.79')$$

valid for every $M = 0, 1, 2, \ldots$.

Remark 5.21. Depending on the arithmetic properties of the parameter a, the residue of $\zeta_{\mathcal{L}}$ at $s = -kD$ may vanish for some values of $k \in \mathbb{N}^*$. This is why one cannot in general specify the exact set of complex dimensions of the a-string $\mathcal{L}$ in the statement of Theorem 5.20.

5.5.2   The Spectrum of the Harmonic String

An interesting example of a non-self-similar fractal string is provided by the harmonic string, introduced and studied in [Lap2, Example 5.4(ii), pp. 171–172] or [Lap3, Remark 2.5, pp. 144–145]. Recall from Equation (3.10) above that it is given by the measure $h = \sum_{j=1}^{\infty} \delta_{\{j\}}$ and hence has lengths $l_j = \frac{1}{j}$ for $j = 1, 2, \ldots$. (Note that this string has infinite total length $\sum_{j=1}^{\infty} \frac{1}{j} = \infty$.) Thus

$$\zeta_h(s) = \zeta(s) \tag{5.80}$$

and

$$\zeta_\nu(s) = \zeta_h(s) \cdot \zeta(s) = (\zeta(s))^2. \tag{5.81}$$

It follows from (5.81) that at the geometric level, $s = 1$ is the only pole of $\zeta_h(s)$, and that it is simple, while at the spectral level, $s = 1$ is the only pole of $\zeta_\nu(s)$ and has multiplicity two.

Since we have, as was noted in [Lap2, p. 171] or [Lap3, p. 144],[9]

$$N_\nu(x) = \sum_{1 \le k \le x} \tau(k), \tag{5.82}$$

where $\tau(k)$ denotes the number of divisors of the integer k, it follows from Equation (5.81) above (and our error estimates) that

$$N_\nu(x) = \sum_{1 \le k \le x} \tau(k) = x \log x + (2\gamma - 1)x + o(x), \tag{5.83}$$

as $x \to \infty$, where γ denotes Euler's constant. This well known formula was recovered in a similar manner in [Lap2–3] and is often referred to as the Dirichlet divisor formula. A long standing open problem—called the Dirichlet divisor problem in the literature [Ti, §12.1, pp. 312–314]—consists of obtaining a much sharper form for the error term on the right-hand side of (5.83).

5.6   Fractal Sprays

We consider fractal sprays, defined in Section 1.4 (following [LapPo3]) and, more generally, in Section 3.3. For simplicity, we assume that $\zeta_B(s) =$

---

[9]Indeed, as is observed in [Lap2–3], the frequencies of the harmonic string consist of the sequence of positive integers $1, 2, 3, \ldots$, with multiplicity respectively equal to $\tau(1), \tau(2), \tau(3), \ldots$. With our notation from Section 3.1 and in view of formulas (3.10) and (3.19), this can be seen as follows: $\nu = h * h = \sum_{m,n=1}^{\infty} \delta_{\{mn\}} = \sum_{k=1}^{\infty} \tau(k)\delta_{\{k\}}$.

$\sum_f w_f f^{-s}$, the spectral zeta function of the basic shape B, has at most finitely many (visible) poles, called the *spectral complex dimensions of B*.[10] Moreover, we restrict our attention to those fractal sprays of a (possibly generalized) fractal string η on B for which no complex dimension of η coincides with a spectral complex dimension of B.

The *Weyl term* at level k associated with the fractal spray of η on B is

$$W_{B,\eta}^{[k]}(x) = \sum_{u:\,\text{pole of }\zeta_B} \operatorname{res}\left(\frac{\zeta_B(s)\zeta_\eta(s)}{(s)_k}x^{s+k-1}; u\right). \tag{5.84}$$

The terms in this sum can be analyzed in the same way as in Section 5.1.1. The result depends on the multiplicity of the spectral complex dimensions. In particular, if these complex dimensions are simple, we obtain the formula

$$W_{B,\eta}^{[k]}(x) = \sum_{u:\,\text{pole of }\zeta_B} \operatorname{res}\left(\zeta_B(s); u\right)\zeta_\eta(u)\frac{x^{u+k-1}}{(u)_k}. \tag{5.85}$$

Recall from Equation (3.35) that the spectral zeta function of this fractal spray is given by

$$\zeta_\nu(s) = \zeta_\eta(s)\cdot\zeta_B(s). \tag{5.86}$$

Therefore, much as was done for fractal strings in Section 5.3.2, we state the following definition:

Definition 5.22. The *spectral operator* (for fractal sprays) is the operator that adds the Weyl term and multiplies each term corresponding to a complex dimension of η (assumed to be simple) by $\zeta_B(\omega)$, and also multiplies the integrand in the error term by $\zeta_B(s)$.

We will focus here on examples of self-similar sprays. However, it should be clear to the reader that non-self-similar fractal sprays—such as those studied, for example, in [LapPo3]—can be treated as well, once their complex dimensions have been analyzed.

Before considering general self-similar sprays in Section 5.6.2, we illustrate our methods applied to fractal sprays by the following example. See also Section 9.4, formulas (9.28), (9.29) and (9.31) for a further example of a fractal spray and of the corresponding Weyl term (in a case where $\zeta_B(s)$ has infinitely many poles). Further, see formulas (9.6) and (9.7) for the simpler Cantor sprays considered in Section 9.2.

---

[10]Thus, 1 is the only spectral complex dimension of the Bernoulli string.

5.6.1   The Sierpinski Drum

The Dirichlet Laplacian on the equilateral triangle T with sides 1 has for eigenvalue spectrum (see [Pi, Note on p. 820 and footnote 1] or [Bér]),

$$\lambda_{m,n} = \frac{16\pi^2}{9}\left(m^2 + mn + n^2\right), \quad m, n = 1, 2, 3, \ldots,$$

and (in view of our convention defining the frequencies as $\sqrt{\lambda_{m,n}}/\pi$) the corresponding spectral zeta function is equal to

$$\zeta_T(s) = \left(\frac{3}{4}\right)^s \sum_{m,n=1}^{\infty} \left(m^2 + mn + n^2\right)^{-s/2}. \tag{5.87}$$

We find the poles and the corresponding residues of this function as follows. The zeta function of the cyclotomic field $\mathbb{Q}[\rho]$, obtained by adjoining a cubic root of unity to the rationals, is given by

$$\zeta_{\mathbb{Q}[\rho]}(s) = \sum_{m=0}^{\infty} \sum_{n=1}^{\infty} \left(m^2 + mn + n^2\right)^{-s}. \tag{5.88}$$

It has a simple pole at $s = 1$ with residue $\pi/(3\sqrt{3})$; see Appendix A, Equation (A.4). It is related to $\zeta_T(s)$ by the equation

$$\zeta_T(s) = \left(\frac{3}{4}\right)^s \zeta_{\mathbb{Q}[\rho]}(s/2) - \left(\frac{3}{4}\right)^s \zeta(s). \tag{5.89}$$

Thus $\zeta_T(s)$ has a simple pole at $s = 2$, with residue $\pi\sqrt{3}/8$, and one at $s = 1$, with residue $-3/4$.

Consider the spray of $\mathcal{L}$ on T obtained by scaling the middle triangle in Figure 5.1 by the self-similar string $\mathcal{L}$ with scaling ratios $r_1 = r_2 = r_3 = \frac{1}{2}$. The boundary of this spray is the classical Sierpinski gasket;[11] see Figure 5.1. We will call this spray the *Sierpinski drum*. Hence, it corresponds to the Dirichlet Laplacian on the infinitely connected bounded open subset of $\mathbb{R}^2$ with boundary the Sierpinski gasket.

In light of Equation (5.86), the spectral zeta function of the Sierpinski drum is given by

$$\zeta_\nu(s) = \zeta_{\mathcal{L}}(s)\zeta_T(s) = \sum_{n=0}^{\infty} 3^n 2^{-ns} \zeta_T(s) = \frac{1}{1 - 3 \cdot 2^{-s}} \zeta_T(s), \tag{5.90}$$

---

[11]Note that $\mathcal{L}$ has dimension $\log_2 3$, which lies between 1 and 2. Hence, strictly speaking, it is not a string in the sense of Section 1.1. Thus, the triangles constituting the Sierpinski gasket would not fit on a line of finite length if they were all aligned, but they do fit inside a bounded region in the plane.

Figure 5.1: The Sierpinski gasket, the boundary of the Sierpinski drum.

and its Weyl term is, by (5.85) along with the comment following (5.89),

$$W_{T,\mathcal{L}}^{[1]}(x) = \frac{\pi\sqrt{3}}{4}x^2 + \frac{3}{2}x. \tag{5.91}$$

From the explicit formula for the geometric counting function of $\mathcal{L}$,

$$N_{\mathcal{L}}(x) = \frac{1}{\log 2}\sum_{n=-\infty}^{\infty}\frac{x^{D+in\mathbf{p}}}{D+in\mathbf{p}} - \frac{1}{2}, \tag{5.92}$$

where $D = \log_2 3$ and $\mathbf{p} = 2\pi/\log 2$, we deduce the explicit formula for the counting function of the frequencies of the Sierpinski drum,

$$N_{T,\mathcal{L}}(x) = \frac{\pi\sqrt{3}}{4}x^2 + x^D g(\log x) + \frac{3}{2}x - \frac{1}{2}\zeta_T(0) + o(1), \tag{5.93}$$

as $x \to \infty$. Here, g is the periodic function, of period $2\pi/\mathbf{p} = \log 2$, given by the Fourier series

$$g(u) = \frac{1}{\log 2}\sum_{n=-\infty}^{\infty}\frac{e^{in\mathbf{p}u}}{D+in\mathbf{p}}\zeta_T(D+in\mathbf{p}), \tag{5.94}$$

where $\zeta_T(s)$ is given by (5.87) or (5.89).

Remark 5.23. It is noteworthy that the spectral zeta function $\zeta_\nu(s)$ of the Sierpinski drum has poles at nonreal values of s. (Specifically, according to (5.90) and the comment following (5.89), the spectral complex dimensions of this fractal spray are located at $s = 2$ and at $s = D + in\mathbf{p}$, $n \in \mathbb{Z}$, with D and $\mathbf{p}$ as above.) This is in contrast to the spectral zeta function of a smooth manifold, which has only real poles; see Theorem B.5 and Remark B.6 of Appendix B.

Theorem 5.24. *The Sierpinski drum has oscillations of order D in its spectrum, where $D = \log_2 3$ is the Minkowski dimension of the Sierpinski gasket.*

More precisely, the asymptotic behavior as $x \to \infty$ of the spectral counting function of the Sierpinski drum is given by the formula (5.93), with g given by (5.94). Further, g is a nonconstant *periodic function, with period $\log 2$.*

Proof. The fact that g is nonconstant follows from Theorem 9.5 below, according to which the Dirichlet series $\zeta_T(s)$ does not have an infinite vertical sequence of zeros in arithmetic progression. It follows that the expansion (5.94), interpreted distributionally (as, for example, in [Sch1, Section VII.1]), has infinitely many nonzero terms.   □

Remark 5.25. In view of [Pi], an entirely analogous result can be obtained for Neumann rather than for Dirichlet boundary conditions. Moreover, in light of [Bér], fractal sprays with more general basic shapes associated with crystallographic groups can be analyzed in the same manner.

Remark 5.26. One can also obtain the counterpart of Theorem 5.24 for the drum with fractal boundary the Sierpinski carpet, or its higher-dimensional analogue, the Menger sponge. (See, for instance, [Man1, Plate 145, pp. 144–145] for a picture of these classical fractals.) In particular, the basic shape of this fractal spray is the unit square or the unit cube, respectively. It is also interesting to study similarly the main example of Brossard and Carmona [BroCa], revisited and extended by Fleckinger and Vassiliev in [FlVa]. Finally, as an exercise, the reader may wish to consider the Cantor spray introduced in Section 1.4; see Figure 1.5. We note that more general Cantor sprays will be studied in Section 9.2 below.

The Sierpinski drum, as well as the examples mentioned in Remark 5.26, is an example of a lattice self-similar spray, in the sense of Theorem 5.27 below. (Its oscillatory period is $p = 2\pi/\log 2$.) Such a self-similar drum has been studied, in particular, by the first author in [Lap2; Lap3, Section 4.4.1] and, in more detail, by Gerling [Ger] and Gerling and Schmidt [GerSc1–2]. We note that the fact that the periodic function g in (5.94) is nonconstant— and thus that the Sierpinski drum has oscillations of order D in its spectrum, as stated in Theorem 5.24 above—was not established in these references. Within our framework, it is a direct consequence of Theorem 9.5, which provides us with a very useful tool for proving such results. See Remark 5.29 below for further references and extensions, which can also be dealt with by our methods.

5.6.2   The Spectrum of a Self-Similar Spray

By entirely analogous methods—based on our explicit formulas and on Theorem 9.5 in Section 9.2 below—we can establish the analogue for self-similar sprays of the results obtained in Section 5.4 for self-similar strings.

Theorem 5.27. *The exact counterpart of the results of Section 5.4 above holds for self-similar sprays (rather than for self-similar strings). (See Remark 5.28 below for more precision.)*

In particular, just as in Section 5.4.3, the spectrum of a lattice spray always has (multiplicatively periodic) oscillations of order D, the dimension of the self-similar string used to define the spray, whereas the spectrum of a nonlattice spray never does[12] *(in agreement with [Lap3, Conjecture 3, p. 163] stated for more general self-similar drums).*

Remark 5.28. By definition, a *self-similar spray* is a fractal spray with basic shape B scaled by a self-similar string $\mathcal{L}$.[13] Moreover, in Theorem 5.27,

---

[12]unless the spectrum of B itself has such oscillations. This is never the case if B is a connected manifold with piecewise smooth boundary. On the other hand, if B has a fractal boundary, it can have such oscillations in its spectrum.

[13]which is allowed to have dimension greater than 1, as was the case for the Sierpinski drum (see footnote 11 at the bottom of page 138).

we assume that $\zeta_B(s)$, the spectral zeta function of B, satisfies hypothesis (P), formulated just before the statement of Theorem 9.5. We further suppose that no complex dimension of $\mathcal{L}$ coincides with the poles of $\zeta_B(s)$. Hypothesis (P) will be satisfied, for instance, when $\zeta_B(s)$ is the spectral zeta function of the Dirichlet (or Neumann) Laplacian on a piecewise smooth bounded domain of $\mathbb{R}^d$ (and the window W is not all of $\mathbb{C}$), provided that the dimension of $\mathcal{L}$ is less than d. (In that case, D is also the dimension of the boundary of the spray.) In particular, it is true for all the examples mentioned in Section 5.6.1 and in the next remark.

Remark 5.29. Theorem 5.27 extends and specifies the corresponding results obtained earlier in [Ger, GerSc1–2, Lap3, Section 4.4.1b; LeVa, vB-Le]. In particular, the statement about the presence of oscillations of order D in the spectrum of lattice sprays seems to be new and cannot be established by means of the Renewal Theorem [Fel, Theorem 2, p. 39], which is used in [Lap3, LeVa]. (See Remark 6.27 for a similar comment in a related situation.) It has been verified, however, in special cases, such as for the main example of [FlVa], extended in [vB-Le].

6

Tubular Neighborhoods
and Minkowski Measurability

In this chapter, we apply our extended distributional explicit formula (Theorem 4.20, derived in Section 4.4.2) to obtain a formula for the volume of the tubular neighborhoods of the boundary of a fractal string. (See Section 6.1.) In Section 6.2, we then deduce from this formula a new criterion for the Minkowski measurability of a fractal string, in terms of its complex dimensions. This completes and extends the earlier criterion obtained in [LapPo1–2].

The results obtained in this chapter provide further insight into the geometric meaning of the notion of complex dimension and suggest new analogies between aspects of fractal and Riemannian geometry. See, in particular, Section 6.1.1.

In Section 6.3, we illustrate our results by discussing several classes of examples. In particular, in Section 6.3.1, we discuss the class of self-similar strings.

Prior to reading this chapter, some readers may find it helpful to briefly consult the beginning of Chapter 1, Section 1.1, where the notions of Minkowski dimension and measurability are defined.

6.1 Explicit Formula for the Volume of a Tubular Neighborhood

Let $\mathcal{L}$ be a standard fractal string given by the sequence of lengths $(l_j)_{j=1}^{\infty}$ and of Minkowski dimension $D \in (0,1)$. Let $\eta = \sum_{j=1}^{\infty} \delta_{\{l_j^{-1}\}}$ be the associated measure.

Given $\varepsilon > 0$, let $V(\varepsilon)$ denote the volume of the ε-neighborhood of the boundary of $\mathcal{L}$, as defined by (1.3). By formula (1.9), the volume is given by

$$V(\varepsilon) = \sum_{j:\, l_j \geq 2\varepsilon} 2\varepsilon + \sum_{j:\, l_j < 2\varepsilon} l_j. \tag{6.1}$$

(See also [LapPo1], [LapPo2, Eq. (3.2)].) We rewrite this expression as

$$V(\varepsilon) = \int_0^{\frac{1}{2\varepsilon}} 2\varepsilon\, \eta(dx) + \int_{\frac{1}{2\varepsilon}}^{\infty} \frac{1}{x}\, \eta(dx)$$
$$= \left\langle \mathcal{P}^{[0]}\eta, v_\varepsilon \right\rangle, \tag{6.2}$$

where $v_\varepsilon(x)$ is the function defined by

$$v_\varepsilon(x) = \begin{cases} 2\varepsilon & \text{for } x \leq (2\varepsilon)^{-1}, \\ 1/x & \text{for } x > (2\varepsilon)^{-1}. \end{cases} \tag{6.3}$$

We assume that $\zeta_{\mathcal{L}}$ has a meromorphic continuation to some neighborhood of $\operatorname{Re} s \geq D$ and that the resulting function satisfies $(\mathbf{H_1})$ and $(\mathbf{H_2})$. Our extended distributional formula (Theorem 4.20) does not apply directly since the test function v_ε is not sufficiently differentiable. Therefore we interpret $V(\varepsilon)$ itself as a distribution.

Theorem 6.1 (The tube formula). *The volume of the (one-sided) tubular neighborhood of radius ε of the boundary of $\mathcal{L}$ is given by the following distributional explicit formula, on test functions in $\mathbf{D}(0,\infty)$, the space of C^∞ functions with compact support contained in $(0,\infty)$:*

$$V(\varepsilon) = \sum_{\omega \in \mathcal{D}_{\mathcal{L}}(W)} \operatorname{res}\left(\frac{\zeta_{\mathcal{L}}(s)(2\varepsilon)^{1-s}}{s(1-s)}; \omega \right) + \{2\varepsilon\zeta_{\mathcal{L}}(0)\} + \mathcal{R}(\varepsilon), \tag{6.4}$$

where the term in braces is only included if $0 \in W \backslash \mathcal{D}_{\mathcal{L}}(W)$, and $\mathcal{R}(\varepsilon)$ is the error term, given by

$$\mathcal{R}(\varepsilon) = \frac{1}{2\pi i} \int_S (2\varepsilon)^{1-s} \zeta_{\mathcal{L}}(s) \frac{ds}{s(1-s)}. \tag{6.5}$$

It is estimated by

$$\mathcal{R}(\varepsilon) = O\left(\varepsilon^{1-\sigma_u}\right), \tag{6.6}$$

as $\varepsilon \to 0^+$. (Here, σ_u is given by Equation (4.15b).)

Moreover, if $\mathcal{L}$ satisfies the stronger hypotheses $(\mathbf{H}_1)$ and $(\mathbf{H}_2')$ with $W = \mathbb{C}$ (rather than $(\mathbf{H}_1)$ and $(\mathbf{H}_2)$), and we apply this formula to a test function supported on a compact subset of $[0, 1/(2A))$, then we have $\mathcal{R}(\varepsilon) \equiv 0$, and the term $2\varepsilon\zeta_{\mathcal{L}}(0)$ is included if $0 \notin \mathcal{D}_{\mathcal{L}}(\mathbb{C})$.

Proof. Let $\varphi(\varepsilon)$ be a test function on $(0, \infty)$. Then, by (6.3),

$$\int_0^\infty \varphi(\varepsilon)v_\varepsilon(x)\,d\varepsilon = \int_0^{\frac{1}{2x}} 2\varepsilon\varphi(\varepsilon)\,d\varepsilon + \frac{1}{x}\int_{\frac{1}{2x}}^\infty \varphi(\varepsilon)\,d\varepsilon \tag{6.7}$$

$$= \varphi_1(x) + \varphi_2(x),$$

where

$$\varphi_1(x) := \int_0^{\frac{1}{2x}} 2\varepsilon\varphi(\varepsilon)\,d\varepsilon \tag{6.8a}$$

and

$$\varphi_2(x) := \frac{1}{x}\int_{\frac{1}{2x}}^\infty \varphi(\varepsilon)\,d\varepsilon. \tag{6.8b}$$

Thus, in light of (6.2) and (6.7),

$$\langle V(\varepsilon), \varphi \rangle = \int_0^\infty \varphi(\varepsilon) \int_0^\infty v_\varepsilon(x)\,\eta(dx)\,d\varepsilon$$

$$= \left\langle \mathcal{P}^{[0]}\eta, \varphi_1 + \varphi_2 \right\rangle.$$

If φ is compactly supported, then the function φ_1 is constant near 0 and vanishes for $x \gg 0$, while the function φ_2 vanishes near 0 and is equal to $\frac{1}{x}\int_0^\infty \varphi(\varepsilon)\,d\varepsilon$ for $x \gg 0$. Hence the explicit formula of Theorem 4.20, the extended distributional formula with error term, applies to φ_1 and to φ_2, with $k = 0$ and $q > \kappa + 1$. Note that

$$\widetilde{\varphi_1}(s) = \frac{2^{1-s}}{s}\widetilde{\varphi}(2 - s) \tag{6.9a}$$

and

$$\widetilde{\varphi_2}(s) = \frac{2^{1-s}}{1 - s}\widetilde{\varphi}(2 - s). \tag{6.9b}$$

The first function (6.9a) has a pole at $s = 0$, and no other poles. If 0 is not a complex dimension, this pole gives the term $2\varepsilon\zeta_{\mathcal{L}}(0)$, to be included if $0 \in W$. Also

$$\tilde{v}_\varepsilon(s) = \frac{1}{s(1-s)}(2\varepsilon)^{1-s}. \tag{6.10}$$

This completes the proof of Equation (6.4), with the error term $\mathcal{R}(\varepsilon)$ given by (6.5). The error estimate (6.6) follows from the first part of Theorem 4.23.

Finally, the fact that $\mathcal{R}(\varepsilon)$ vanishes identically when $\mathcal{L}$ satisfies the stronger hypothesis ($\mathbf{H}_2'$) follows from the corresponding explicit formula without error term given in Theorem 4.21.   □

If the complex dimension ω (and hence also $\bar{\omega}$) is simple, then the associated local term in formula (6.4) is equal to

$$\text{res}\,(\zeta_{\mathcal{L}};\omega)\,\frac{(2\varepsilon)^{1-\omega}}{\omega(1-\omega)} = \text{res}\,(\zeta_{\mathcal{L}};\omega)\left(\frac{(2\varepsilon)^{1-\omega}}{\omega} + \frac{(2\varepsilon)^{1-\omega}}{1-\omega}\right). \tag{6.11}$$

In particular, when $\mathcal{L}$ is a self-similar string, this will be the case of all the complex dimensions on the vertical line $\text{Re}\,s = D$; see Theorem 2.13. We thus obtain the following corollary of Theorem 6.1:

Corollary 6.2. *If in Theorem 6.1, we assume in addition that all the visible complex dimensions of $\mathcal{L}$ are simple, and that 0 is not a complex dimension, then the sum over the complex dimensions in the 'tube formula' (6.4) becomes*

$$V(\varepsilon) = \sum_{\omega \in \mathcal{D}_{\mathcal{L}}(W)} \text{res}\,(\zeta_{\mathcal{L}};\omega)\,\frac{(2\varepsilon)^{1-\omega}}{\omega(1-\omega)} + \{2\varepsilon\zeta_{\mathcal{L}}(0)\} + \mathcal{R}(\varepsilon)$$

$$= \sum_{\omega \in \mathcal{D}_{\mathcal{L}}(W)} \text{res}\,(\zeta_{\mathcal{L}};\omega)\,\frac{(2\varepsilon)^{1-\omega}}{\omega} + \sum_{\omega \in \mathcal{D}_{\mathcal{L}}(W)} \text{res}\,(\zeta_{\mathcal{L}};\omega)\,\frac{(2\varepsilon)^{1-\omega}}{1-\omega} \tag{6.12}$$

$$+ \{2\varepsilon\zeta_{\mathcal{L}}(0)\} + \mathcal{R}(\varepsilon),$$

where the error term $\mathcal{R}(\varepsilon)$ is given by (6.5) and is estimated by (6.6).

Moreover, as in the statement of Theorem 6.1, $\mathcal{R}(\varepsilon) \equiv 0$ if $\mathcal{L}$ satisfies ($\mathbf{H}_1$) and ($\mathbf{H}_2'$) (rather than ($\mathbf{H}_1$) and ($\mathbf{H}_2$)).

Remark 6.3. In view of (4.15b) and (4.21), we have $\sigma_u \leq D$, and so estimate (6.6) implies that $\mathcal{R}(\varepsilon) = O\left(\varepsilon^{1-D}\right)$, as $\varepsilon \to 0^+$, both in Theorem 6.1 and Corollary 6.2. We point out that under the stronger assumptions of Theorem 6.12 below, we actually have the better estimate $\mathcal{R}(\varepsilon) = o\left(\varepsilon^{1-D}\right)$, as $\varepsilon \to 0^+$. (This follows from Theorem 4.24 and from the second part of Theorem 4.23.)

Remark 6.4. Since a self-similar string satisfies hypothesis $(\mathbf{H'_2})$, we can let $W = \mathbb{C}$, as was seen at the beginning of Section 5.4. We refer to Section 6.3.1 below for the precise form of the corresponding tube formula (6.4) in the lattice and nonlattice cases.

6.1.1 Analogy with Riemannian Geometry

There is an interesting analogy between formula (6.4) and H. Weyl's formula [Wey3] for the volume of the tubular ε-neighborhood of a compact, n-dimensional Riemannian submanifold of Euclidean space $\mathbb{R}^d$. Indeed, by [BergGo, Theorem 6.9.9, p. 235], this volume is given by the formula

$$V_M(\varepsilon) = a_0 \varepsilon^{d-n} + a_1 \varepsilon^{d-n+1} + \cdots + a_n \varepsilon^d. \tag{6.13}$$

When $V_M(\varepsilon)$ is the volume of the two-sided ε-neighborhood, the odd-numbered coefficients $a_1, a_3, \ldots$ all vanish, and for $j = 0, 1, 2, \ldots$, a_{2j} is given by

$$a_{2j} = \frac{1}{d - n + 2j} \int_M K_{2j} \delta, \tag{6.14}$$

where K_{2j} is a universal j-th degree polynomial in the curvature tensor of M, and δ is the canonical density on M (see [BergGo, Proposition 6.6.1, p. 214]). Thus, formula (6.13) is a polynomial in ε whose coefficients (6.14) are expressed in terms of the Weyl curvatures in different (integer) dimensions.

When $V_M(\varepsilon)$ is the volume of the one-sided ε-neighborhood (which corresponds more closely to our definition of $V(\varepsilon)$), then the counterpart of (6.13) still holds. In that case, the odd coefficients do not necessarily vanish. We refer to Sections 6.6–6.9, and especially Section 6.9.8 and Theorem 6.9.9 of [BergGo], and also to [Gra], for a much more detailed discussion and various helpful examples.

In the explicit formula (6.4), $V(\varepsilon)$ is expressed as an expansion in $\varepsilon^{1-\omega}$, where ω ranges over the visible complex dimensions. This suggests that the complex dimensions of fractal strings and the associated residues[1] could have a direct geometric interpretation.

Remark 6.5. The Weyl curvatures occurring in Weyl's tube formula were shown in [Wey3] to be intrinsic to the submanifold. One of them is closely tied to the notion of Euler characteristic and hence to the Gauss–Bonnet Formula. It is pointed out in [BergGo, 6.9.8, p. 235] that Chern's classical work [Chern1–2] on a higher-dimensional analogue of the Gauss–Bonnet formula (and on characteristic classes) was influenced by the geometric

---

[1]or, more generally, the Laurent expansions at these poles, when they have a higher multiplicity.

interpretation of Weyl's tube formula given in [Wey3].[2] Additional related information can be found, for example, in [CheeMüS1,Ko] and the relevant references therein.

Remark 6.6. Further references closely related to or extending in various directions Weyl's tube formula (or the earlier Steiner's formula [Stein] for convex bodies in Euclidean space [Fed2, Theorem 3.2.35]) include [Mink, Bl,Fed1–2, Ban,CheeMüS2,Gra,Ful–2, Mil], where a more detailed history and description of this formula can be found. We point out, in particular, the recent book by Gray entitled *Tubes* [Gra], as well as the extension in the context of geometric measure theory [Fed2] of the notion of Weyl's curvatures obtained by Federer in his paper on curvature measures [Fed1]. The more recent paper [Fu2], and the relevant references therein, gives a further generalization and interpretation of [Fed1].

6.2   Minkowski Measurability and Complex Dimensions

In Theorem 6.12 of this section, we obtain a new criterion for the Minkowski measurability of the boundary of a fractal string $\mathcal{L}$, expressed in terms of the complex dimensions of $\mathcal{L}$. This completes and extends (under our present assumptions) the earlier criterion obtained by the first author and C. Pomerance in [LapPo1–2]. We will also comment in Remarks 6.15 and 6.16 on the relationship between our new criterion and that obtained previously in [LapPo1–2].

Before stating and proving our main result, Theorem 6.12, we need to establish several technical lemmas (Lemmas 6.7, 6.8 and 6.10). At this point, the reader primarily interested in the results—rather than their proof—may wish to move ahead directly to the statement of Theorem 6.12.

Let $\mathcal{L}$ be an ordinary fractal string, or more generally, a (local) positive measure. We assume that $\zeta_{\mathcal{L}}$ satisfies $(\mathbf{H_1})$ and $(\mathbf{H_2})$. In particular, $s = D$ is a pole of $\zeta_{\mathcal{L}}$, and the singularities on the line $\mathrm{Re}\, s = D$ are poles of $\zeta_{\mathcal{L}}$.

Lemma 6.7. *Let $\mathcal{L}$ be an ordinary fractal string with lengths l_j ($j = 1, 2, \dots$). Then $N_{\mathcal{L}}(x) = O\left(x^D\right)$ as $x \to \infty$ if and only if $l_j = O\left(j^{-1/D}\right)$ as $j \to \infty$.*

Proof. If $N_{\mathcal{L}}(x) \leq C \cdot x^D$, then for $x = l_j^{-1}$ we find $j \leq C \cdot l_j^{-D}$. Thus $l_j \leq (C/j)^{1/D}$.

---

[2]The fact that Weyl's tube formula influenced this aspect of his work in [Chern1–2] was confirmed by S.-S. Chern during a recent conversation with the first author on the occasion of the symposium given in his honor at the University of California, Berkeley (and at the Mathematical Sciences Research Institute (MSRI)). (See also [Chern3].)

On the other hand, given $x > 0$, if $l_j \leq C \cdot j^{-1/D}$ for all $j = 1, 2, \ldots$, then $l_j^{-1} \geq x$ for $j \geq (Cx)^D$. Thus $N_{\mathcal{L}}(x) \leq (Cx)^D$. $\qquad\square$

Lemma 6.8. *Let $\mathcal{L}$ be a generalized fractal string given by a (local) positive measure η. If $N_{\mathcal{L}}(x) = O\left(x^D\right)$ as $x \to \infty$ or if $V(\varepsilon) = O\left(\varepsilon^{1-D}\right)$ as $\varepsilon \to 0^+$, then $\zeta_{\mathcal{L}}$ has a simple pole at D.*

Proof. First, recall from Remark 1.10 or from [Pos] or [Wid] that D is a singularity of $\zeta_{\mathcal{L}}$ because the measure η associated with $\mathcal{L}$ is positive. Suppose $N_{\mathcal{L}}(x) \leq C \cdot x^D$ and $N_{\mathcal{L}}(x) = 0$ for $x \leq x_0$. Then for $s > D$,

$$\zeta_{\mathcal{L}}(s) = s \int_0^\infty N_{\mathcal{L}}(x) x^{-s-1}\, dx \leq \frac{Cs}{s-D} x_0^{D-s}.$$

It follows that the singularity at D is at most a simple pole. Since, by assumption, $\zeta_{\mathcal{L}}$ has a meromorphic extension to a neighborhood of D, it follows that D is a simple pole of $\zeta_{\mathcal{L}}$.

The second part follows from the first part since

$$V(\varepsilon) = 2\varepsilon N_{\mathcal{L}}\left(\frac{1}{2\varepsilon}\right) + \sum_{j:\, l_j < 2\varepsilon} l_j = O\left(\varepsilon^{1-D}\right) \quad (\text{as } \varepsilon \to 0^+)$$

implies that $N_{\mathcal{L}}(x) = O\left(x^D\right)$ as $x \to \infty$. $\qquad\square$

The following example shows that the function $(s - D)\zeta_{\mathcal{L}}(s)$ can have a finite and positive limit as $s \to D^+$, even though $N_{\mathcal{L}}(x)$ is not of order x^D as $x \to \infty$ and $V(\varepsilon)$ is not of order ε^{1-D} as $\varepsilon \to 0^+$. This does not contradict our explicit formulas because the vertical line $\operatorname{Re} s = D$ is a natural boundary for the analytic continuation of the geometric zeta function of the fractal string considered in this example.

Example 6.9. Let

$$\zeta_{\mathcal{L}}(s) = \sum_{j=1}^\infty j e^{Dj^2} e^{-j^2 s}. \tag{6.15}$$

That is, we consider the generalized fractal string with lengths e^{-j^2}, each repeated with multiplicity $j e^{Dj^2}$, for $0 < D < 1$. (We obtain an ordinary fractal string with the same behavior if we choose for the multiplicity the integer nearest to $j e^{Dj^2}$.) Thus

$$N_{\mathcal{L}}(x) = \sum_{j=1}^n j e^{Dj^2} \geq n e^{Dn^2}$$

for $e^{n^2} \leq x < e^{(n+1)^2}$. For $x = e^{n^2}$, $N_{\mathcal{L}}(x) \geq x^D \sqrt{\log x}$, and hence $N_{\mathcal{L}}(x)$ is not of order x^D as $x \to \infty$.

On the other hand, $(s - D)\zeta_{\mathcal{L}}(s)$ has limit $1/2$ as $s \to D^+$, as we now show. For $t > 0$, we have

$$t\zeta_{\mathcal{L}}(D + t) = t \sum_{j=1}^{\infty} je^{-j^2 t}$$

$$= \sum_{j=1}^{\infty} j\sqrt{t}e^{-(j\sqrt{t})^2}\sqrt{t}.$$

This is the Riemann sum for the integral $\int_0^{\infty} xe^{-x^2}\, dx$ (with mesh width $\sqrt{t}$), which has value $1/2$.

We conclude, in particular, that the vertical line $\operatorname{Re} s = D$ is a natural boundary for the analytic continuation of $\zeta_{\mathcal{L}}$.

Lemma 6.10. *Let $\mathcal{L}$ be a generalized fractal string, given by a (local) positive measure η. If the pole of $\zeta_{\mathcal{L}}$ at D is of order $m \geq 1$, then any pole at $D + it$ (with $t \in \mathbb{R}$) is of order at most m.*

Proof. Let $\operatorname{Re} s = \sigma > D$. Since $|\zeta_{\mathcal{L}}(s)| \leq \zeta_{\mathcal{L}}(\sigma)$, we deduce that the function $(\sigma - D)^m \zeta_{\mathcal{L}}(s)$ is bounded as $\sigma \to D^+$. $\square$

Remark 6.11. It follows from the $m = 1$ case of Lemma 6.10 that if D is a simple complex dimension of $\mathcal{L}$, then every other complex dimension of $\mathcal{L}$ on the vertical line $\operatorname{Re} s = D$ is also simple. For example, this is consistent with the fact—established in Theorem 2.13—that for a lattice self-similar string $\mathcal{L}$, D and all the complex dimensions above D (namely, $D + in\mathbf{p}$, $n \in \mathbb{Z}$, where $\mathbf{p}$ is the oscillatory period of $\mathcal{L}$), are simple.

The next result—which follows in particular from the tube formula given in Theorem 6.1 (in conjunction with Theorem 4.24)—extends and puts in a more conceptual framework the criterion for Minkowski measurability of fractal strings obtained by M. L. Lapidus and C. Pomerance in [LapPo1] and [LapPo2, Theorem 2.2, p. 46]. Note that statement (i) below does not appear in [LapPo2].[3] We refer to Remarks 6.14–6.19 below for further comments about our Minkowski measurability criterion and about its relationship with earlier work, in particular [LapPo1-2].

Recall from Section 1.1 that $\mathcal{L}$ is Minkowski measurable if the limit $\lim_{\varepsilon \to 0^+} V(\varepsilon)\varepsilon^{D-1}$ exists and lies in $(0, \infty)$. Then, necessarily, D coincides with the Minkowski dimension of $\mathcal{L}$.

Theorem 6.12 (Criterion for Minkowski measurability). *Let $\mathcal{L}$ be an ordinary fractal string such that $(\mathbf{H}_1)$ and $(\mathbf{H}_2)$ are satisfied for a screen passing between the vertical line $\operatorname{Re} s = D$ and all the complex dimensions of $\mathcal{L}$ with real part strictly less than D. Then the following are equivalent:*

---

[3]On the other hand, the hypotheses about $\zeta_{\mathcal{L}}$ made here are not assumed in [LapPo2]. See Remark 6.15.

(i) *The only complex dimension with real part D is D itself, and D is simple.*

(ii) $N_{\mathcal{L}}(x) \sim c \cdot x^D$ *(i.e., $x^{-D}N_{\mathcal{L}}(x) \to c$) as $x \to \infty$, for some positive constant c.*

(iii) *The boundary of $\mathcal{L}$ is Minkowski measurable.*

Moreover, if these conditions are satisfied, then

$$\mathcal{M} = 2^{1-D}\frac{c}{1-D} = 2^{1-D}\frac{\text{res}\,(\zeta_{\mathcal{L}}(s); D)}{D(1-D)} \tag{6.16}$$

is the Minkowski content of the boundary of $\mathcal{L}$.

Proof. First, (i) implies (ii) by Theorem 4.12 and (iii) by Theorem 6.1, choosing a screen such that only D is visible and using Theorems 4.23 and 4.24.

Assume (ii). Then $\zeta_{\mathcal{L}}$ has only simple poles on the line $\mathrm{Re}\,s = D$, by Lemmas 6.8 and 6.10. Let $\{D + i\gamma_n\}$ be the (finite or infinite) sequence of these poles. By Theorem 4.12, we have that

$$N_{\mathcal{L}}(x) = \sum_n a_n x^{D+i\gamma_n} + o\left(x^D\right), \qquad \text{as } x \to \infty,$$

where $a_n = \text{res}\left(s^{-1}\zeta_{\mathcal{L}}(s); D + i\gamma_n\right)$. Hence

$$\sum_n a_n x^{D+i\gamma_n} - c \cdot x^D \to 0, \qquad \text{as } x \to \infty.$$

By the Uniqueness Theorem for almost periodic functions (see [Sch1, Section VI.9.6, p. 208]), we conclude that $a_n = 0$ for $\gamma_n \neq 0$. This implies (i).

To deduce (i) from (iii), we reason similarly, using $V(\varepsilon)$ instead of $N_{\mathcal{L}}(x)$, and letting $a_n = \text{res}\left((s(1-s))^{-1}\zeta_{\mathcal{L}}(s); D + i\gamma_n\right)$. $\qquad\square$

Remark 6.13. Lemmas 6.7, 6.8 and 6.10 above are needed to ensure that $N_{\mathcal{L}}(x) = x^D \cdot g(x) + o\left(x^D\right)$ as $x \to \infty$ (or $V(\varepsilon) = \varepsilon^{1-D}g_1(\varepsilon) + o\left(\varepsilon^{1-D}\right)$ as $\varepsilon \to 0^+$), where g and g_1 are multiplicatively almost periodic functions. The proof would be simpler if we could split the sum

$$\sum_{\mathrm{Re}\,\omega = D} \text{res}\left(\frac{\zeta_{\mathcal{L}}(s)x^{s+k-1}}{(s)_k}; \omega\right) \tag{6.17}$$

into the different subsums of the type

$$\sum_{\mathrm{Re}\,\omega = D} a_{\omega,n} x^{\omega+k-1}(\log x)^n,$$

for fixed n, which would arise when $\mathcal{L}$ has multiple complex dimensions. Then we would simply apply the Uniqueness Theorem for almost periodic functions to each of these sums. But such a decomposition may not be possible in general, since the series (6.17) is only conditionally convergent. However, we do think that such a decomposition of the sums in our explicit formulas is possible.

Remark 6.14. Let $(l_j)_{j=1}^{\infty}$ denote the sequence of lengths of $\mathcal{L}$. Then condition (ii) is equivalent to the following condition:

(ii′) $l_j \sim L \cdot j^{-1/D}$ (i.e., $j^{1/D} l_j \to L$) as $j \to \infty$, for some positive constant $L > 0$.

Further, the constants c in (ii) and L in (ii′) are connected by $c = L^D$.

Remark 6.15. The criterion for Minkowski measurability that was obtained in [LapPo2, Theorem 2.2, p. 46] is the following:

> Let $\mathcal{L} = (l_j)_{j=1}^{\infty}$ be an (arbitrary) ordinary fractal string of Minkowski dimension $D \in (0, 1)$. Then the boundary of $\mathcal{L}$ is Minkowski measurable if and only if (ii) above holds for some $c > 0$ (or equivalently, if and only if condition (ii′) from Remark 6.14 holds for some $L > 0$).

Further, in that case, the Minkowski content of $\mathcal{L}$ is given by

$$\mathcal{M} = 2^{1-D} \frac{c}{1-D} = 2^{1-D} \frac{L^D}{1-D}. \tag{6.18}$$

Note that in [LapPo1–2], $\zeta_{\mathcal{L}}$ is not required to admit a meromorphic extension to a neighborhood of $\mathrm{Re}\, s = D$ or to satisfy suitable growth conditions. On the other hand, under the hypotheses of Theorem 6.12, our present theory enables us to introduce the new criterion (i), expressed in terms of the notion of complex dimension. The latter criterion gives a rather clear and intuitive geometric meaning to the notion of Minkowski measurability in the present context of fractal strings. It also provides one more geometric interpretation of the notion of complex dimension. (See also the next remark.)

The following comment may help the reader, as it did the writers, to develop further intuition for the notion of complex dimension and the associated oscillatory phenomena (in the geometry). It will also be very useful in Chapter 7, when we reformulate (and extend) the inverse spectral problem for fractal strings studied in [LapMa1–2].

Remark 6.16 (Dimensions above D and geometric oscillations).   The Minkowski measurability of $\mathcal{L}$ (condition (iii) in Theorem 6.12) means, heuristically, that the volume of small tubular neighborhoods does not oscillate. Similarly, condition (ii) (or equivalently, (ii′), by Remark 6.14) says that the sequence of lengths of $\mathcal{L}$ does not oscillate (asymptotically) either; that is, (ii) and (iii) can be interpreted as corresponding to the absence of oscillations of order D in the geometry of $\mathcal{L}$. Therefore, in some sense (provided that D is simple), Theorem 6.12 says that the absence of geometric oscillations of order D in $\mathcal{L}$ is equivalent to the absence of nonreal complex dimensions of $\mathcal{L}$ above D. Note that $\mathcal{L}$ could still have oscillations of lower order.

As will be discussed in more detail below, for a self-similar string $\mathcal{L}$, this fact is illustrated rather clearly by the lattice vs. nonlattice dichotomy which, by Theorem 2.13, corresponds precisely to the existence vs. the absence of (nonreal) complex dimensions above D and hence to the non-Minkowski measurability vs. the Minkowski measurability of $\mathcal{L}$.

Remark 6.17. As was pointed out in Example 4.25 (see also Remark 5.14), for certain fractal strings, and in particular, for certain nonlattice strings, we cannot choose a screen as in Theorem 6.12, and the proof of our criterion for Minkowski measurability does not apply to such strings. However, by the work of M. L. Lapidus and K. J. Falconer ([Lap3, Section 4.4.1b] and [Fa3]), nonlattice strings are always Minkowski measurable. (See Remark 6.27 below.) We show in Section 6.3.1 how we can also recover this result within our framework.

Remark 6.18. A different proof (more of a dynamical systems than of an analytical nature) of the Minkowski measurability criterion of [LapPol–2] recalled in Remark 6.15 above was later obtained in [Fa3]. Moreover, as will be further discussed in the next remark in relationship with our present work, the Minkowski measurability criterion of [LapPol–2] (together with the notion of Minkowski measurability) was extended to a large class of gauge functions (going beyond the traditional power functions) in [HeLap1; HeLap2, Theorem 2.5 and §4.1].

Remark 6.19. Our present approach enables us to analyze in more detail the effect on the geometry (or the spectrum) of fractal strings due to certain of the gauge functions (other than the usual power functions) involved in the definition of the generalized Minkowski content studied by C. Q. He and the first author in [HeLap1–2]. It also enables us to deal with, for example, gauge functions of the form of a power function times a multiplicatively periodic function, as for lattice self-similar strings, which are not within the scope of the theory developed in [HeLap2]. This is worked out in Section 6.3.1 below.

6.3   Examples

In this section, we discuss two classes of examples, namely, the self-similar strings and the a-string, in order to illustrate our results from Sections 6.1 and 6.2.

6.3.1 Self-Similar Strings

Let $\mathcal{L}$ be a self-similar string, with boundary $\partial\mathcal{L}$ of Minkowski dimension D, as studied in Chapter 2.

Nonlattice Case

If $\mathcal{L}$ is a nonlattice string, then, by Theorem 2.13, D is the only complex dimension located on the line $\mathrm{Re}\, s = D$; it is simple (still by Theorem 2.13) and hence, by Corollary 6.2, we have

$$
\begin{aligned}
V(\varepsilon) &= \frac{(2\varepsilon)^{1-D}}{D(1-D)}\,\mathrm{res}\,(\zeta_{\mathcal{L}}(s); D) + \sum_{\mathrm{Re}\,\omega < D} \mathrm{res}\left(\frac{\zeta_{\mathcal{L}}(s)(2\varepsilon)^{1-s}}{s(1-s)}; \omega \right) \\
&\qquad + \{2\varepsilon\zeta_{\mathcal{L}}(0)\} + \mathcal{R}(\varepsilon) \\
&= \mathcal{M} \cdot \varepsilon^{1-D} + o\left(\varepsilon^{1-D}\right),
\end{aligned}
\tag{6.19}
$$

as $\varepsilon \to 0^+$. It follows that $\mathcal{L}$ is Minkowski measurable (in agreement with Theorem 6.12), with Minkowski content $\mathcal{M}$ given by (6.16):

$$
\mathcal{M} = \mathrm{res}\,(\zeta_{\mathcal{L}}(s); D)\, \frac{2^{1-D}}{D(1-D)}.
\tag{6.20}
$$

For instance, if $\mathcal{L}$ is the golden string defined in Section 2.2.4, then (6.20) holds with $D \approx .77921$.

By Remark 5.14, this analysis is not valid if there is no screen passing between $\mathrm{Re}\, s = D$ and the complex dimensions strictly to the left of this line, as in Example 4.25. In that case, we apply formula (5.49) of Section 5.4.2, along with Theorem 2.33. It remains to estimate

$$
\sum_{D-\delta/2 < \mathrm{Re}\,\omega < D} \mathrm{res}\,(\zeta_{\mathcal{L}}; \omega)\, \frac{(2\varepsilon)^{1-\omega}}{\omega(1-\omega)},
\tag{6.21}
$$

for small positive δ, and this is done in Theorem 4.11. The latter result implies that this sum is $o\left(x^D\right)$, as $x \to +\infty$, from which we deduce that (6.19) holds for an arbitrary nonlattice string.

Note that the hypotheses of Theorem 4.11 are satisfied because the sum (6.21) is absolutely convergent. Indeed, by Theorem 2.33, $\mathrm{res}\,(\zeta_{\mathcal{L}}; \omega)$ is uniformly bounded. Hence (6.21) can be compared to $\sum_{D-\delta/2 < \mathrm{Re}\,\omega < D} \omega^{-2}$, which converges by the density estimate (2.25).

We summarize this discussion in the following theorem (see Remarks 6.26 and 6.27 below):

Theorem 6.20. *Every nonlattice string is Minkowski measurable. Further, the volume $V(\varepsilon)$ of the tubular neighborhoods of $\mathcal{L}$ is given by the formulas in (6.19), with $\mathcal{M}$, the Minkowski content of $\mathcal{L}$, given by (6.20).*

Lattice Case

If $\mathcal{L}$ is a lattice string, then the complex dimensions located on the vertical line $\operatorname{Re} s = D$ are all simple and exactly of the form $D + in\mathbf{p}$ $(n \in \mathbb{Z})$, where $\mathbf{p} = 2\pi / \log r^{-1}$ is the oscillatory period of $\mathcal{L}$ and r is its multiplicative generator. (See Theorem 2.13 and Definition 2.10.)

We will establish, in particular, the following result:

Theorem 6.21. *Lattice strings are never Minkowski measurable and always have multiplicatively periodic oscillations of order D in their geometry. Moreover, when $\mathcal{L}$ is normalized as in Remark 2.5, the volume $V(\varepsilon)$ of the tubular neighborhoods of $\mathcal{L}$ is given by the pointwise formulas in (6.29), where the nonconstant periodic function G (of period 1) is given by (6.26).*

If, in addition, the complex dimensions of $\mathcal{L}$ are all simple (and $\mathcal{L}$ is normalized), then $V(\varepsilon)$ is given by (6.30), where $G_1 = G$ and the nonconstant periodic function G_u (also of period 1) is given by (6.31) for $u = 1, \ldots, q$.

We first consider the generalized Cantor string with dimension D and oscillatory period $\mathbf{p}$, introduced in Section 3.1.1 and studied in Chapter 8. This string has lengths $1, a^{-1}, a^{-2}, \ldots$, with multiplicity $1, b, b^2, \ldots$, where $a = e^{2\pi/\mathbf{p}}$ is the inverse of the multiplicative generator r, and $b = a^D$. Thus, a, b and r are related to D and $\mathbf{p}$ by

$$a = r^{-1} = e^{\frac{2\pi}{\mathbf{p}}}, \qquad \mathbf{p} = \frac{2\pi}{\log a},$$
$$b = a^D, \qquad D = \log_a b. \tag{6.22}$$

The geometric zeta function of this generalized Cantor string is

$$\zeta_{D,\mathbf{p}}(s) = \frac{1}{1 - b \cdot a^{-s}},$$

with residue $1/\log a$ at D. In order for $\mathcal{L}$ to be an ordinary fractal string and for $V(\varepsilon)$ to be geometrically meaningful, we assume in the following that $a > b > 1$ and that b is an integer. However, our computation is formally valid without these hypotheses.

In view of Corollary 6.2, we have that

$$V(\varepsilon) = \frac{1}{\log a} \sum_{n \in \mathbb{Z}} \frac{(2\varepsilon)^{1-D-in\mathbf{p}}}{(D + in\mathbf{p})(1 - D - in\mathbf{p})} - \frac{2\varepsilon}{b - 1}, \tag{6.23}$$

for $0 < \varepsilon < a/2$. On the other hand, by formula (1.9),

$$V(\varepsilon) = \sum_{k=0}^{[\log_a \frac{1}{2\varepsilon}]} 2\varepsilon \cdot b^k + \sum_{k=1+[\log_a \frac{1}{2\varepsilon}]}^{\infty} b^k a^{-k}. \tag{6.24}$$

By a direct computation, we obtain

$$V(\varepsilon) = \frac{1}{\log a}(2\varepsilon)^{1-D} G\left(\log_a (2\varepsilon)^{-1}\right) - \frac{2\varepsilon}{b-1}, \qquad (6.25)$$

for $0 < \varepsilon < a/2$, where G is the periodic function (of period 1) given by

$$G(x) = \sum_{n \in \mathbb{Z}} \frac{e^{2\pi i n x}}{(D+in\mathbf{p})(1-D-in\mathbf{p})}$$

$$= r^D \log r^{-1}\left(\frac{r^{D\{x\}}}{1-r^D} + \frac{r^{(D-1)\{x\}}}{r^{D-1}-1}\right), \qquad (6.26)$$

for $x > -1$. Clearly, G is nonconstant since all its Fourier coefficients are nonzero. Further, G is a positive function that is bounded away from zero and infinity. In fact, $G(x)$ attains its maximum

$$r^D \log r^{-1} \frac{r^{D-1} - r^D}{(1-r^D)(r^{D-1}-1)} \qquad (6.27a)$$

at the endpoints $x = 0$ and $x = 1$, and its minimum

$$r^D \log r^{-1} \frac{(1-D)^{D-1}(1-r^D)^{D-1}}{D^D(r^{D-1}-1)^D} \qquad (6.27b)$$

at the point $x > 0$ such that $r^x = \frac{1-D}{D}\frac{1-r^D}{r^{D-1}-1}$.

In view of (6.25) and definitions (1.6a) and (1.6b), it follows from (6.27a) and (6.27b) that the upper and lower Minkowski contents of $\mathcal{L}$ are given by

$$\mathcal{M}^* = \frac{2^{1-D}}{\log a} r^D \log r^{-1} \frac{r^{D-1} - r^D}{(1-r^D)(r^{D-1}-1)}$$

$$= 2^{1-D} \frac{a-1}{(a-b)(b-1)}, \qquad (6.28a)$$

and

$$\mathcal{M}_* = \frac{2^{1-D}}{\log a} r^D \log r^{-1} \frac{(1-D)^{D-1}(1-r^D)^{D-1}}{D^D(r^{D-1}-1)^D}$$

$$= 2^{1-D} \frac{D^{-D}(1-D)^{-(1-D)}}{(a-b)^D(b-1)^{1-D}}, \qquad (6.28b)$$

respectively, where $r = 1/a$ and $D = \log_a b$ (so that $r^D = 1/b$ and hence $r^{D-1} = a/b$). Note that $0 < \mathcal{M}_* < \mathcal{M}^* < \infty$ since G is nonconstant and bounded away from zero and infinity. It follows that $\mathcal{L}$ is not Minkowski measurable.

For an arbitrary lattice string $\mathcal{L}$, normalized for simplicity as in Remark 2.5, we denote by $\mathbf{p}$ the oscillatory period of $\mathcal{L}$ and by r its multiplicative generator. These two quantities are related by $\mathbf{p} = 2\pi/\log r^{-1}$ and $r = e^{-2\pi/\mathbf{p}}$. We choose a number $\Theta < D$ such that the first line of complex dimensions to the left of D lies to the left of $\operatorname{Re} s = \Theta$. Then, by the above computation for the generalized Cantor string,

$$
\begin{aligned}
V(\varepsilon) &= \operatorname{res}\left(\zeta_{\mathcal{L}}(s); D\right) \sum_{n \in \mathbb{Z}} \frac{(2\varepsilon)^{1-D-in\mathbf{p}}}{(D+in\mathbf{p})(1-D-in\mathbf{p})} \\
&\quad + \sum_{\operatorname{Re}\omega < D} \operatorname{res}\left(\frac{\zeta_{\mathcal{L}}(s)(2\varepsilon)^{1-s}}{s(1-s)}; \omega\right) + 2\varepsilon\zeta_{\mathcal{L}}(0) \\
&= \operatorname{res}\left(\zeta_{\mathcal{L}}(s); D\right)(2\varepsilon)^{1-D} G\left(\log_r(2\varepsilon)\right) + o\left(\varepsilon^{1-\Theta}\right), \quad \text{as } \varepsilon \to 0^+,
\end{aligned}
\tag{6.29}
$$

where G is given by (6.26) and $\operatorname{res}\left(\zeta_{\mathcal{L}}(s); D\right)$ is given by (2.29). Since the periodic function G is nonconstant, it follows that $\mathcal{L}$ is not Minkowski measurable, in agreement with Theorems 6.12 and 2.13.

Note that formula (6.29) converges pointwise since there are only finitely many lines of complex dimensions, and the denominators of the terms are quadratic in ω.

We stress that given the special structure of the complex dimensions of a lattice string (see Theorem 2.13), we can rewrite the first equality of Equation (6.29) in a much more explicit manner. Indeed, continue to assume for notational simplicity that $\mathcal{L}$ is normalized as in Remark 2.5, so that it has first length 1, and the last statement of the lattice case of Theorem 2.13 applies. (See Remark 6.22 below for the general case.) Then, for instance, in the notation of Theorem 2.13 and in the case where the complex dimensions ω_u are all simple (as in Examples 6.23 and 6.24 below), we have[4]

$$
V(\varepsilon) = \sum_{u=1}^{q} \operatorname{res}\left(\zeta_{\mathcal{L}}(s); \omega_u\right) (2\varepsilon)^{1-\omega_u} G_u\left(\log_r(2\varepsilon)\right) + 2\varepsilon\zeta_{\mathcal{L}}(0),
\tag{6.30}
$$

where $\omega_1 = D$, $G_1 = G$, as in (6.26), and for $u = 1, \ldots, q$, the function G_u is periodic (of period 1) given by

$$
G_u(x) = \sum_{n \in \mathbb{Z}} \frac{e^{2\pi in x}}{(\omega_u + in\mathbf{p})(1 - \omega_u - in\mathbf{p})}.
\tag{6.31}
$$

---

[4]Recall from (2.28) that $\mathcal{D} = \{\omega_u + in\mathbf{p} : n \in \mathbb{Z}, u = 1, \ldots, q\}$ and from the last statement of the lattice case of Theorem 2.13 that for each fixed $u = 1, \ldots, q$, we have $\operatorname{res}\left(\zeta_{\mathcal{L}}(s); \omega_u + in\mathbf{p}\right) = \operatorname{res}\left(\zeta_{\mathcal{L}}(s); \omega_u\right)$, for all $n \in \mathbb{Z}$. (Since $\mathcal{L}$ is normalized, this common value is given in Equation (2.35), with the numerator on the right-hand side of (2.35) set equal to one.)

Clearly, G_u is nonconstant and, as above, it is bounded away from zero and from infinity. Further, according to Remark 2.14, $\mathrm{res}\,(\zeta_{\mathcal{L}}(s); \omega_u)$ is given by (2.35) with $n = 0$.

Remark 6.22. If the lattice string $\mathcal{L}$ is not normalized, then an additional phase factor is introduced. More precisely, the numerator of the n-th term in the Fourier series expansion of G (respectively, G_u) in (6.26) (respectively, (6.31)) is multiplied by $(L(1-R))^{i n \mathbf{p}}$, where $L(1-R)$ is the first length of $\mathcal{L}$. See the comment following Equation (2.34), along with Remark 2.14.

Example 6.23 (The Cantor string). The Cantor string is a special case of the generalized Cantor string discussed above, with parameters $a = 3$ and $b = 2$. It is a lattice string, and it was studied in Chapter 1 and in Section 2.2.1. In view of (6.25) and the discussion in Section 1.1.2, we have

$$V(\varepsilon) = \frac{1}{\log 3}(2\varepsilon)^{1-D}G\left(\log_3(2\varepsilon)^{-1}\right) - 2\varepsilon,$$

where $D = \log_3 2$, $r = 1/3$, and G is the (nonconstant) periodic function given by (6.26), with $\mathbf{p} = 2\pi/\log 3$. Further, G is bounded away from zero and infinity, as explained above (in (6.27a) and 6.27b)). Moreover, by specializing (6.28) to $a = 3$ and $b = 2$, we have

$$\mathcal{M}^* = 2^{2-D} \qquad \text{and} \qquad \mathcal{M}_* = 2^{1-D}D^{-D}(1-D)^{-(1-D)}, \qquad (6.32)$$

in agreement with the result of [LapPo2, Theorem 4.6, p. 65] recalled in Section 1.1.2. In particular, we recover the fact that the Cantor string is not Minkowski measurable.

Example 6.24 (The Fibonacci string). Recall from Section 2.2.2 that this is a lattice string with two lines of complex dimensions. (See Figure 2.3 on page 31.) Since all these complex dimensions are simple, the tube formula (6.29) becomes

$$V(\varepsilon) = \frac{2+\phi}{5\log 2} \sum_{n \in \mathbb{Z}} \frac{(2\varepsilon)^{1-D-in\mathbf{p}}}{(D+in\mathbf{p})(1-D-in\mathbf{p})} - 2\varepsilon$$

$$+ \frac{3-\phi}{5\log 2} \sum_{n \in \mathbb{Z}} \frac{(2\varepsilon)^{1+D-i(n+\frac{1}{2})\mathbf{p}}}{(-D+i(n+\frac{1}{2})\mathbf{p})(1+D-i(n+\frac{1}{2})\mathbf{p})}, \qquad (6.33)$$

where $D = \log_2 \phi$, $\mathbf{p} = 2\pi/\log 2$ and $0 < \varepsilon < 1$.

Example 6.25 (A lattice string with multiple poles). An example of a lattice string $\mathcal{L}$ with multiple poles was considered in Section 2.2.3. Since $\mathcal{L}$ has one (discrete) line of complex dimensions above D, $\omega = D + in\mathbf{p}$ ($n \in \mathbb{Z}$, $D = \log_2 3$ and $\mathbf{p} = 2\pi/\log 3$), and another line of double poles $\omega = \frac{1}{2}i\mathbf{p} + in\mathbf{p}$ ($n \in \mathbb{Z}$) (see Figure 2.4 on page 32), the tube formula (6.29) becomes

$$V(\varepsilon) = \frac{4}{9\log 3} \sum_{n\in\mathbb{Z}} \frac{(2\varepsilon)^{1-D-in\mathbf{p}}}{(D+in\mathbf{p})(1-D-in\mathbf{p})} - \frac{\varepsilon}{2}$$

$$+ \frac{1}{9(\log 3)^2} \sum_{n\in\mathbb{Z}} \frac{(2\varepsilon)^{1-i(n+\frac{1}{2})\mathbf{p}}}{i(n+\frac{1}{2})\mathbf{p}(1-i(n+\frac{1}{2})\mathbf{p})}$$

$$\cdot \left(5\log 3 + \frac{3i(2n+1)\mathbf{p}-1}{i(n+\frac{1}{2})\mathbf{p}(1-i(n+\frac{1}{2})\mathbf{p})} \right) \qquad (6.34)$$

$$+ \frac{1}{3\log 3} \sum_{n\in\mathbb{Z}} \frac{(2\varepsilon)^{1-i(n+\frac{1}{2})\mathbf{p}} \log_3(2\varepsilon)^{-1}}{i(n+\frac{1}{2})\mathbf{p}(1-i(n+\frac{1}{2})\mathbf{p})},$$

as $\varepsilon \to 0^+$.

Remark 6.26. Theorems 6.20 and 6.21 above establish completely the geometric part of [Lap3, Conjecture 3, p. 163] in the one-dimensional case (that is, for self-similar strings rather than for self-similar drums). When specialized to one dimension ($d = 1$), the latter stated, in particular, that nonlattice self-similar strings are Minkowski measurable whereas lattice strings are not (because they have oscillations of order D in their geometry). It was motivated in part by the work of Lalley in [Lal1–3] on various geometric counting functions associated with self-similar sets and by works of the first author in [Lap1–2] as well as of Lapidus and Pomerance in [LapPo1–2]. (See especially the Minkowski measurability criterion in [LapPo1–2], recalled in Remark 6.15 above, and the example of the Cantor set studied in [LapPo2, Example 4.5]. Also see the main example of [BroCa], revisited in [FlVa].) We note, however, that our results supplement significantly the geometric aspects of [Lap3, Conjecture 3], as is explained toward the end of the next remark.

Remark 6.27. We should point out that the fact (stated in Theorem 6.20) that nonlattice strings are Minkowski measurable was already known in the literature, as was noted earlier in this book. Indeed, it was first observed by Lapidus in [Lap3, Section 4.4.1b] and later recovered independently by Falconer in [Fa3]. The method used in both [Lap3] and [Fa3] relies on a suitable use of the Renewal Theorem from probability theory [Fel, Theorem 2,

p. 39], much as was done earlier by Lalley [Lal1] in a related context.[5,6,7]
In the lattice case, the results of [Lap3] and of [Fa3] (also based on the
Renewal Theorem) yield the existence of a (multiplicatively) periodic func-
tion G in the leading term of $\varepsilon^{-(1-D)}V(\varepsilon)$ as $\varepsilon \to 0^+$. However, as was
noted in [Lap3], one cannot expect to deduce from the Renewal Theorem
alone (or from a Tauberian-type argument) that G is nonconstant (as was
conjectured in [Lap3]) and hence that a lattice string is not Minkowski measurable. As was alluded to at the end of the previous remark, our present
results go further (than in either [Lap3] or [Fa3]) in several directions:

(i) We provide an explicit formula for the volume of the tubular neigh-
borhoods, $V(\varepsilon)$, valid (pointwise) for all $0 < \varepsilon < 1$ and expressed in terms
of the complex dimensions of $\mathcal{L}$. This formula is especially transparent
when $\mathcal{L}$ is a lattice string. (See formula (6.29) and, when $\mathcal{L}$ has simple
complex dimensions, formula (6.30)).

(ii) In the lattice case, we show that the periodic function G is nonconstant and hence that a lattice string is never Minkowski measurable. We
also obtain the Fourier series expansion of G (see formula (6.26)).

(iii) Our results provide a new intuitive understanding of the dichotomy
lattice vs. nonlattice and of its consequence for the Minkowski measurability of self-similar strings. Namely, the presence of nonreal complex dimensions of real part D characterizes lattice strings and explains the fact that
such strings are not Minkowski measurable (since they have oscillations of
order D in their geometry). Analogously, nonlattice strings are Minkowski
measurable because they do not have nonreal complex dimensions above D.
(See Theorems 2.13, 6.1 and 6.12 above; also see the related discussion in
Section 10.3 below.)

6.3.2 The a-String

Let $\mathcal{L}$ be the a-string studied in Section 5.5.1. Recall from Theorem 5.20
that $\mathcal{L}$ has Minkowski dimension $D = \frac{1}{a+1}$ and that its complex dimensions
are all simple, real, and located at D and possibly at $-D, -2D, \dots$. Also,
the residue of $\zeta_{\mathcal{L}}$ at D is equal to a^D. Hence, by Corollary 6.2, we have, for

---

[5]Recall from Remark 6.18 that the main goal of [Fa3] was to obtain an alternative
proof of the Minkowski measurability criterion of [LapPo2, Theorem 2.1] for fractal
strings.

[6]Shortly after the preliminary version of [Lap3] was written, a similar method—
based on the Renewal Theorem—was used by Kigami and the first author in [KiLap]
to obtain a Weyl-type formula for the eigenvalue distribution of Laplacians on finitely
ramified self-similar fractals, thereby proving (and specifying) for this class of fractals
Conjecture 5 of [Lap3, p. 190] for self-similar drums with fractal membrane (rather than
with fractal boundary, as in [Lap3, Conjecture 3]).

[7]A more general version of the Renewal Theorem that is well suited for this or related
situations was later obtained in [LeVa].

every fixed integer $M \geq 1$,

$$V(\varepsilon) = \frac{(2\varepsilon)^{1-D}}{D(1-D)} a^D - \sum_{k=1}^{M} \frac{(2\varepsilon)^{1+kD}}{kD(1+kD)} \operatorname{res}(\zeta_{\mathcal{L}}; -kD) + O\left(\varepsilon^{1+(M+1)D}\right),$$

(6.35)

as $\varepsilon \to 0^+$.

In particular, it follows that $\mathcal{L}$ is Minkowski measurable with Minkowski content

$$\mathcal{M} = \frac{2^{1-D}}{D(1-D)} a^D,$$

(6.36)

as was shown by a direct computation in [Lap1, Appendix C, pp. 523–524] (and later reproved more conceptually in [LapPo2, Example 4.3, pp. 64–65] and then in [HeLap2]). Since D is simple and is the only pole of $\zeta_{\mathcal{L}}$ above D, this is in agreement with Theorem 6.12. We stress, however, that our tube formula (6.35) gives much more precise information about $V(\varepsilon)$.

Remark 6.28. All the results of this chapter can be extended to (suitable) fractal sprays, in the sense of [LapPo3] and Section 1.4. The main difficulty consists in finding a formula for the (inner) tubular neighborhood of the underlying basic shape $B \subset \mathbb{R}^d$. (See formulas (6.13) and (6.14) for smooth manifolds B, and (10.1) and (10.3) for two examples of a fractal basic shape.) In particular, under appropriate assumptions, tube formulas analogous to formula (6.4) of Theorem 6.1 and a criterion for Minkowski measurability much like Theorem 6.12 can be obtained in such a setting.

In the important special case of self-similar sprays (in the sense of Section 5.6), a counterpart of Theorems 6.20 and 6.21 can also be obtained in the nonlattice and lattice case, respectively. For example, the Sierpinski drum (or rather, its boundary, the Sierpinski gasket, see Figure 5.1 on page 139) is not Minkowski measurable (as was stated in [Lap2–3]) because it is a lattice spray, whereas (suitable) nonlattice self-similar sprays are always Minkowski measurable (as was shown in [Lap3], see also [LeVa], by means of the Renewal Theorem). Again, even in those situations, our tube formulas give more precise information than was previously available in the literature.

7

The Riemann Hypothesis, Inverse Spectral Problems and Oscillatory Phenomena

In this chapter, we provide a geometric reformulation of the Riemann Hypothesis in terms of a natural inverse spectral problem for fractal strings. After stating this inverse problem in Section 7.1, we show in Section 7.2 that its solution is equivalent to the nonexistence of critical zeros of the Riemann zeta function on a given vertical line. This was done earlier in [LapMa1–2], but now we use the point of view of complex dimensions and the explicit formulas of Chapter 4. Then, in Section 7.3, we extend this characterization to a large class of zeta functions, including all the number-theoretic zeta functions for which the extended Riemann Hypothesis is expected to hold.

7.1 The Inverse Spectral Problem

In this chapter, we study the question of whether one can deduce geometric information about a fractal string from information about its spectrum. In other words, we consider the question *"Can one hear the shape of a fractal drum?"* In Section 5.3.2, we introduced the spectral operator, which allows us to formalize questions of this type as the question of the invertibility of this operator.

We give here two examples of such questions. Namely,

Can one hear the dimension of a fractal string?

and

Can one hear whether a fractal string is Minkowski measurable?

In [LapPo2, Theorem 2.4, p. 47], the first question is answered in the affirmative, and in [LapMa2] (announced in [LapMa1]) it is shown that a positive answer to the second inverse problem is equivalent to the fact that $\zeta(s)$ has no zeros on the line $\operatorname{Re} s = D$. Thus, Lapidus and Maier obtained a reformulation of the Riemann Hypothesis in terms of a natural inverse spectral problem for (standard) fractal strings. (See Figure 7.1 on the opposite page; also see [Lap2, Figure 3.1, p. 165] or [Lap3, Figure 2.1, p. 143] and the related comments therein.)

The inverse spectral problem studied in [LapMa2] is the following:

(S)   Let a standard fractal string with Minkowski dimension $D \in (0,1)$ be given. If this string has no oscillations of order D in its spectrum, does it follow that it is Minkowski measurable?

See [LapMa2] and below for a precise formulation. In [LapMa2], condition (ii) of Theorem 6.12 was used to characterize the Minkowski measurability of a fractal string. In our present situation, we will use the equivalent condition (i) of Theorem 6.12 involving the complex dimensions above D.

Remark 7.1. Intuitively, the inverse spectral problem (S) can be interpreted as follows (see Remark 6.16): If an ordinary fractal string $\mathcal{L}$ of Minkowski dimension D has no oscillations of order D in its frequency spectrum, does it follow that it has no oscillations of order D in its geometry (i.e., by Theorem 6.12, that D is simple and $\mathcal{L}$ has no other complex dimensions on the vertical line $\operatorname{Re} s = D$)? By contraposition, we obtain the equivalent formulation: If $\mathcal{L}$ has oscillations of order D in its geometry, does it follow that it has oscillations of order D in its spectrum?

The inverse spectral problem is in the spirit of questions raised in the spectral theory of smooth manifolds; see Appendix B. Let Ω be a bounded open subset of $\mathbb{R}$, with boundary $\partial\Omega$ of Minkowski dimension $D \in [0,1]$. That is, Ω is an ordinary fractal string $\mathcal{L} = (l_j)_{j=1}^{\infty}$. Then a counterpart of estimate (B.12) was obtained in [LapPo1–2], where the first connections between the vibrations of fractal strings and the Riemann zeta function $\zeta = \zeta(s)$ were also established. This provided, in particular, a resolution of the (modified) Weyl–Berry Conjecture for fractal strings (from [Lap1]). More precisely, [LapPo2, Corollary 2.3, p. 46] states that if $\partial\Omega$ is Minkowski measurable in dimension $D \in (0,1)$ with Minkowski content $\mathcal{M}(D;\partial\Omega)$, then an analogue (with $d = 1$ and $d - 1$ replaced by D) of (B.12) holds for the Dirichlet Laplacian on Ω; namely, we have the following pointwise asymptotic formula:

$$N_\nu(x) = W(x) - g_D \mathcal{M}(D;\partial\Omega)x^D + o\left(x^D\right), \tag{7.1}$$

as $x \to \infty$, where

$$W(x) = \operatorname{vol}_1(\Omega)x$$

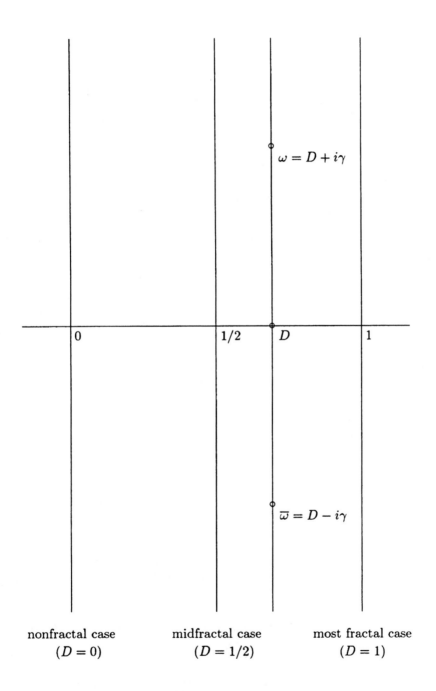

Figure 7.1: The critical strip for $\zeta(s)$: $0 \leq \operatorname{Re} s \leq 1$.

is the Weyl term and

$$g_D = 2^{-(1-D)}(1 - D)\left(-\zeta(D)\right).$$

(Since $\zeta(D) < 0$ for D in the critical interval $(0, 1)$, we have $g_D > 0$.)
Note that (7.1) implies that $N_\nu(x)$ admits a monotonic asymptotic second
term.[1] The converse of this result (namely, the corresponding inverse spectral problem for fractal strings), was shown in [LapMa1-2] to be closely
connected with the Riemann Hypothesis. More precisely, Theorems 2.3
and 2.4 in [LapMa2, p. 20] show, in particular, that for a fixed $D \in (0, 1)$,
the existence of a monotonic asymptotic second term for $N_\nu(x)$ (with the
pointwise error term $o\left(x^D\right)$ replaced by $O\left(x^D \log^{-(1+\varepsilon)} x\right)$, for some $\varepsilon > 0$)
always implies that $\partial\Omega$ is Minkowski measurable if and only if $\zeta = \zeta(s)$ does
not vanish on the vertical line $\operatorname{Re} s = D$.

Remark 7.2. On the basis of [LapPo2, Theorem 2.1] (and its proof), the
authors of [LapPo1-2] have raised the question of finding a suitable notion
of complex dimension that would extend, in particular, the usual notion of
real fractal (i.e., Minkowski) dimension and would provide (when $d = 1$) a
new interpretation for the critical strip $0 \le \operatorname{Re} s \le 1$ of the Riemann zeta
function $\zeta = \zeta(s)$. See [LapPo1, p. 347] and [LapPo2, Section 4.4.b, p. 67],
along with, for example, [Lap2, Remark 2.2 and Figure 3.1, pp. 142–143]
and Figure 7.1. The later work in [LapMa1-2] used this intuition and also
began to corroborate it (see, e.g., [LapMa2, Section 3.3 and Remark 3.21(c),
(d), p. 32]), while the present work provides a rigorous theory for the notion
of complex dimension.

Remark 7.3. Finally, we note that (still for fractal strings) the following
conditions were shown to be equivalent in [LapPo2, Theorem 2.4, p. 47]:

(a) $0 < \mathcal{M}_*(D; \partial\Omega) \le \mathcal{M}^*(D; \partial\Omega) < \infty$;

(b) $0 < \liminf_{j\to\infty} l_j j^{-1/D} \le \limsup_{j\to\infty} l_j j^{-1/D} < \infty$;

(c) $0 < \delta_* \le \delta^* < \infty$,

where δ_* (respectively, δ^*) denotes the lower (respectively, upper) limit
of $x^{-D}\left(W(x) - N_\nu(X)\right)$ as $x \to \infty$.[2] In conjunction with the aforementioned results [LapPo2, Theorems 2.1 and 2.2], this shows that the existence of an oscillatory second term of order D for $N_\nu(x)$ (i.e., by definition,

---

[1] The characterization of Minkowski measurability obtained in [LapPo2, Theorem 2.2,
p. 46] played a key role in deriving this result. Recall that the criterion of [LapPo1-2]
(namely, $l_j \sim Lj^{-1/D}$, for some $L > 0$) was extended in Chapter 6 and interpreted in
terms of the notion of complex dimension; see especially Theorem 6.12 and Remark 6.15
above.

[2] Observe that by the Minkowski measurability criterion of [LapPo2, Theorem 2.2],
(a) holds with strict inequalities if and only if (b) does. Further observe that the fact
that condition (a) (or, equivalently, (b)) holds with strict inequalities means that the
geometry of the boundary $\partial\Omega$ (i.e., of the fractal string $\mathcal{L}$) has oscillations of order D.

$0 < \delta_* < \delta^* < \infty)$ implies that (a) and (b) hold with strict inequalities, and hence that $\partial\Omega$ is not Minkowski measurable. On the other hand, the results of [LapMa1–2] (specifically, [LapMa2, Theorem 3.16, p. 28]) show that the converse is not true; namely, $\partial\Omega$ may be non-Minkowski measurable (with strict inequalities in (a) and (b)) whereas $N_\nu(x)$ admits a monotonic asymptotic second term of order D (i.e., $0 < \delta_* = \delta^* < \infty$). The fact that for Cantor-type sets (or strings), the spectral operator is invertible (in the terminology of Section 5.3.2) and hence this phenomenon cannot occur, is established in Chapter 8 and is used in Chapter 9 below. In turn, the results of Chapter 9 can be used to show that a lattice string always has oscillations of order D in its spectrum, and hence satisfies (c) as well as conditions (a) and (b) with strict inequalities; see Remark 5.16 in Section 5.4.3. Further, by using the results of Section 9.2, an entirely analogous statement can be shown to hold for a large class of (possibly generalized) lattice self-similar sprays (as in Section 5.6).

7.2 Complex Dimensions of Fractal Strings and the Riemann Hypothesis

Our explicit formulas shed new light on the results and methods of M. L. Lapidus and H. Maier in [LapMa2]. We will use condition (i) of Theorem 6.12 to characterize the Minkowski measurability of a fractal string in order to recover (and extend) the results of [LapMa2] by focusing on the fractal string η introduced in [LapMa2, §3.3], as well as on its continuous analogue μ.

Consider the generalized fractal string

$$\mu(dx) = \left(x^{D-1} + \beta x^{D-1+i\gamma} + \overline{\beta} x^{D-1-i\gamma}\right) dx, \tag{7.2}$$

supported on $[1,\infty)$,[3] where $\gamma \in \mathbb{R}$, $D \in (0,1)$ and $\beta \in \mathbb{C}$, $|\beta| \leq 1/2$, so that μ is a positive measure. Then

$$\zeta_\mu(s) = \frac{1}{s-D} + \frac{\beta}{s-D-i\gamma} + \frac{\overline{\beta}}{s-D+i\gamma}, \tag{7.3}$$

and the complex dimensions of μ are D, $D+i\gamma$ and $D-i\gamma$. By Theorem 6.12, this string is not Minkowski measurable. By Theorem 4.8, applied for $k = 1$,

---

[3]More precisely, we have

$$\mu(dx) = \left(x^{D-1} + \beta x^{D-1+i\gamma} + \overline{\beta} x^{D-1-i\gamma}\right) \mathbf{1}_{(1,\infty)} \, dx,$$

where $\mathbf{1}_{(1,\infty)}$ denotes the characteristic function of the interval $(1,\infty)$.

we have the following pointwise equality, for all $x \geq 1$:

$$N_\mu(x) = \frac{x^D}{D} + \beta \frac{x^{D+i\gamma}}{D+i\gamma} + \overline{\beta} \frac{x^{D-i\gamma}}{D-i\gamma} + \zeta_\mu(0). \qquad (7.4)$$

(This formula could also be obtained by a direct computation.)

For the frequency counting function, we cannot apply Theorem 4.8 to obtain a pointwise formula at level 1, hence we have to interpret $N_\nu(x)$ as a distribution. Thus, by Theorems 4.12 and 4.23, we have, in a distributional sense,

$$N_\nu(x) = \zeta_\mu(1)x + \zeta(D)\frac{x^D}{D}$$
$$+ \beta\zeta(D+i\gamma)\frac{x^{D+i\gamma}}{D+i\gamma} + \overline{\beta}\zeta(D-i\gamma)\frac{x^{D-i\gamma}}{D-i\gamma} + O(1), \qquad (7.5)$$

as $x \to \infty$.

Now, if $\zeta(D+i\gamma)$, and hence also $\zeta(D-i\gamma)$, vanishes, then

$$N_\nu(x) = \zeta_\mu(1)x + \zeta(D)\frac{x^D}{D} + O(1); \qquad (7.6)$$

thus $N_\nu(x)$ has no oscillatory terms. Therefore, μ provides a counterexample to problem (S).

Remark 7.4. Instead of using the distributional explicit formula, we can apply the pointwise formula, Theorem 4.4, at level $k = 2$. Then we obtain an integrated version of $N_\nu(x)$ (with $(s)_2 = s(s+1)$, as in Equation (4.10)):

$$N_\nu^{[2]}(x) = \zeta_\mu(1)\frac{x^2}{2} + \zeta(D)\frac{x^{D+1}}{(D)_2} + \beta\zeta(D+i\gamma)\frac{x^{D+1+i\gamma}}{(D+i\gamma)_2}$$
$$+ \overline{\beta}\zeta(D-i\gamma)\frac{x^{D+1-i\gamma}}{(D-i\gamma)_2} + O(x^{1+\varepsilon}), \qquad (7.7)$$

as $x \to \infty$, for all $\varepsilon > 0$. (In this formula, $O(x^{1+\varepsilon})$ has the usual pointwise meaning, as opposed to $O(1)$ in the above formula, which must be interpreted distributionally.) Consequently, if $\zeta(s)$ vanishes at $s = D \pm i\gamma$, the spectrum has no oscillations of order $D + 1$ at level 2.

Since the function given by (7.4) is strictly increasing, we can define an ordinary fractal string η by the property

$$N_\eta(x) = [N_\mu(x)], \quad \text{for every } x > 0. \qquad (7.8)$$

Thus η is the ordinary fractal string $\eta = \sum_{j=1}^{\infty} \delta_{\{l_j^{-1}\}}$, with $l_j > 0$ defined uniquely by $N_\mu(l_j^{-1}) = j$, for each $j = 1, 2, \ldots$. Then we have

$$\zeta_\eta(s) = \zeta_\mu(s) + \zeta_{\eta-\mu}(s).$$

The function $\zeta_{\eta-\mu}(s)$ is holomorphic for $\operatorname{Re} s > 0$. Hence, η and μ have the same complex dimensions, namely D, $D+i\gamma$ and $D-i\gamma$, which are all simple. Since η has nonreal complex dimensions above D, it follows from the above Minkowski measurability criterion (Theorem 6.12) that η is not Minkowski measurable: It has oscillations of order D in its geometry. Thus we recover the counterexample of [LapMa2] to the inverse spectral problem.

We now consider the converse of the above question. That is, we want to show that if ζ does not vanish on the vertical line $\operatorname{Re} s = D$, then the inverse spectral problem (S) in dimension D always has an affirmative answer for a suitable class of generalized fractal strings. Indeed, if η is an arbitrary generalized fractal string of dimension $D \in (0, 1)$ satisfying $(\mathbf{H_1})$ and $(\mathbf{H_2})$, its oscillations in dimension D are described by its complex dimensions with real part D. We choose a screen S to the left of $\operatorname{Re} s = D$ so that only the complex dimensions with real part D are visible. Hence we need to assume that η allows such a screen (see Examples 4.25 and 4.26). Then, by Theorems 4.12 and 4.23,

$$N_\eta(x) = \sum_{\operatorname{Re} \omega = D} \operatorname{res}\left(\frac{\zeta_\eta(s)x^s}{s}; \omega\right) + o\left(x^D\right) \tag{7.9}$$

and

$$N_\nu(x) = \zeta_\eta(1)x + \sum_{\operatorname{Re} \omega = D} \operatorname{res}\left(\frac{\zeta_\eta(s)\zeta(s)x^s}{s}; \omega\right) + o\left(x^D\right), \tag{7.10}$$

as $x \to \infty$. (These formulas have to be interpreted distributionally.) If ζ has no zeros on the line $\operatorname{Re} s = D$, all terms in the second series (which represents an almost periodic function, in an extended sense) remain, and we obtain a positive answer to the inverse spectral problem considered in [LapMa2].

Remark 7.5. In [LapMa1–2], the converse was established by using the Wiener–Ikehara(–Landau) Tauberian Theorem [Pos, Section 27, pp. 109–112] rather than an explicit formula. Accordingly, the assumption made on the fractal string was on the error term beyond the asymptotic second term in the spectral counting function $N_\nu(x)$, rather than on the existence of a suitable screen. (See [LapMa2, Theorem 2.3 and 3.2, pp. 20 and 23] and the discussion preceding Remark 7.2.)

We summarize the above discussion in the following theorem:

Theorem 7.6. *For a given D in the critical interval $(0, 1)$, the inverse spectral problem (S) in dimension D—suitably interpreted as above—has a positive answer if and only if $\zeta(s)$ does not have any zero on the vertical line $\operatorname{Re} s = D$.*

Since $\zeta(s)$ has zeros on the critical line $\text{Re}\, s = 1/2$, we also obtain

Corollary 7.7. *The inverse spectral problem* (S) *is not true in the mid-fractal case* (*i.e., when $D = 1/2$, see Figure 7.1*).

On the other hand, it is true for every $D \in (0,1)$, $D \neq 1/2$, if and only if the Riemann Hypothesis holds. In the terminology of Section 5.3.2, the spectral operator is invertible for all fractal strings of dimension $D \neq 1/2$ if and only if the Riemann Hypothesis holds. In that case, the inverse spectral operator is given as in Remark 5.8 above.

Remark 7.8. The first part of the above corollary, stating that (S) is not true in the midfractal case, does not necessarily have a counterpart for the general zeta functions $\zeta_B(s)$ considered in Section 7.3 below.

Remark 7.9. Close inspection of the proof of Theorem 7.6 reveals the following important fact. In the statement of Theorem 7.6 and of Corollary 7.7, we can replace arbitrary fractal strings of dimension D by the particular fractal string η of dimension D (and with additional complex dimensions $D \pm i\gamma$) defined by Equation (7.8). A similar comment holds for the results obtained in Section 7.3, for fractal sprays.

7.3   Fractal Sprays and the Generalized Riemann Hypothesis

Instead of considering fractal strings, we consider other fractal sprays on the basic shape B, as defined in Section 1.4. The spectral zeta function of the fractal spray of $\mathcal{L}$ on B is

$$\zeta_\nu(s) = \zeta_{\mathcal{L}}(s) \cdot \zeta_B(s), \qquad (7.11)$$

where

$$\zeta_B(s) = \sum_\lambda \lambda^{-s/2}. \qquad (7.12)$$

Here, λ runs over the normalized eigenvalues of the positive Laplacian on B, with Dirichlet boundary conditions (and hence, the frequencies of B are the numbers $f = \sqrt{\lambda}$).

Example 7.10. The frequencies of the fundamental domain of the lattice $\mathbb{Z}^m$ in $\mathbb{R}^m$ ($m \in \mathbb{N}^*$), with identification of opposite sides, are described by the classical Epstein zeta functions $\zeta_m(s) = \sum_{v \in \mathbb{Z}^m - \{0\}} \|v\|^{-s}$, where $\|v\|$ is the Euclidean norm. These functions satisfy a functional equation, relating $\zeta_m(s)$ and $\zeta_m(m - s)$; see [Te, Theorem 1, p. 59] or Section A.4 of Appendix A. Note that $\zeta_1(s) = 2\zeta(s)$.

In general, we do not need to assume the existence of B as a subset of $\mathbb{R}^m$.[4] Indeed, for our theory to apply, we only need to assume that $\zeta_B(s) = \sum_f w_f f^{-s}$ is a zeta function that satisfies ($\mathbf{H_1}$) and ($\mathbf{H_2}$) for some suitable screen S, as in Section 7.2. If the coefficients w_f—to be thought of as the (complex) multiplicities of f—are real, we use the above generalized fractal string μ (with complex dimensions D, $D + i\gamma$ and $D - i\gamma$) to test whether ζ_B has zeros at $D \pm i\gamma$, by applying the analogue of formula (7.5). If the coefficients w_f are not real—such as for a Dirichlet L-series associated with a complex, nonreal character (see, e.g., [Lan] or Appendix A.2 below)—we cannot assume that the zeros come in complex conjugate pairs. However, if the window of ζ_B is symmetric with respect to the real axis (and ω is not a zero when $\overline{\omega}$ is a pole of ζ_B), we replace $\zeta_B(s)$ by the symmetrized Dirichlet series $\zeta_B(s)\overline{\zeta_B(\overline{s})}$, the zeros of which come in complex conjugate pairs. We can then go through exactly the same reasoning as above to characterize the absence of zeros of $\zeta_B(s)$ on the line $\operatorname{Re} s = D$ in terms of an inverse spectral problem for fractal sprays.

We summarize the above discussion in the following theorem:

Theorem 7.11. *Let $\zeta_B(s)$ be a Dirichlet series (or integral) satisfying hypotheses ($\mathbf{H_1}$) and ($\mathbf{H_2}$) for a suitable screen S (and the associated window W). Let $D \in W \cap \mathbb{R}$. Then, if $\zeta_B(s)$ has real coefficients (or is associated with a real measure), the exact counterpart of the inverse spectral problem (S) for the corresponding generalized fractal spray is true for this value of D if and only if $\zeta_B(s)$ does not have any zero on the vertical line $\operatorname{Re} s = D$.*

Furthermore, in the general case when $\zeta_B(s)$ is not real-valued on the real axis, then, provided that W is symmetric with respect to the real axis (and ω is not a zero when $\overline{\omega}$ is a pole of $\zeta_B(s)$), the same criterion as above holds.

Remark 7.12. We note that, in general, $\zeta_B(s)$ need not satisfy a functional equation or an Euler product.

We thus obtain a criterion for the absence of zeros on vertical lines within the critical strip for all Epstein zeta functions [Te], and for all the number-theoretic zeta functions for which the generalized Riemann Hypothesis is expected to hold: for example, the Dedekind zeta functions and more generally, the Hecke L-series associated with an algebraic number field [Lan, ParSh1-2]; also, the zeta functions of algebraic varieties over a finite field [ParSh1, Chapter 4, §1]. In all these cases, we may choose the window W to be all of $\mathbb{C}$. (We refer to Appendix A for a brief introduction to the abovementioned zeta functions.)

---

[4]In other words, we consider virtual basic shapes and the associated generalized fractal sprays.

Remark 7.13. In the case of zeta functions of varieties over a finite field, the Weyl term takes a special form. Consequently, the inverse spectral problem must be interpreted suitably. We refer to Section 9.4 below for a detailed discussion of the special case of curves over a finite field.

8

Generalized Cantor Strings and their Oscillations

In this chapter, we analyze the oscillations in the geometry and the spectrum of the simplest type of generalized self-similar fractal strings. The complex dimensions of these so-called generalized Cantor strings lie on just one vertical line $D + inp$ ($n \in \mathbb{Z}$), for some $D \in (0,1)$ and $\mathbf{p} > 0$. We construct such a generalized Cantor string for any choice of D and $\mathbf{p}$.

In Chapter 9, we will apply the results of the present chapter and our explicit formulas from Chapter 4 to prove the theorem that suitable Dirichlet series with positive coefficients have no infinite vertical sequence of critical zeros in arithmetic progression.

8.1 The Geometry of a Generalized Cantor String

Definition 8.1. For real numbers $a > 1$ and $b > 0$, the *generalized Cantor string* (with parameters a and b) is the string

$$\sum_{n=0}^{\infty} b^n \, \delta_{\{a^{-n}\}}. \tag{8.1}$$

That is, it is the string with lengths $1, a^{-1}, a^{-2}, \ldots$, repeated with (possibly noninteger) multiplicity $1, b, b^2, \ldots$.

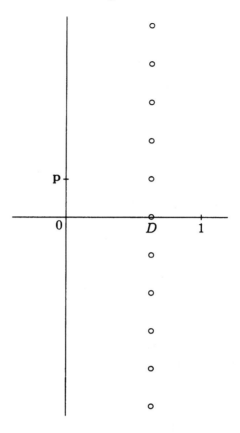

Figure 8.1: The complex dimensions of the generalized Cantor string with parameters a and b. Here, $D = \log_a b$ and $\mathbf{p} = 2\pi/\log a$.

The geometric zeta function of this string is

$$\zeta_{D,\mathbf{p}}(s) = \frac{1}{1 - b \cdot a^{-s}}, \tag{8.2}$$

where $D = \log_a b$ is the dimension and $\mathbf{p} = 2\pi/\log a$ is the oscillatory period of the string.

One deduces from (8.2) that, indeed, the complex dimensions are located at the points $\omega = D + in\mathbf{p}$, $n \in \mathbb{Z}$. Hence, they lie on a single vertical line, $\operatorname{Re} s = D$, and their imaginary parts form a doubly infinite arithmetic progression; see Figure 8.1. Further, the poles are simple, with residue $1/\log a$.

Remark also that the pairs $(D, \mathbf{p})$ and (a, b) determine each other: D and $\mathbf{p}$ are given above in terms of a and b, and conversely, $a = e^{2\pi/\mathbf{p}}$ and $b = a^D$. (Note that if $a > b > 1$, then $D \in (0, 1)$ as in our usual framework.

However, we need to consider here for later use in Chapter 9 the more general case when D is an arbitrary positive real number.)

The geometric counting function is easy to compute. There are $1 + b + b^2 + \cdots + b^n$ lengths less than x, with $n = [\log_a x]$. Thus

$$N_{D,\mathbf{p}}(x) = \frac{b^{n+1} - 1}{b - 1} = \frac{b}{b - 1} b^{[\log_a x]} - \frac{1}{b - 1}. \tag{8.3}$$

Much as in Section 5.4.1, and using the Fourier series of $u \mapsto b^{-\{u\}}$ (see formula (1.13)), we obtain

$$N_{D,\mathbf{p}}(x) = \frac{x^D}{\log a} \sum_{n \in \mathbf{Z}} \frac{x^{in\mathbf{p}}}{D + in\mathbf{p}} - \frac{1}{b - 1}. \tag{8.4}$$

Indeed, this coincides with the expression given by the explicit formula of Theorem 4.8.

More generally, for the k-th integrated counting function, the pointwise explicit formula without error term, Theorem 4.8, yields for every $k \geq 1$ (with $(s)_k$ given by Equation (4.10)):

$$\begin{aligned}
N_{D,\mathbf{p}}^{[k]}(x) = &\frac{x^{D+k-1}}{\log a} \sum_{n \in \mathbf{Z}} \frac{x^{in\mathbf{p}}}{(D + in\mathbf{p})_k} \\
&+ \frac{1}{(k-1)!} \sum_{j=0}^{k-1} \binom{k-1}{j} (-1)^j \frac{x^{k-1-j}}{1 - b \cdot a^j}.
\end{aligned} \tag{8.5}$$

This formula could also be derived by integration from (8.4), keeping in mind that the constants of integration are fixed by the condition that $N_{D,\mathbf{p}}^{[k]}(0) = 0$ for $k \geq 1$. It is not an easy matter, though, to compute the second sum over j in this way.

In view of formula (8.3), we clearly have the following result:

Theorem 8.2. *The counting function of the lengths of a generalized Cantor string is monotonic, locally constant, with jumps of b^n at the points $x = a^n$ $(n \in \mathbf{N})$.*

8.2   The Spectrum of a Generalized Cantor String

Theorem 4.4, the pointwise explicit formula with error term, gives us the following expansion for the spectral counting function:

$$
N_{\nu,D,\mathbf{p}}^{[k]}(x) = \frac{a}{(a-b)}\frac{x^k}{k!} + \frac{x^{D+k-1}}{\log a}\sum_{n\in\mathbb{Z}}\frac{\zeta(D+in\mathbf{p})}{(D+in\mathbf{p})_k}x^{in\mathbf{p}}
$$

$$
+ \frac{1}{(k-1)!}\sum_{j=0}^{k-1}\binom{k-1}{j}(-1)^j\frac{x^{k-1-j}}{1-b\cdot a^j}\zeta(-j) + R_\nu(x),
$$

$$
(8.6)
$$

for $k \geq 2$. For $k = 1$, $\mathcal{P}^{[1]}\nu$ is given by the same formula, as a distribution. This follows from Theorem 4.12, the distributional formula with error term. (See Remark 2.35, in conjunction with Remark 5.9, which explains why our explicit formulas can be applied in this situation to obtain (8.6).)

There is an important difference in the spectrum between integral and nonintegral values of a. In the next section, we will study the first case, when a is integral, and in Section 8.2.2, we will study the case when a is nonintegral.

8.2.1   Integral Cantor Strings: a-adic Analysis of the Geometric and Spectral Oscillations

Definition 8.3. The generalized Cantor string constructed with the parameters a, b, is called *integral* if $a \in \mathbb{N}$, $a \geq 2$, and otherwise it is said to be *nonintegral*.

Remark 8.4. When b is integral, $b \in \mathbb{N}^*$ (i.e., when the multiplicities are positive integers),[1] the Cantor string is an ordinary fractal string, and we will then call it geometrically realizable, or simply geometric, when no confusion can arise with our usage of 'geometric' in opposition to 'spectral'.

For example, the (ternary) Cantor string—studied, in particular, in the papers [LapPo1; LapPo2, Example 4.5; Lap2–3] as well as in Chapter 1 and Sections 2.2.1, 5.4 and 6.3.1 of this book—is a generalized Cantor string with parameters $a = 3$ and $b = 2$. Hence, it is both an integral Cantor string (since $a = 3$ is an integer) and an ordinary geometric fractal string (since $b = 2$ is an integer).

Remark 8.5. It may be helpful to the reader to see the precise connection between our present setting and that of Chapter 2 where we considered arbitrary self-similar strings. Assume that b is an integer ($b = 2, 3, \ldots$), so that $\eta = \sum_{n=0}^{\infty} b^n \delta_{\{a^{-n}\}}$ is geometrically realizable. Then, in the language

---

[1] and when $b < a$ (so that $D = \log_a b$ lies in $(0,1)$).

of Chapter 2, η is an (ordinary, lattice) self-similar string with b scaling ratios that are all equal to a^{-1} (with $a > b$): $r_1 = r_2 = \cdots = r_b = a^{-1}$. Thus, in particular, its multiplicative generator is a^{-1} and $k_1 = \cdots = k_b = 1$, in the notation of Definition 2.10. The associated polynomial equation (2.26) is of degree 1, and takes the form $bz = 1$, $z = a^{-\omega}$. Therefore, we recover the fact that the complex dimensions of η lie on a single vertical line. Indeed, they are the points $\omega = D + in\mathbf{p}$, with $n \in \mathbb{Z}$, $D = \log_a b$ and $\mathbf{p} = 2\pi/\log a$.

The following result will enable us to determine very precisely the nature of the jumps of the spectral counting function of an integral Cantor string (see Corollary 8.7 below, along with its proof).

Recall from Theorem 1.13 that

$$N_{\nu,D,\mathbf{p}}(x) = N_{D,\mathbf{p}}(x) + N_{D,\mathbf{p}}(x/2) + N_{D,\mathbf{p}}(x/3) + \dots . \qquad (8.7)$$

Thus, by a direct computation, we derive the spectral counterpart of (8.3).

Theorem 8.6. *Let x be a positive real number and let*

$$x = \sum_{k \in \mathbb{Z}} x_k a^k$$

be the expansion of x in base a. Then the spectral counting function, $N_\nu(x)$, of the integral Cantor string with parameters a and b is given by the following formula:

$$N_\nu(x) = \frac{a}{a-b}x - \frac{1}{a-b}\left(a \sum_{k=-\infty}^{-1} x_k a^k + b \sum_{k=0}^{\infty} x_k b^k \right). \qquad (8.8)$$

Proof. Observe that the digits of x form a finite sequence to the left in the sense that $x_k = 0$ for $k < 0$ with $|k|$ sufficiently large. The expansion of x allows us to obtain a simple formula for the integer part of $a^{-n}x$:

$$[a^{-n}x] = \left[\sum_{k \in \mathbb{Z}} x_k a^{k-n} \right] = \sum_{k=n}^{\infty} x_k a^{k-n}.$$

Thus we can compute the counting function of the frequencies as follows:

$$N_\nu(x) = \sum_{n=0}^{\infty} [a^{-n}x] \, b^n$$

$$= \sum_{n=0}^{\infty} \sum_{k=n}^{\infty} x_k a^{k-n} b^n$$

$$= \sum_{k=0}^{\infty} x_k \sum_{n=0}^{k} a^{k-n} b^n$$

$$= \sum_{k=0}^{\infty} x_k \frac{a^{k+1} - b^{k+1}}{a - b}$$

$$= \frac{a}{a-b} x - \frac{1}{a-b} \left(a \sum_{k=-\infty}^{-1} x_k a^k + b \sum_{k=0}^{\infty} x_k b^k \right).$$

This yields the required result.   □

Corollary 8.7. *The counting function of the frequencies, $N_\nu(x)$, of an integral Cantor string jumps by*

$$\frac{b^{n+1} - 1}{b - 1}$$

at integral values of x that are divisible by a^n and not by a^{n+1} ($n \in \mathbb{N}$).

Proof. Let x be exactly divisible by a^n. Then, there are two ways to represent x in base a:

$$x_+ = \sum_{k=n}^{\infty} x_k a^k \quad \text{and} \quad x_- = \sum_{k=n+1}^{\infty} x_k a^k + (x_n - 1) a^n + \sum_{k=-\infty}^{n-1} (a-1) a^k.$$

Note that the n-th digit, x_n, is different from 0. By the above formula (8.7) for N_ν, we find the two values

$$N_\nu(x_+) = \frac{a}{a-b} x - \frac{1}{a-b} \left(b x_n b^n + b \sum_{k=n+1}^{\infty} x_k b^k \right)$$

and

$$N_\nu(x_-) = \frac{a}{a-b} x - \frac{1}{a-b} \left(a \sum_{k=-\infty}^{-1} (a-1) a^k + b \sum_{k=0}^{n-1} (a-1) b^k \right.$$
$$\left. + b(x_n - 1) b^n + b \sum_{k=n+1}^{\infty} x_k b^k \right).$$

These values are the limit of $N_\nu(t)$ when t approaches x from above and from below, respectively. Hence we conclude that, at the point x, N_ν jumps by

$$\frac{a}{a-b} \sum_{k=-\infty}^{-1} (a-1)a^k + \frac{b}{a-b} \sum_{k=0}^{n-1} (a-1)b^k - \frac{b}{a-b}b^n = \frac{b^{n+1}-1}{b-1},$$

as was to be proved. $\square$

Remark 8.8. This greatly improves some of the results of [LapPo2, Example 4.5, pp. 65–67], dealing with the usual Cantor string.

8.2.2 Nonintegral Cantor Strings: Analysis of the Jumps in the Spectral Counting Function

In this section, we study the oscillations in the spectrum of nonintegral Cantor strings. This is the most important case for us. In applications, we need to choose the oscillatory period $\mathbf{p}$ to be large. Since $a = e^{2\pi/\mathbf{p}}$, this means that a is a nonintegral real number, a little greater than 1.

Theorem 8.9. *The spectral counting function $N_\nu(x)$ jumps by at least b^n at $x = a^n$ ($n \in \mathbb{N}$). Hence $N_\nu(x)$ has jumps of order D at such points.*[2]

Proof. We will prove a slightly more general statement: Let m be the smallest positive integer such that a^m is an integer, or let $m = \infty$ when a^m is an integer for $m = 0$ only. Then N_ν jumps by

$$b^n + b^{n-m} + b^{n-2m} + \cdots + b^{n-qm} \qquad (n - qm \geq 0 > n - (q+1)m),$$

at $x = a^n$. (The case $m = 1$ corresponds to integral Cantor strings and has been treated in more detail in the previous section. When $m = \infty$, the sum contains only the term b^n.)

The frequencies are the numbers $k \cdot a^j$, with multiplicity b^j (with $k \in \mathbb{N}^*$, $j \in \mathbb{N}$). Thus the frequency a^n is found for all possible choices of k and j such that $ka^j = a^n$. We find $j = n, n - m, \ldots, n - qm$ as possible choices, and the corresponding multiplicities add up to a positive jump of

$$b^n + b^{n-m} + \cdots + b^{n-qm}.$$

This concludes the proof of the theorem. $\square$

Remark 8.10. We see that when passing from the geometry to the spectrum of a generalized Cantor string, the oscillations remain of the same order. Thus, in a sense, the spectral operator (defined in Section 5.3.2 above) is invertible when restricted to the class of generalized Cantor strings. This observation will play an important role in Chapter 9.

---

[2]Indeed, if $x = a^n$, then $b^n = x^D$, since $D = \log_a b$.

9

The Critical Zeros of Zeta Functions

As we saw in the previous chapter, the complex dimensions of a generalized Cantor string form an arithmetic progression $\{D + in\mathbf{p}\}_{n \in \mathbf{Z}}$ (for $D \in (0,1)$ and $\mathbf{p} > 0$). In this chapter, we use this fact to study arithmetic progressions of critical zeros of zeta functions.

By combining our explicit formulas with the analysis of the oscillations in the geometry and the spectrum of generalized Cantor strings carried out in Chapter 8, we show, in particular, that the Riemann zeta function—and other zeta functions from a large class of Dirichlet series not necessarily satisfying a functional equation—does not have an infinite sequence of critical zeros forming an (almost) arithmetic progression. (See Theorems 9.1, 9.5 and 9.9.) C. R. Putnam [Pu1–2] was the first to obtain such a result in the special case of the Riemann zeta function. We have found this result independently, first for the Riemann and Epstein zeta functions [Lap-vF4], and then in the generality presented here [Lap-vF2–3], which is natural in our framework, and for which Putnam's method does not work.

In Section 9.1, we present our proof first for $\zeta(s)$, the Riemann zeta function, and then indicate in Section 9.2.1 the changes necessary to obtain the general theorem. In Section 9.3, we combine a special case of Theorem 9.9 with methods from algebraic number theory to extend some of our results to Hecke L-series. Finally, in Section 9.4, we discuss a situation where our argument does not apply, and where the conclusion of Theorem 9.9 does not hold, the case of the zeta function of a curve over a finite field.

As will be clear to the reader, the tools developed in Chapter 4, namely, our distributional formula (Theorem 4.12) along with the corresponding distributional error estimate (Theorem 4.23), once again play an essential role in our proof of the main results of this chapter, Theorems 9.1, 9.5 and 9.9.

9.1 The Riemann Zeta Function: No Critical Zeros in an Arithmetic Progression

Let $\mathcal{L}$ be a generalized Cantor string with lengths a^{-n}, repeated with multiplicity b^n $(1 < b < a)$. As before, we denote the geometric and spectral counting function by $N_{\mathcal{L}}(x)$ and $N_\nu(x)$, respectively. In the previous chapter, we have computed $N_{\mathcal{L}}(x)$ and $N_\nu(x)$ by a direct calculation. In this chapter, we compute them again, but this time by using our explicit formulas, in order to obtain the desired connection with the zeros of the Riemann zeta function.

Theorem 4.8, applied to $N_{\mathcal{L}}(x)$, yields

$$N_{\mathcal{L}}(x) = \frac{1}{\log a} \sum_{n=-\infty}^{\infty} \frac{x^{D+in\mathbf{p}}}{D + in\mathbf{p}} - \frac{1}{b-1}, \tag{9.1}$$

where $\mathbf{p} = \frac{2\pi}{\log a}$ and D is determined by $a^D = b$; that is, $D = \log_a b$.

We have to interpret $N_\nu(x)$ as a distribution. We choose a screen to the left of $\operatorname{Re} s = 0$. Theorem 4.12 in conjunction with Theorem 4.23 yields

$$N_\nu(x) = \frac{a}{a-b}x + \frac{1}{\log a} \sum_{n=-\infty}^{\infty} \zeta(D+in\mathbf{p})\frac{x^{D+in\mathbf{p}}}{D + in\mathbf{p}} + O(1), \tag{9.2}$$

as $x \to \infty$. (See Remarks 2.35 and 5.9.)

Theorem 9.1. *Let $0 < D < 1$ and $\mathbf{p} > 0$ be given. Then there exists an integer $n \neq 0$ such that $\zeta(D+in\mathbf{p}) \neq 0$. That is, the Riemann zeta function does not have an infinite sequence of critical zeros forming an arithmetic progression.*

Remark 9.2. Note that the theorem implies that $\zeta(D + in\mathbf{p}) \neq 0$ for infinitely many integers n. Indeed, if $\zeta(D + in\mathbf{p})$ were to be nonzero for only finitely many $n \in \mathbb{Z}$, say for $|n| < M$, then we would obtain a contradiction by applying the theorem to D and $M\mathbf{p}$ instead of to D and $\mathbf{p}$. In the next section, we also obtain information about the density of the set $\{n \colon \zeta(D + in\mathbf{p}) \neq 0\}$ in $\mathbb{Z}$. The same remark applies to Theorem 9.5 below.

Remark 9.3. A priori, our result implies that ζ does not have a vertical infinite sequence of critical zeros in infinite arithmetic progression. However, since ζ satisfies a functional equation relating $\zeta(s)$ and $\zeta(1 - s)$, and $\zeta(s)$ does not vanish for $\operatorname{Re} s \geq 1$, it follows immediately that every infinite sequence of zeros on one line must be vertical, say on the line $\operatorname{Re} s = D$, with $D \in (0, 1)$.

The analogue of this remark applies to all the natural arithmetic zeta functions since they also satisfy a functional equation and have a convergent Euler product to the right of the critical strip. This is the case, for example, of the Hecke L-series of number fields, as well as of the more general L-series, to which the results of Section 9.2 or 9.3 can be applied.

Proof of Theorem 9.1. Assume that $\zeta(D + inp) = 0$ for all $n \neq 0$. Let $a = e^{2\pi/p}$ and $b = a^D$. The generalized Cantor string with these parameters has complex dimensions at all the points $D + inp$ ($n \in \mathbb{Z}$); see Section 8.1 above. Equation (9.2), the explicit formula for the frequencies, becomes very simple because all the terms corresponding to x^{D+inp}, $n \neq 0$, disappear since $\zeta(D + inp) = 0$. The resulting formula reads

$$N_\nu(x) = \frac{a}{a-b}x + \zeta(D)\frac{x^D}{D\log a} + o\left(x^D\right), \qquad \text{as } x \to \infty, \qquad (9.3)$$

as a distribution on $(0, \infty)$. We see that the frequencies of this Cantor string have no oscillations of order D. On the other hand, we have seen in Theorem 8.9 that $N_\nu(x)$ jumps by at least b^q at a^q, for each $q \in \mathbb{N}$. But this means that N_ν has oscillations of order D. Since this is a contradiction, we conclude that $\zeta(D + inp) \neq 0$ for some integer $n \neq 0$. $\qquad \square$

Alternate proof of Theorem 9.1. Note that it suffices to prove the theorem for $D \geq \frac{1}{2}$ because of the functional equation satisfied by ζ. (See, for example, [Ti, Chapter II].) Now, let $D > \frac{1}{4}$ and $p > 0$ be such that $\zeta(D + inp) \neq 0$ for at most finitely many values of n. Put $a = e^{2\pi/p}$ and $b = a^D$ and consider the same generalized Cantor string as above. Instead of applying the distributional explicit formula at level 1 to obtain a formula for $N_\nu(x)$, we can apply the pointwise explicit formula at level 2, as we now explain.

We choose $S(t) = \sigma_0 + it$, for some value of σ_0 strictly between $-\frac{1}{2}$ and 0. Then hypotheses $(\mathbf{H_1})$ and $(\mathbf{H_2})$ are satisfied with $\kappa = \frac{1}{2} - \sigma_0$. (See Section 5.2.2.) By Theorem 4.4, applied for $k = 2$, the integrated counting function of the frequencies is given by

$$N_\nu^{[2]}(x) = \frac{a}{(a-b)}\frac{x^2}{2} + \frac{x^{D+1}}{\log a}\sum_{n\in\mathbb{Z}}\frac{\zeta(D+inp)}{(D+inp)_2}x^{inp}$$
$$+ \frac{x}{1-b}\zeta(0) + O\left(x^{\sigma_0+1}\right), \qquad (9.4)$$

as $x \to \infty$. Here, $(s)_2 = s(s+1)$, as in Equation (4.10). The sum over n in formula (9.4) is finite.

Much as in [In, Theorem C, p. 35], we deduce from Equation (9.4) by a Tauberian argument that

$$N_\nu(x) = N_\nu^{[1]}(x) = \frac{a}{a-b}x + \sum_{n\in\mathbb{Z}}\frac{x^{D+inp}}{\log a}\frac{\zeta(D+inp)}{D+inp} + O\left(x^{\frac{\sigma_0+1}{2}}\right), \quad (9.5)$$

as $x \to \infty$. Again, the sum over n in formula (9.5) is finite, hence continuous. Choosing $\sigma_0 > -\frac{1}{2}$ such that $(\sigma_0 + 1)/2 < D$, we deduce that the frequencies of this Cantor string have no jumps of order D. But this contradicts Theorem 8.9. □

In order to obtain the analogue of this result for a broader class of zeta functions, we have to use our distributional explicit formula (Theorems 4.12 and 4.23) in Section 9.2 below, as in the first proof of Theorem 9.1 presented above, rather than our pointwise formula (Theorem 4.4) as in the alternative proof of this theorem. This was one of our main original motivations for developing the distributional theory of explicit formulas, as presented in Section 4.4.3 and Theorem 4.23.

As was alluded to in Remark 8.10, the following corollary captures the essence of our method of proof in this context (compare Corollary 7.7 above and review Section 5.3.2, where we have defined the spectral and inverse spectral operators for fractal strings):

Corollary 9.4. *The spectral operator is invertible when restricted to the class of generalized Cantor strings.*

9.2   Extension to Other Zeta Functions

The above (distributional) argument generalizes naturally to a large subclass of the class of zeta functions introduced at the end of Chapter 7. As was mentioned in the introduction, this subclass includes all Epstein zeta functions and all Dedekind zeta functions of algebraic number fields (see [Te] and [Lan], along with Appendix A). By a refinement of the argument, we also obtain an estimate for the density of nonzeros in arithmetic progressions, in Section 9.2.1 below.

We now precisely state the assumptions on the zeta functions ζ_B for which our results apply:

(P)   Given a sequence w_f of positive coefficients associated with a sequence of positive real numbers f, let $\zeta_B(s) = \sum_f w_f f^{-s}$ be the corresponding (generalized) Dirichlet series. Assume that, for some screen S, this series has only finitely many poles contained in the associated window W, and that it satisfies the hypotheses $(\mathbf{H_1})$ and $(\mathbf{H_2})$ of Equations (4.16) and (4.17).

Recall that $(\mathbf{H_1})$ and $(\mathbf{H_2})$ mean that the meromorphic continuation of ζ_B grows polynomially along horizontal lines and along the vertical direction of the screen. Under these assumptions, we prove

Theorem 9.5. *Let ζ_B be a zeta function satisfying the above hypothesis (P). Let $\mathbf{p} > 0$ be arbitrary and let $D \in W \cap \mathbb{R}$ be such that the vertical*

line Re $s = D$ *lies entirely within* W. *Then there exists an integer* $n \neq 0$ *such that* $\zeta_B(D + in\mathbf{p}) \neq 0$. *That is,* $\zeta_B(s)$ *does not have an infinite vertical sequence of zeros forming an arithmetic progression within the window* W.

Before proving Theorem 9.5, we indicate how to associate a generalized fractal spray to ζ_B. First of all, we view the Dirichlet series ζ_B as the spectral zeta function of a virtual basic shape B. By definition, the 'frequency' f has multiplicity w_f; see Section 3.3. Then, given D and $\mathbf{p}$ as in Theorem 9.5, let $\mathcal{L}$ be the generalized Cantor string of dimension D and oscillatory period $\mathbf{p}$. Note that here, in contrast to Section 9.1, we do not restrict D to be between 0 and 1. Next, consider the generalized Cantor spray of $\mathcal{L}$ with virtual basic shape B. Recall from Section 3.3, Equation (3.35), that the spectral zeta function of this spray is given by

$$\zeta_\nu(s) = \zeta_B(s) \cdot \zeta_{\mathcal{L}}(s).$$

We have to interpret $N_\nu(x)$ as a distribution. Theorem 4.12 in conjunction with Theorem 4.23 yields the explicit formula for the frequencies

$$N_\nu(x) = W_{B,\mathcal{L}}^{[1]}(x) + \frac{1}{\log a} \sum_{n=-\infty}^{\infty} \zeta_B(D + in\mathbf{p}) \frac{x^{D+in\mathbf{p}}}{D + in\mathbf{p}} + O(x^{\sigma_u}), \quad (9.6)$$

as $x \to \infty$, where

$$W_{B,\mathcal{L}}^{[1]}(x) = \sum_{u: \text{ pole of } \zeta_B} \operatorname{res}\left(\frac{\zeta_B(s)\zeta_{\mathcal{L}}(s)}{s} x^s; u\right) \quad (9.7)$$

is the Weyl term (see Section 5.6, formula (5.84), along with Remarks 2.35 and 5.9).

Proof of Theorem 9.5. Assume that $\zeta_B(D + in\mathbf{p}) = 0$ for all $n \neq 0$. As above, let $\mathcal{L}$ be the generalized Cantor string with $a = e^{2\pi/\mathbf{p}}$ and $b = a^D$. Then the generalized Cantor spray of $\mathcal{L}$ on B has complex dimensions at all the points $D + in\mathbf{p}$ ($n \in \mathbb{Z}$), but since $\zeta_B(D + in\mathbf{p}) = 0$ for all $n \in \mathbb{Z}\backslash\{0\}$, Equation (9.6) simplifies, just as in the proof of Theorem 9.1:

$$N_\nu(x) = W_{B,\mathcal{L}}^{[1]}(x) + \zeta_B(D)\frac{x^D}{D\log a} + o(x^D), \quad \text{as } x \to \infty, \quad (9.8)$$

as a distribution on $(0, \infty)$. We see that the frequencies of this Cantor spray have no oscillations of order D.

On the other hand, the counting function of the frequencies is related to the geometric counting function by

$$N_\nu(x) = \sum_f w_f N_{\mathcal{L}}\left(\frac{x}{f}\right). \quad (9.9)$$

Since $N_{\mathcal{L}}$ has jumps, the positivity of w_f guarantees that N_ν has jumps too, at the points $x = f \cdot a^n$, where f runs through the frequencies of B and $n = 0, 1, \ldots$. Thus we obtain the analogue of Theorem 8.9: N_ν has jumps of order D. But this means that N_ν has oscillations of order D. Since this is a contradiction, we conclude that $\zeta_B(D + in\mathbf{p}) \neq 0$ for some integer $n \neq 0$. $\square$

Note that Remark 9.2 applies, so that we can immediately conclude that there are infinitely many integers n such that $\zeta_B(D + in\mathbf{p}) \neq 0$.

The following corollary is the exact analogue of Corollary 9.4 in the present more general setting. See Definition 5.22 in Section 5.6 for the definition of the spectral operator for fractal sprays.

Corollary 9.6. *The spectral operator is invertible when restricted to the class of all generalized Cantor sprays, with virtual basic shape B defined by a zeta function ζ_B satisfying hypothesis (P).*

9.2.1   Density of Nonzeros on Vertical Lines

By using a refinement of the above argument, we can obtain a lower bound for the density of points where $\zeta_B(D + in\mathbf{p})$ is nonzero.

Let ζ_B be a zeta function satisfying the above hypothesis (P). Let $\rho \geq 0$ be such that

$$\zeta_B(D + it) = O\left(t^\rho\right), \quad \text{as } |t| \to \infty. \tag{9.10}$$

Theorem 9.7. *Let $\delta > 0$ and assume that $\rho < 1$, where ρ is the exponent of Equation (9.10). Then, for infinitely many values of T, tending to infinity, the set*

$$\{n \in \mathbb{Z} : |n| \leq T, \ |\zeta_B(D + in\mathbf{p})| \neq 0\} \tag{9.11}$$

contains more than $T^{1-\rho-\delta}$ elements.

Proof. Suppose that the set defined by (9.11) contains fewer than $T^{1-\rho-\delta}$ elements, for all sufficiently large T. Let $0 < n_1 < n_2 < n_3 < \ldots$ be the sequence of positive elements of (9.11) (for $T = \infty$). Then we have that $n_j \geq j^{1/(1-\rho-\delta)}$, except possibly for the first few integers n_j. Moreover, if n is not in the sequence, then $\zeta_B(D + in\mathbf{p}) = 0$. Thus,

$$\begin{aligned}
N_\nu(x) = W_{B,\mathcal{L}}^{[1]}(x) &+ \frac{x^D}{D \log a} \zeta_B(D) \\
&+ \frac{1}{\log a} \sum_{j=1}^{\infty} \frac{x^{D \pm i n_j \mathbf{p}}}{D \pm i n_j \mathbf{p}} \zeta_B(D \pm i n_j \mathbf{p}) + o\left(x^D\right),
\end{aligned} \tag{9.12}$$

as $x \to \infty$. The j-th term in this series is bounded by a constant times $n_j^{\rho-1}$, hence the series is absolutely convergent. Therefore its sum is a continuous function. But this is a contradiction, since N_ν has jumps of order D.   □

For the Riemann zeta function $\zeta(s)$, we thus obtain, choosing $D = 1/2$ and $\rho = 1/6$ (see [Ti, Theorems 5.5 and 5.12]):[1]

Corollary 9.8. *For every* **p** > 0 *and* $\delta > 0$,

$$\# \left\{ n \in \mathbb{Z} \colon |n| \le T, \ \zeta \left(\tfrac{1}{2} + in\mathbf{p} \right) \ne 0 \right\} \ge T^{5/6-\delta},$$

for infinitely many values of T, tending to infinity.

9.2.2 Almost Arithmetic Progressions of Zeros

By a classical result from Fourier theory, a Fourier series $\sum_{n \in \mathbb{Z}} a_n e^{inx}$ does not have jump discontinuities if $a_n = o(n^{-1})$ as $|n| \to \infty$, see [Zyg, Theorem 9.6, p. 108] or [Ru1, §5.6.9, p. 118]. Since this result is formulated in terms of the derivative of this Fourier series (i.e., for measures), we formulate the following argument on level $k = 0$.

Theorem 9.9. *Let ζ_B satisfy hypothesis (P) above. Then there do not exist $D \in W \cap \mathbb{R}$ and* **p** > 0 *such that $\zeta_B(D + in\mathbf{p}) \to 0$ as $|n| \to \infty$.*[2]

Proof. By Theorems 4.12 and 4.23 applied at level $k = 0$, we obtain the analogue of formula (9.6):

$$\nu = W_{B,\mathcal{L}}^{[0]}(x) \, dx + \frac{1}{\log a} \sum_{n \in \mathbb{Z}} x^{D-1+in\mathbf{p}} \zeta_B(D + in\mathbf{p}) \, dx + o \left(x^{D-1} \, dx \right),$$

$$(9.13)$$

as $x \to \infty$. Here, $W_{B,\mathcal{L}}^{[0]}(x)$ is the distributional derivative of the Weyl term, given by

$$W_{B,\mathcal{L}}^{[0]}(x) = \sum_{u: \text{ pole of } \zeta_B} \text{res} \left(\zeta_B(s) \zeta_\eta(s) x^{s-1}; u \right). \qquad (9.14)$$

If $\zeta_B(D + in\mathbf{p}) \to 0$ as $|n| \to \infty$, then, by the classical result cited above, this measure is continuous. That is, it does not have atoms.

On the other hand, by Equation (9.9), N_ν has jumps of order D. Since $N_\nu(x) = \int_0^x \nu(dx)$, this shows that ν has atoms. This contradiction shows that $\zeta_B(D + in\mathbf{p})$ does not converge to 0 as $|n| \to \infty$.   □

---

[1]By [Ti, Theorem 5.18, p. 99], one could even take $\rho = 27/164$. Further, if the Lindelöf Hypothesis holds, then one can take $\rho = 0$.

[2]We assume as in Theorem 9.5 that the line $\text{Re } s = D$ lies within the window W.

9.3   Extension to L-Series

By using some well known results from algebraic number theory [Lan, ParSh1–2], we can also obtain information about the zeros of certain Dirichlet series with complex (rather than positive) coefficients. For example, we can deduce that given any Hecke L-series (see Appendix A, Section A.2) associated with an algebraic number field (and with a complex-valued character), its critical zeros do not form an arithmetic progression. In particular, this is true of any Dirichlet L-series.[3,4] This statement is a simple corollary of the special case of our results for Dedekind zeta functions (which have positive coefficients) and of class field theory.

We now precisely state the resulting theorem. Further extensions of Theorem 9.1 are possible using higher-dimensional representations (as, for example, in [Lan, ParSh1–2, RudSar]) rather than one-dimensional representations (i.e., characters), but we omit this discussion here for simplicity of exposition.

Theorem 9.10. *Let K be a number field and let χ_0 be a character of a generalized ideal class group of K. Then the associated Hecke L-series, $L(s, \chi_0)$, has no infinite sequence of critical zeros forming an arithmetic progression.*

Proof. Let L be the class field associated with this ideal class group, and let ζ_L be the Dedekind zeta function of L. This is the Hecke L-series associated with the trivial character. Let χ run over the characters of the ideal class group. According to a well known result from class field theory (see Equation (A.10) in Appendix A or, for example, [Lan, Chapter XII, Theorem 1, p. 230] and [ParSh2, Chapter 2, Theorem 2.24, p. 106]), we have

$$\zeta_L(s) = \prod_\chi L(s, \chi). \tag{9.15}$$

We deduce from (9.15) that if the factor $L(s, \chi_0)$ has an infinite sequence of zeros in an arithmetic progression, then so does $\zeta_L(s)$. But this Dirichlet series has positive coefficients. Hence, by Theorem 9.5, it does not have such a sequence of zeros. It follows that $L(s, \chi_0)$ does not have such a sequence of zeros either.   □

Remark 9.11. The reader unfamiliar with the terminology used in the proof of Theorem 9.10 may wish to assume that $K = \mathbb{Q}$, the field of rational numbers. Then $L(s, \chi_0)$ is nothing but an ordinary Dirichlet L-series

---

[3]We are grateful to Ofer Gabber for suggesting to us this extension of our results [Gab].

[4]After learning about our work in [Lap-vF1–4], Mark Watkins has obtained a similar extension of our results in the case of Dirichlet L-series, for arithmetic progressions of zeros of finite length. (See [Wa].)

(see [Da,Lan; Ser, §VI.3] or Example A.2 in Appendix A). We note that the corresponding special case of formula (9.15) is established in [Da, Chapter 6] or [Ser, §VI.3.4].

9.4   Zeta Functions of Curves Over Finite Fields

The zeta function of a curve over a finite field is periodic with a purely imaginary period. If the curve is not rational, the zeta function has zeros, and, by periodicity, each zero gives rise to a shifted arithmetic progression of zeros. Moreover, depending on the curve, the zeta function can have a real zero, and then it has a vertical arithmetic progression of zeros starting on the real axis. So the conclusion of Theorem 9.5 does not hold in this case. However, this situation is very similar to that considered in Section 9.2: The associated measure is positive, and the zeta function has an Euler product and satisfies a functional equation. The main goal of this section is to explain why our proof does not go through in this situation. As we will show, the key reason is that the Weyl term itself has large jumps (see the comment following Equation (9.31)).

The interested reader can find additional examples and further information about the theory in [vF1, vL-vdG, Wei2], as well as, for instance, in [ParSh1, Chapter 4, §1].

Let $\mathbb{F}_q$ be the finite field with q elements, where q is a power of a prime number. Let C be a curve defined over $\mathbb{F}_q$. We denote the function field of C by $\mathbb{F}_q(C)$. Thus, $\mathbb{F}_q(C)$ is the field of algebraic functions from C to $\mathbb{P}^1$ ($\mathbb{F}_q$), the projective line over $\mathbb{F}_q$. We view C as embedded in projective space, with homogeneous coordinates $(x_0 : \cdots : x_n)$. The affine part of C is then given as the subset where $x_0 \neq 0$, and this gives rise to the affine coordinate ring $R = \mathbb{F}_q[C]$, the ring of functions that have no poles in the affine part of C. Note that the field of fractions of R is $\mathbb{F}_q(C)$.

The ring R has ideals. For an ideal $\mathfrak{a}$ of R, the ring $R/\mathfrak{a}$ is a finite-dimensional vector space over $\mathbb{F}_q$, and we denote its dimension by $\deg \mathfrak{a}$, the *degree* of $\mathfrak{a}$. Ideals have a unique factorization into prime ideals, and this fact is expressed by the equality

$$\sum_{\mathfrak{a}} q^{-s \deg \mathfrak{a}} = \prod_{\mathfrak{p}} \frac{1}{1 - q^{-s \deg \mathfrak{p}}}, \qquad (9.16)$$

where $\mathfrak{a}$ (respectively, $\mathfrak{p}$) runs over all ideals (respectively, prime ideals) of R. This function of the complex variable s is the (incomplete) zeta function associated with C. It is incomplete, because the factors corresponding to the points of C with $x_0 = 0$ are missing. Also, C may have singularities.

Example 9.12 (The projective line). If $C = \mathbb{P}^1$, that is, if the curve has genus $g = 0$, then $R = \mathbb{F}_q[X]$, the ring of polynomials over $\mathbb{F}_q$. In this case, every ideal of R is generated by a single polynomial. The number of ideals of degree n is equal to the number of monic polynomials of degree n, which is q^n. Hence

$$\sum_{\mathfrak{a}} q^{-s \deg \mathfrak{a}} = \sum_{n=0}^{\infty} q^n q^{-ns} = \frac{1}{1 - q^{1-s}}.$$

There is only one point at infinity, and the corresponding factor in the Euler product is

$$\frac{1}{1 - q^{-s}}.$$

Thus, the zeta function of $\mathbb{P}^1$ over $\mathbb{F}_q$ is

$$\zeta_{\mathbb{P}^1(\mathbb{F}_q)}(s) = q^{-s} \frac{1}{(1 - q^{-s})(1 - q^{1-s})}, \tag{9.17}$$

where the factor q^{-s} has been inserted in order that this function satisfies the functional equation

$$\zeta_{\mathbb{P}^1(\mathbb{F}_q)}(1 - s) = \zeta_{\mathbb{P}^1(\mathbb{F}_q)}(s). \tag{9.18}$$

Finally, we note that the logarithmic derivative of this zeta function is the generating function of the number of points of $\mathbb{P}^1$ with values in algebraic extensions $\mathbb{F}_{q^n}$ of $\mathbb{F}_q$, for $n = 1, 2, \dots$. Indeed, $\mathbb{P}^1(\mathbb{F}_{q^n})$ contains $q^n + 1$ points, and

$$-\frac{\zeta'_{\mathbb{P}^1(\mathbb{F}_q)}}{\zeta_{\mathbb{P}^1(\mathbb{F}_q)}}(s) = \log q \left(1 + \frac{q^{-s}}{1 - q^{-s}} + \frac{q^{1-s}}{1 - q^{1-s}} \right)$$
$$= \log q \left(1 + \sum_{n=1}^{\infty} (q^n + 1) q^{-ns} \right). \tag{9.19}$$

This result holds for the logarithmic derivative of the zeta function of an arbitrary curve over a finite field, as the reader may check after the subsequent discussion. We do not need this fact here, but it was the original motivation for introducing these functions. (See [Weil–3].)

Again let C be a curve over $\mathbb{F}_q$, of genus g. We describe a way to obtain the (completed) zeta function of C without first having to choose an affine part of C. The field $\mathbb{F}_q(C)$ has valuations. Associated with a valuation v, we have a residue class field $\mathbb{F}_v$, which is a finite extension of $\mathbb{F}_q$. The degree of this extension is called $\deg v$, the *degree* of v. A formal sum of valuations

$$\mathfrak{D} = \sum_v m_v v, \quad \text{with } m_v \in \mathbb{Z}, \tag{9.20}$$

with only finitely many nonzero coefficients m_v, is called a *divisor* of C. A valuation will also be called a *prime divisor*. Thus, divisors form a group, the free group generated by the prime divisors. The *degree* of a divisor is $\deg \mathfrak{D} = \sum_v m_v \deg v$. A divisor is *positive*, $\mathfrak{D} \geq 0$, if $m_v \geq 0$ for all v.

Recall that g is the genus of C. The zeta function of C is defined by

$$\zeta_C(s) = q^{-s(1-g)} \sum_{\mathfrak{D} \geq 0} q^{-s \deg \mathfrak{D}}. \tag{9.21}$$

Since, by definition, the factorization of a divisor into prime divisors is unique, we have the Euler product

$$\zeta_C(s) = q^{-s(1-g)} \prod_v \frac{1}{1 - q^{-s \deg v}}. \tag{9.22}$$

It is known that $\zeta_C(s)$ is a rational function of q^{-s}, of the form

$$\zeta_C(s) = x^{1-g} \frac{P(x)}{(1-x)(1-qx)} \qquad (x = q^{-s}), \tag{9.23}$$

where

$$P(x) = p_0 + p_1 x + p_2 x^2 + \cdots + p_{2g} x^{2g}$$

has degree $2g$, integer coefficients, and $p_0 = 1$. It can be shown that

$$P(1) = p_0 + p_1 + p_2 + \cdots + p_{2g}$$

is the *class number* of C; i.e., the number of divisor classes up to linear equivalence. Note that

$$P(1) = -\operatorname{res}(\zeta_C; 0)(q-1) \log q,$$

and that $\operatorname{res}(\zeta_C; 0) = -\operatorname{res}(\zeta_C; 1)$.

The function $\zeta_C(s)$ satisfies the functional equation

$$\zeta_C(s) = \zeta_C(1 - s), \tag{9.24}$$

which, in terms of P, means that

$$P\left(\frac{1}{x}\right) = q^g x^{-2g} P\left(\frac{x}{q}\right).$$

In other words, $p_{2g-j} = q^{g-j} p_j$ for $0 \leq j \leq 2g$.

By (9.23), ζ_C is periodic with period $2\pi i / \log q$. The poles of ζ_C are simple and located at $1 + 2k\pi i / \log q$ and $2k\pi i / \log q$ ($k \in \mathbb{Z}$). It is also known that the zeros of ζ_C have real part $1/2$. In other words, the zeros of $P(x)$ have absolute value $q^{-1/2}$. This is the Riemann Hypothesis for curves over a finite field, established by André Weil in [Weil-3]; see also, for example, [Roq, Step, Bom] or [ParSh1, Chapter 4, §1] and the relevant references therein. But we do not use this fact in the sequel.

Using the Dirichlet series for $\zeta_{\mathbb{P}^1}$ over $\mathbb{F}_q$,

$$\frac{x}{(1-x)(1-qx)} = \sum_{k=0}^{\infty} \frac{q^k - 1}{q - 1} x^k,$$

we find successively (with $x = q^{-s}$),

$$
\begin{aligned}
\zeta_C(s) &= \sum_{j=-g}^{g} p_{j+g} x^j \sum_{k=0}^{\infty} \frac{q^k - 1}{q - 1} x^k \\
&= \sum_{n=1-g}^{\infty} x^n \sum_{\substack{j=-g \\ j \le n}}^{g} p_{j+g} \frac{q^{n-j} - 1}{q - 1} \\
&= \sum_{n=1-g}^{g-1} x^n \sum_{j=-g}^{n} p_{j+g} \frac{q^{n-j} - 1}{q - 1} + \sum_{n=g}^{\infty} x^n \sum_{j=-g}^{g} p_{j+g} \frac{q^{n-j} - 1}{q - 1} \\
&= \sum_{n=1-g}^{g-1} x^n \sum_{j=-g}^{n} p_{j+g} \frac{q^{n-j} - 1}{q - 1} + P(1) \sum_{n=g}^{\infty} x^n \frac{q^n - 1}{q - 1},
\end{aligned}
$$

(9.25)

where in the last equality we have used that $\sum_{j=-g}^{g} p_{j+g} = P(1)$ and $\sum_{j=-g}^{g} p_{j+g} q^{-j} = q^g P(1/q) = P(1)$. Note that for $g = 0$, the first sum is to be interpreted as 0.

We interpret this zeta function as the spectral zeta function of a (virtual) basic shape B. Thus, the frequencies of B are

$$q^n, \text{ with multiplicity } P(1) \frac{q^n - 1}{q - 1}, \text{ when } n \ge g, \qquad (9.26)$$

and the small frequencies are

$$q^n, \text{ with multiplicity } \sum_{j=-g}^{n} p_{j+g} \frac{q^{n-j} - 1}{q - 1}, \text{ when } 1 - g \le n \le g - 1. \qquad (9.27)$$

Now, let η be a (generalized) fractal string, and consider the fractal spray of η on B. Its frequencies are counted by

$$N_\nu(x) = \sum_{n=1-g}^{g-1} N_\eta\left(\frac{x}{q^n}\right) \sum_{j=-g}^{n} p_{j+g} \frac{q^{n-j} - 1}{q - 1} + P(1) \sum_{n=g}^{\infty} N_\eta\left(\frac{x}{q^n}\right) \frac{q^n - 1}{q - 1}. \qquad (9.28)$$

The first term in the asymptotic expansion of $N_\nu(x)$ is the Weyl term. If $\eta = \mathcal{L}$ is an ordinary fractal string of dimension D with lengths l_j, we can find the corresponding Weyl term as follows:

$$
\sum_{n=0}^{\infty} N_{\mathcal{L}}\left(\frac{x}{q^n}\right) \frac{q^n - 1}{q - 1} = \sum_{n=0}^{\infty} \sum_{j:\, l_j^{-1} \leq xq^{-n}} \frac{q^n - 1}{q - 1}
$$

$$
= \sum_{j:\, l_j^{-1} \leq x} \sum_{n=0}^{[\log_q xl_j]} \frac{q^n - 1}{q - 1}
$$

$$
= \sum_{j:\, l_j^{-1} \leq x} \frac{q^{[\log_q xl_j]+1} - 1}{(q-1)^2} - \frac{[\log_q xl_j] + 1}{q - 1}
$$

$$
= x \sum_{j=1}^{\infty} l_j \frac{q^{1-\{\log_q xl_j\}} - 1}{(q-1)^2} + O\left(x^D \log x\right),
$$

as $x \to \infty$. The first sum is multiplicatively periodic with period q. The other terms in formula (9.28), namely

$$
\sum_{n=1-g}^{g-1} N_{\mathcal{L}}\left(\frac{x}{q^n}\right) \sum_{j=-g}^{n} p_{j+g} \frac{q^{n-j} - 1}{q - 1} - P(1) \sum_{n=0}^{g-1} N_\eta\left(\frac{x}{q^n}\right) \frac{q^n - 1}{q - 1},
$$

add up to $O\left(x^D \log x\right)$. Thus, the Weyl term is given by

$$
W(x) = x \cdot P(1) \sum_{j=1}^{\infty} l_j \frac{q^{1-\{\log_q xl_j\}}}{(q-1)^2} = x \cdot G\left(\log_q x\right), \tag{9.29}
$$

where G is the periodic function (of period 1) defined by

$$
G(u) = P(1) \sum_{j=1}^{\infty} l_j \frac{q^{1-\{u+\log_q l_j\}}}{(q-1)^2}. \tag{9.30}
$$

We see that G is bounded in terms of the total length L of $\mathcal{L}$:

$$
L \frac{P(1)}{(q-1)^2} \leq G(u) \leq L \frac{qP(1)}{(q-1)^2}.
$$

Moreover, we see that the Weyl term has jumps of order x at the points $x = q^m l_1^{-1}$, for $m = 0, 1, 2, \ldots$. By our explicit formulas, we also immediately find the Fourier expansion of the Weyl term, by applying formula (5.85) for $k = 1$:

$$
W(x) = \operatorname{res}\left(\zeta_C; 1\right) \sum_{n \in \mathbb{Z}} \zeta_{\mathcal{L}}\left(1 + n\frac{2\pi i}{\log q}\right) \frac{x^{1+2n\pi i/\log q}}{1 + 2n\pi i/\log q}. \tag{9.31}
$$

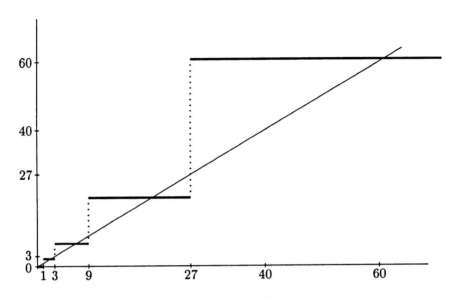

Figure 9.1: The Weyl term of the ordinary Cantor spray on $\mathbb{P}^1(\mathbb{F}_3)$, viewed additively.

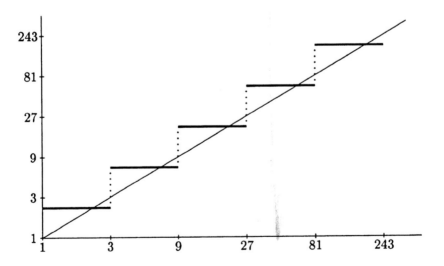

Figure 9.2: The Weyl term of the ordinary Cantor spray on $\mathbb{P}^1(\mathbb{F}_3)$, viewed multiplicatively.

We apply this to the generalized Cantor string $\eta = \sum_{m=0}^{\infty} b^m \delta_{\{q^m\}}$ with parameters $a = q$ and $b > 0$, so that the oscillatory period is $\mathbf{p} = 2\pi/\log q$. (See Figures 9.1 and 9.2 for a graph of the Weyl term associated with the spray of the ordinary Cantor string on the basic shape $\mathbb{P}^1(\mathbb{F}_3)$.) We deduce that at $x = q^m$, $m = 0, 1, 2, \ldots$, the counting function $N_\nu(x)$ jumps by

$$\sum_{n=0}^{m} b^{m-n} \frac{q^n - 1}{q - 1} = \frac{1}{q-1} \left(\frac{q^{m+1} - b^{m+1}}{q - b} - \frac{b^{m+1} - 1}{b - 1} \right),$$

up to jumps of order b^m, caused by the terms with $n \leq g - 1$. These latter jumps are smaller since $b < q$. We see that the jump is of order x, and not of order x^D. Therefore, we cannot conclude that there must be oscillatory terms of order D in the explicit formula for $N_\nu(x)$. Moreover, in contrast to the situation in Sections 9.1 and 9.2, we see that the jump of $N_\nu(x)$ is caused by the terms for large values of n, and not by the first terms in the sum for $N_\nu(x)$.

We illustrate this discussion by two examples (see [vF1]).

Example 9.13. The curve given by $y^2 = x^3 - x$ over $\mathbb{F}_3$ has genus 1. Its zeta function is

$$\frac{1 + 3^{1-2s}}{(1 - 3^{-s})(1 - 3^{1-s})} = 1 + 4 \sum_{n=1}^{\infty} 3^{-ns} \frac{3^n - 1}{2},$$

and the Weyl term associated with the string $\mathcal{L} = (l_j)_{j=1}^{\infty}$ is

$$W(x) = 4x \sum_{j=1}^{\infty} l_j \frac{3^{1 - \{\log_3 x l_j\}}}{4}.$$

Example 9.14. The curve given by $x^3 y + y^3 z + z^3 x = 0$ over $\mathbb{F}_2$ has genus 3. Its zeta function is

$$2^{2s} \frac{1 + 5 \cdot 2^{-3s} + 8 \cdot 2^{-6s}}{(1 - 2^{-s})(1 - 2^{1-s})}$$

$$= 2^{2s} + 3 \cdot 2^s + 7 + 20 \cdot 2^{-s} + 46 \cdot 2^{-2s} + 14 \sum_{n=3}^{\infty} 2^{-ns} (2^n - 1),$$

and the Weyl term associated with the string $\mathcal{L} = (l_j)_{j=1}^{\infty}$ is

$$W(x) = 14 x \sum_{j=1}^{\infty} l_j 2^{1 - \{\log_2 x l_j\}}.$$

We close this section with an example of a Dirichlet series that is a rational function of q^{-s}, has a functional equation and an Euler product, but does not satisfy the analogue of the Riemann Hypothesis.

Example 9.15. Consider the zeta function

$$\zeta_X(s) = \frac{1 - 5x + 5x^2}{(1 - x)(1 - 5x)}, \qquad x = 5^{-s}.$$

As a Dirichlet series,

$$\zeta_X(s) = 1 + \sum_{n=1}^{\infty} \frac{5^n - 1}{4} x^n, \qquad x = 5^{-s}.$$

One checks that this function satisfies the functional equation (9.24). Moreover, the numerator factors as $1 - 5x + 5x^2 = (1 - (3 - \phi)x)(1 - (2 + \phi)x)$, where ϕ is the golden ratio. Since $3 - \phi < \sqrt{5} < 2 + \phi$, the zeros of $\zeta_X(s)$ do not have real part $1/2$. The logarithmic derivative of $\zeta_X(s)$ is given by

$$-\frac{\zeta_X'(s)}{\zeta_X(s)} = (\log 5) \sum_{n=1}^{\infty} x^n \left(1 + 5^n - (3 - \phi)^n - (2 + \phi)^n \right).$$

Let a_n be $(3 - \phi)^n + (2 + \phi)^n$, for $n = 0, 1, 2, \ldots$. One checks that $a_{n+2} = 5(a_{n+1} - a_n)$ and $a_0 = 2$, $a_1 = 5$. Thus $a_2 = 15$, $a_3 = 50$, $a_4 = 175, \ldots$. If we want to write $\zeta_X(s)$ as an Euler product of the form

$$\zeta_X(s) = \prod_{n=1}^{\infty} (1 - x^n)^{-v_n},$$

where v_n stands for the number of valuations of degree n, then we find the following formula for v_n:

$$\sum_{d \mid n} d v_d = 1 + 5^n - (3 - \phi)^n - (2 + \phi)^n = 1 + 5^n - a_n.$$

From this, v_n can be successively computed: $v_1 = 1$, $v_2 = 5$, $v_3 = 25$, $v_4 = 110$, $v_5 = 500, \ldots$. In particular, $v_n > 0$ for all $n \geq 1$.

The Weyl term associated with the string $\mathcal{L} = (l_j)_{j=1}^{\infty}$ is

$$W(x) = x \sum_{j=1}^{\infty} l_j \frac{5^{1 - \{\log_5 x l_j\}}}{16},$$

just as in the case of a curve C. However, ζ_X is not the zeta function of a curve, since otherwise it would satisfy the Riemann Hypothesis.

10
Concluding Comments

In this chapter, we make several suggestions for the direction of future research related to, and naturally extending in various ways, the theory developed in this book. In several places, we also provide some additional background material that may be helpful to the reader.

In Section 10.1, we formulate general conjectures about the zeros of Dirichlet series, going beyond our results in Chapter 9, concerning zeros in infinite arithmetic progressions. We also give examples showing the necessity of the assumptions that we are led to make in the resulting 'irrationality conjectures'.

In Section 10.2, we propose a new definition of fractality, involving the notion of complex dimension. Namely, a given set is 'fractal' if it has at least one nonreal complex dimension (with positive real part). We illustrate this definition by means of two examples, and in Section 10.2.1, we compare our definition with other definitions of fractality that have previously been suggested in the literature. This enables us, in particular, to resolve several paradoxes concerning the fractality of certain geometric objects, such as the a-string and the Devil's staircase (i.e., the graph of the Cantor function).

In Section 10.3, we explore some of the relationships between fractality (as defined in Section 10.2) and self-similarity. In particular, we discuss the geometric aspects of this relationship for self-similar fractal drums, and propose a natural interpretation (expressed in terms of complex dimensions) of the lattice–nonlattice dichotomy for higher-dimensional self-similar sets. We illustrate our conjecture by discussing the example of the (von Koch) snowflake curve, a lattice fractal of which we determine the complex dimensions. We also point out in Section 10.3, as well as in Section 10.4, some of the challenges associated with the extension of our theory of complex dimensions to higher-dimensional objects, such as self-similar fractal drums.

In Section 10.4.1, we discuss the problem of describing the spectrum of a fractal drum, as initially formulated in the Weyl–Berry Conjecture. In Section 10.4.2, we consider this problem for the important special case of self-similar drums, including the snowflake drum and its natural generalizations. In Section 10.4.3, we briefly examine the question of understanding the spectrum of a fractal drum in terms of the periodic orbits of a dynamical system naturally associated to it. In general, and for example, for the snowflake drum, this is a very difficult problem, the formulation of which is even in a preliminary phase. We focus here on the case of a self-similar string for which more can be said at this point.

We close this chapter with a very speculative section, Section 10.5, in which we raise the question of whether a suitable homological interpretation can be found for the complex dimensions of a fractal. We conclude by making a suggestion for such a theory of complex homology.

10.1   Conjectures about Zeros of Dirichlet Series

We state and briefly discuss several conjectures that could be tackled, or at least receive geometric meaning, within our framework.

We begin by formulating a natural conjecture regarding the vertical distribution of the zeros of the Riemann zeta function $\zeta = \zeta(s)$:

Conjecture 10.1 (Irrationality Conjecture). *The imaginary parts of the critical zeros of ζ on one vertical line in the upper half-plane are rationally independent. In particular, the critical zeros of ζ are all simple.*

We refer, for example, to the paper by Odlyzko and te Riele [Od-tR] and the relevant references therein for numerical and theoretical evidence in support of this conjecture. (Concerning the numerical evidence in support of the simplicity of the zeros, see also, e.g., [Od1–2] and [vL-tR-W], where additional relevant information can be found.)

Several comments are in order:

(i) Because $\zeta(s) = \sum_{n=1}^{\infty} n^{-s}$ is defined by a Dirichlet series with real coefficients, its zeros come in complex conjugate pairs. Thus, in the statement of Conjecture 10.1, only the zeros of $\zeta(s)$ with positive imaginary part are considered, say, those on a given vertical half-line $\operatorname{Re} s = D$, $\operatorname{Im} s > 0$. A similar convention is assumed, most often implicitly, in the statement of all the conjectures discussed in the present section. Indeed, the more general zeta functions considered below also have complex conjugate pairs of zeros because they are defined by Dirichlet series with real coefficients.

(ii) We have formulated Conjecture 10.1 in such a way that it is independent of the truth of the Riemann Hypothesis. This will enable us, in particular, to extend this conjecture to more general zeta functions, without undue restrictions from the perspective of our present theory. On the other hand, in the literature on this subject, the Riemann Hypothesis is often assumed. That is, the critical zeros of $\zeta(s)$ are assumed to be of the form $1/2 \pm i\gamma_n$, $n = 1, 2, \ldots$, with γ_n real and positive. Conjecture 10.1 then asserts that the sequence $\{\gamma_n\}_{n=1}^{\infty}$ is linearly independent over the rationals, or, equivalently, over the ring of rational integers.

(iii) In the statement of Conjecture 10.1, one could even replace 'rational independence' by 'algebraic independence'. We prefer not to do so here, however, because it is less clear how to interpret the resulting statement in the framework of fractal strings. We also point out that Conjecture 10.1 could be recast in the language of fractal strings as the question of invertibility of the spectral operator when restricted to a certain class of fractal strings (much as was done in Corollary 9.4 in a related context).

(iv) Conjecture 10.1 implies, in particular, that $\zeta(s)$ does not have an infinite vertical sequence of critical zeros in arithmetic progression, as was first shown by Putnam in [Pu1–2] and was reproved by a different method in Theorem 9.1 above. It also implies that (nontrivial) finite vertical arithmetic progressions of zeros of $\zeta(s)$ do not exist. We note that even though it is much less general than Conjecture 10.1, the latter statement would provide a significant step toward that conjecture. It is not too difficult, at least in principle, to adapt our methods of proof, presented in Sections 8.2.2 and 9.1, to try to establish it. However, significant additional technical difficulties still need to be overcome in order to be able to analyze the nature of the oscillations in the frequency spectrum of the resulting fractal strings. (But see [Wa] for a partial result.)

We have used above the important example of the Riemann zeta function as a motivation for stating and exploring further conjectures regarding a broader class of Dirichlet series. The reader will recall that, in Chapter 9, we were able to show that many zeta functions do not have an infinite vertical arithmetic progression of critical zeros by considering generalized Cantor sprays instead of Cantor strings. However, both the zeros and the poles of the zeta function of a curve over a finite field are periodically distributed along vertical lines. (See Section 9.4 above.) Hence, it seems that some restrictions on the multiplicity or the number of the poles or zeros are necessary in order to avoid obvious counterexamples, such as Examples 10.6 and 10.7 below. We therefore make the following assumptions. (In the following, all the poles and zeros are counted according to their multiplicity.)

Let $\zeta_B(s)$ be a Dirichlet series with positive coefficients satisfying hypothesis (P), formulated on page 184. At this point, we invite the reader to review the statement of this hypothesis, which involves the existence of a window W for $\zeta_B(s)$. Note that according to hypothesis (P), $\zeta_B(s)$ is required neither to satisfy a functional equation nor to have an Euler product.

Moreover, we recall that we use throughout this section the convention described in comment (i) above. Thus, according to this convention, only the zeros of $\zeta_B(s)$ with positive imaginary part should be taken into account in the statement of Conjectures 10.3–10.5 and 10.8 below.

We first show by means of an example that we need to assume positivity and discreteness of the measure.

Example 10.2. Let $\zeta_0, \zeta_1, \ldots, \zeta_n$ be $n+1$ Dirichlet series. For example, ζ_0 is the Riemann zeta function, and $\zeta_1, \ldots, \zeta_n$ are L-series associated with n independent nontrivial characters (see Appendix A, Section A.2). Then, for any choice of coefficients $c_1, \ldots, c_n$, the function

$$\zeta_0(s) + c_1\zeta_1(s) + \cdots + c_n\zeta_n(s)$$

has a simple pole at $s = 1$. By suitably choosing the coefficients, one can arrange for this function to have a sequence of zeros in an arithmetic progression of length n. However, according to the following conjectures, the resulting Dirichlet series will not have real positive coefficients.

If one chooses $\zeta_0, \zeta_1, \ldots, \zeta_n$ to be completed zeta functions (see Section A.3 of Appendix A), associated with real-valued characters, we can choose the coefficients $c_1, \ldots, c_n$ to be real and small, so that the resulting Dirichlet integral is associated with a positive measure and has a sequence of zeros in an arithmetic progression of length n. However, in that case, it is no longer a Dirichlet series, associated with a discrete measure.

In analogy with the Riemann zeta function, we formulate the following conjectures:

Conjecture 10.3. *If ζ_B has at most one pole in a window, then it does not have a vertical arithmetic progression of zeros of length two in this window.*

More generally, if ζ_B has one pole, then it does not have two zeros with positive imaginary part on one vertical line in this window, the imaginary parts of which are rationally dependent.

Conjecture 10.4. *If ζ_B has n poles in a window, then it can have a vertical arithmetic progression of zeros of length at most n in this window.*

More generally, if ζ_B has n poles in a window, then it can have at most n zeros with positive imaginary part on one vertical line in this window, the imaginary parts of which are rationally dependent.

The next conjecture is the exact counterpart in this context of Conjecture 10.1 concerning the Riemann zeta function.

Conjecture 10.5 (General Irrationality Conjecture). *If ζ_B has one pole in a window, then the imaginary parts of the zeros of ζ_B in the upper halfplane on a given vertical line in this window are rationally independent. In particular, the zeros of ζ_B are all simple in this window.*

Note that since $\zeta(s)$ is a particular instance of the zeta functions that we consider, this conjecture implies Conjecture 10.1.

Before stating our last conjecture, we provide two examples showing that there are some further restrictions on what may be expected to hold.

Example 10.6. The function

$$\prod_{k=1}^{n} \zeta\left(\frac{1}{2} + \frac{s}{k}\right)$$

shows that a function with n poles can have n zeros in arithmetic progression. Indeed, let $1/2 + i\gamma$ be a critical zero of $\zeta(s)$, with $\gamma > 0$. Then the values $s = ki\gamma$, for $k = 1, \ldots, n$, are zeros of the above function in the upper half-plane.

Example 10.7. Let $\zeta_K(s)$ be the zeta function of an algebraic number field K. As is well known, this function has zeros with real part $1/2$. Then the function

$$\zeta\left(\frac{1}{2} + s\right) + c_1 \zeta_K\left(\frac{1}{2} + c_2 s\right),$$

with a suitable choice of the real constants c_1 and $c_2 > 0$, shows that a Dirichlet series with two poles can have a double zero.

Conjecture 10.8. *If ζ_B has at most n poles in a window, then its zeros have multiplicity at most n in this window.*

10.2 A New Definition of Fractality

Our work shows that the complex dimensions of a fractal string contain important geometric information; see especially Section 6.1.1 and Theorems 2.13, 6.1 and 6.12, along with Chapter 5 and Section 6.3. Therefore,

> We propose to define 'fractality' as *the presence of at least one nonreal complex dimension with positive real part.*

Note that nonreal complex dimensions come in conjugate pairs ω, $\overline{\omega}$, since $\zeta_{\mathcal{L}}(s)$ is real-valued for $s \in \mathbb{R}$. Hence a fractal set, in this definition, has at least two nonreal complex dimensions.

For example, given $a > 0$, the a-string—viewed as the complement in $(0, 1)$ of the sequence j^{-a} ($j = 1, 2, 3, \ldots$)—is not 'fractal', since $\zeta_{\mathcal{L}}$ has a meromorphic continuation to $\operatorname{Re} s \geq 0$, with only one pole at $D = 1/(1+a)$. (See Theorem 5.20 in Section 5.5.1 for a complete analysis of the complex dimensions of this string and compare [Lap1, Example 5.1 and Appendix C].) By definition, the same statement applies to the boundary of $\mathcal{L}$, which is equal to the compact set $\{1, 2^{-a}, 3^{-a}, \ldots, 0\}$. This is reassuring because, as is mentioned in [Fa2, p. 45],

> *No one would regard this set, with all but one of its points iso-*
> *lated, as a fractal, yet it has fractional box dimension.*[1]

Moreover, by Corollary 2.16, a self-similar string (and thus a self-similar set in $\mathbb{R}$ other than a single interval) is always fractal, since it has infinitely many complex dimensions with positive real part.

For a higher-dimensional fractal set, it becomes a difficult problem to define its complex dimensions. We plan to address the question of defining the complex dimensions of higher-dimensional fractal sets in a subsequent work, but for the moment, we will regard the exponents of ε in the asymptotic expansion of the volume of the (inner) tubular neighborhoods as the complex codimensions.

Consider, for example, the Devil's staircase (i.e., the graph of the Cantor function), which is a one-dimensional set. (See Figure 10.1 or [Man1, Plate 83, p. 83], along with Remark 10.9 below.) It is not fractal in the traditional sense, because its topological and Hausdorff dimensions coincide. (See Section 10.2.1 below.) However, it is at least as irregular as the Cantor set. We quote from Mandelbrot's book [Man1, p. 82]:

> *One would love to call the present curve a fractal, but to achieve*
> *this goal we would have to define fractals less stringently, on the*
> *basis of notions other than D alone.*[2]

The volume $V(\varepsilon)$ of the inner tubular neighborhoods of the Devil's staircase (Figure 10.2) is approximated by

$$2\varepsilon - \varepsilon^{2-\Theta}\left(1 - \frac{\pi}{4}\right)\left(\left(\frac{1}{2}\right)^{\{x\}} + \left(\frac{3}{2}\right)^{\{x\}}\right), \tag{10.1}$$

where $x = \log_3 \varepsilon^{-1}$. By formula (1.13) on page 12, we obtain the expression

$$2\varepsilon^{2-1} + \frac{\log 3}{2}\left(1 - \frac{\pi}{4}\right)\sum_{n=-\infty}^{\infty}\frac{\varepsilon^{2-\Theta-in\mathbf{p}}}{(\log 2 + 2\pi in)(-\log\frac{3}{2} + 2\pi in)} \tag{10.2}$$

for this approximate volume, where $\Theta = \log_3 2$ and $\mathbf{p} = 2\pi/\log 3$.[3] Thus the complex dimensions of the Devil's staircase are 1 and $\Theta + in\mathbf{p}$, $n \in \mathbb{Z}$, and they are all simple. (See Figure 10.3.)[4] Hence the Devil's staircase is fractal in our new sense.

---

[1] That is, noninteger Minkowski dimension.

[2] In the notation of Mandelbrot, D denotes the Hausdorff dimension.

[3] *Caution:* Even though the complex dimensions of the Devil's staircase are precisely determined by formula (10.2), the residues corresponding to the poles other than 1 in (10.2) are not necessarily accurate.

[4] For the nonexpert, it may be worth pointing out that the Devil's staircase is self-affine but not self-similar. Thus our result does not contradict Conjecture 10.13 stated for the complex dimensions of self-similar fractals in Section 10.3 below.

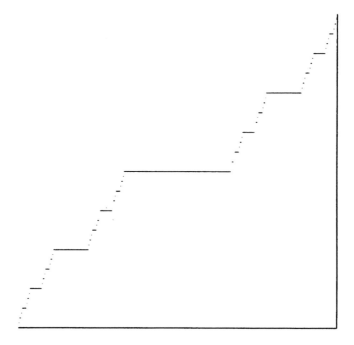

Figure 10.1: The Devil's staircase.

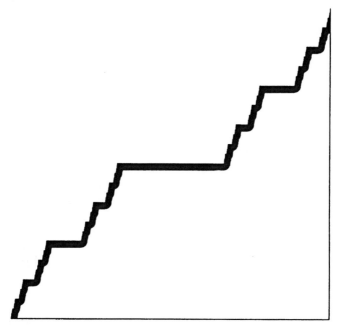

Figure 10.2: The inner ε-neighborhood of the Devil's staircase.

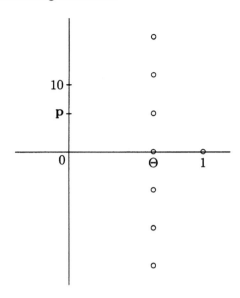

Figure 10.3: The complex dimensions of the Devil's staircase. $\Theta = \log_3 2$ and $\mathbf{p} = 2\pi/\log 3$.

Remark 10.9. Recall that the Devil's staircase is defined as the graph of the Cantor function $\mathcal{C}$. The latter is a nondecreasing continuous function on $[0, 1]$ such that $\mathcal{C}(0) = 0$ and $\mathcal{C}(1) = 1$. Further, $\mathcal{C}$ is constant on each interval in the complement of the ternary Cantor set in $[0, 1]$. In fact, $\mathcal{C}$ is nothing but the primitive of the natural (Θ-dimensional) Hausdorff measure on the Cantor set. (See, for example, [Coh, p. 55 and pp. 22–24] or [ReSi1, pp. 20–23].)

Remark 10.10. Observe that $\Theta = \log_3 2$ is the dimension of the Cantor set, which is the set where $\mathcal{C}$ is increasing (by infinitesimal amounts), while 1 is the dimension of the complement of the Cantor set in $[0, 1]$, the open set which is composed of the open intervals on which $\mathcal{C}$ is constant. Further note that in formula (10.1) (or (10.2)), the leading term 2ε can be interpreted as follows: The Devil's staircase is rectifiable with finite length equal to 2. This well known fact (see, e.g., [Man1, p. 82]) can be understood intuitively by adding up the infinitesimal steps of the Devil's staircase. This amounts to projecting the Devil's staircase onto the horizontal and vertical axes in Figure 10.1 and adding up the resulting lengths.

As we see from the above discussion, our new definition of fractality enables us to resolve several paradoxes connected with the previous definitions of fractality, based on a single real number, either the Hausdorff or the Minkowski dimension.

It goes without saying that any definition of fractality should be taken with a grain of salt. Indeed, it is difficult to imagine that a single definition can capture the multitude of forms taken by this elusive notion. Without any doubt, the present definition will not escape this difficulty. However, it may help gain insight into certain aspects of this multifaceted notion, especially those associated with various forms of oscillatory behavior. We quote from the introduction of Falconer's book [Fa2, p. xx]:

> *My personal feeling is that the definition of a 'fractal' should be regarded in the same way as the biologist regards the definition of 'life'. There is no hard and fast definition, but just a list of properties characteristic of a living thing, such as the ability to reproduce or to move or to exist to some extent independently of the environment. Most living things have most of the characteristics on the list, though there are living objects that are exceptions to each of them.*

According to Conjecture 10.13 in Section 10.3 below, any (nontrivial) self-similar set in $\mathbb{R}^d$ is fractal in our new sense. This is in agreement with the other definitions of fractality discussed in the next subsection. We point out that when $d = 1$, Conjecture 10.13 has been proved in Theorem 2.13 above.

We note that more refined information about fractal geometries can be obtained by taking into account the structure of the complex dimensions (and the associated residues). In particular, we believe that the (conjectured) almost periodicity of the complex dimensions—along with the presence of geometric oscillations of order D or of order arbitrarily close to but less than D—is an important feature of self-similar geometries.[5] In view of Remark 10.15 below, an analogous statement should hold for approximately self-similar fractals as well.

10.2.1   Comparison with Other Definitions of Fractality

The classical definition of fractality, as stated by Mandelbrot in [Man1, p. 15], is the following:

> *A fractal is by definition a set for which the Hausdorff–Besicovitch dimension strictly exceeds the topological dimension.*

In other words, if H denotes the Hausdorff dimension of a set $F \subset \mathbb{R}^d$ and T denotes its topological dimension, then F is fractal, according to Mandelbrot, if and only if $H > T$.[6] Mandelbrot discusses in detail in [Man1, Chap-

---

[5]This almost periodicity was established in the case of self-similar strings in Chapter 2.

[6]Recall that T is a nonnegative integer and that we always have $H \geq T$; see, e.g., [HurWa] and [Rog].

ter 3] and elsewhere in his book the implications as well as the limitations of his definition. (Also see the introduction of [Man3].) In the case of the Devil's staircase F, one has $H(= D) = T = 1$ because F is rectifiable; see, for example, [Man1, p. 82] and [Fed2, Theorem 3.2.39, p. 275]. On the other hand, we have seen that F has complex dimensions with positive real part.

Other possible definitions of fractality have been introduced in the literature, involving various notions of (real) fractal dimensions, such as the packing dimension P [Tr2] (which is in some sense dual to the Hausdorff dimension) and the Minkowski(–Bouligand) (or box) dimension D. For the former notion, the example of the Devil's staircase still poses a problem since we also have $P = H = D = T = 1$ (because $H \leq P \leq D$ by Remark 1.3), while for the latter notion, the above example of the a-string[7] with boundary $F = \{1, 2^{-a}, 3^{-a}, \ldots, 0\}$ shows that one can have $D = (a+1)^{-1} > T = 0$ for a set F that one would not want to call fractal. On the other hand, we showed in Theorem 5.20 that F does not have any nonreal complex dimension. (Note that since F is countable, we have $H = P = 0$ in that case; see, for example, [Fa2], [Tr2] or [Lap1, Example 5.1].)[8]

10.2.2 Possible Connections with the Notion of Lacunarity

In his book [Man1, Chapter X], Mandelbrot suggests to complement the notion of (real) fractal dimension by that of 'lacunarity', which is aimed at better taking into account the texture of a fractal. (See [Man1, Section X.34] and, for a more quantitative approach, [Man2], along with the relevant references therein, including [BedFi].)

We quote from the introduction of [Man2, pp. 16–17]:

> *Fractal lacunarity is an aspect of "texture" that is dominated by the sizes of the largest open components of the complement, which are perceived as "holes" or "lacunas."* ...
>
> *Lacunarity is very small when a fractal is nearly translation invariant, being made of "diffuse" clumps separated by "very small" empty lacunas, and lacunarity is high when this set is made of "tight" clumps separated by "large" empty gaps or lacunas.*

Our present theory of complex dimensions may help shed some new light on this somewhat elusive notion of lacunarity, especially in the one-

---

[7]already known to Bouligand [Bou] in a different terminology.

[8]A dichotomy fractal vs. nonfractal—based on D rather than H—was used in [Lap1] (and in later papers, such as [Lap2–3, LapPo2, LapMa2]) for pragmatic reasons in the context of drums with fractal boundary; it was not intended to provide a definition of fractality.

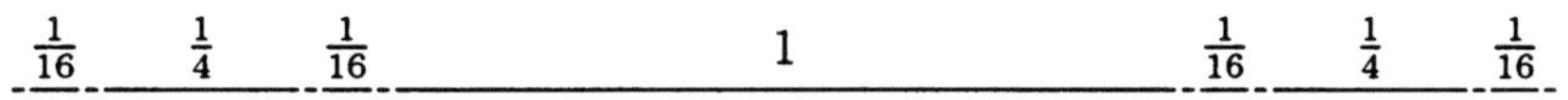

$$\frac{1}{16} \qquad \frac{1}{4} \qquad \frac{1}{16} \qquad\qquad 1 \qquad\qquad \frac{1}{16} \qquad \frac{1}{4} \qquad \frac{1}{16}$$

Figure 10.4: A lacunary Cantor string, with $a = 4$, $b = 2$ ($D = 1/2$, $\mathbf{p} = \pi/\log 2$ and $k = 1$).

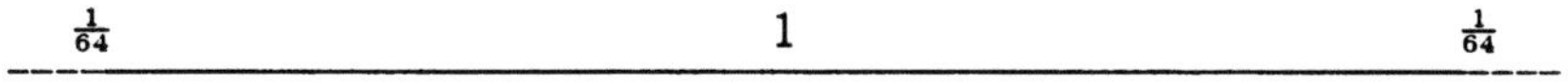

$$\frac{1}{64} \qquad\qquad\qquad 1 \qquad\qquad\qquad \frac{1}{64}$$

Figure 10.5: A less lacunary Cantor string, with $a = 64$, $b = 8$ ($D = 1/2$, $\mathbf{p} = \pi/(3\log 2)$ and $k = 3$).

dimensional situation. In particular, we hope to explain in more detail elsewhere how one of the main examples discussed in [Man1, Section X.34] and [Man2]—namely, in our notation, the sequence of Cantor sets (or 'Cantor dusts') $\eta_k = \sum_{j=0}^{\infty} 2^{jk}\delta_{\{4^{jk}\}}$ ($k \geq 1$)[9] represented in [Man2, Figure 1, pp. 18–19] and discussed in [Man2, Section 2.2, p. 21]—can be understood in terms of the complex dimensions of the associated Cantor strings. More specifically, for each fixed $k \geq 1$, the complex dimensions of the Cantor set η_k are $1/2 + in\pi/(k\log 2)$ ($n \in \mathbb{Z}$) and the corresponding residues are independent of n and equal to $1/(2k\log 2)$. (See Figure 8.1 in Chapter 8 above, with the choice of parameters $a = 4^k$ and $b = 2^k = \sqrt{a}$.) Therefore, even though these Cantor strings have the same real dimension $D = 1/2$, they have very different sets of complex dimensions (and residues), which accounts for their different lacunarities (in the language of [Man1–2]).

In particular, as $k \uparrow \infty$, the oscillatory period $\mathbf{p}_k = \pi/(k\log 2)$ decreases to 0, so the complex dimensions become denser and denser on the vertical line $\mathrm{Re}\, s = 1/2$, with smaller and smaller residues. Since the imaginary parts of the complex dimensions vanish in the limit, the nonreal complex dimensions eventually disappear, as might be expected intuitively for the homogeneous limiting set, which has zero lacunarity according to [Man2]; see Figure 10.5. Note that as $k \uparrow \infty$, the gaps (also called holes or lacunas in [Man1–2]) in the complement of the Cantor set defined by η_k become smaller and smaller, while the Cantor sets themselves appear to be more and more translation invariant or homogeneous and therefore "less and less fractal" or "lacunary." (See the discussion in Section 2.1, page 21, of [Man2].)

In contrast, as $k \downarrow 0$ (a limiting case that is not considered in [Man1–2] because it does not correspond to a geometric Cantor set but that is

---

[9]In ordinary language, in the j-th stage of its construction, the complement in $(0,1)$ of the Cantor set η_k consists of b^j open intervals of length a^{-j}, with $a = 4^k$ and $b = 2^k = \sqrt{a}$.

compatible with our notion of generalized Cantor strings, see Chapter 8),
$\mathbf{p}_k = \pi/(k\log 2)$ increases to infinity and so do the associated residues
$1/(2k\log 2)$. Therefore, the complex dimensions become sparser and sparser
on the vertical line $\operatorname{Re} s = 1/2$, but with larger and larger residues. This
is in agreement with the intuition that the (generalized) Cantor string η_k
is more and more lacunary (i.e., has larger and larger gaps or holes, see
Figure 10.4) as the real number k decreases to 0.

Remark 10.11. Recall that in Chapter 6, we have obtained precise tube
formulas for the volume of the ε-neighborhoods of a fractal string $\mathcal{L}$; these
formulas are expressed in terms of the complex codimensions of $\mathcal{L}$ and
of the associated residues (or principal parts) of $\zeta_{\mathcal{L}}$. This is of interest in
the present context because in [Man1] and especially in [Man2], lacunarity
has also been linked heuristically to the (possibly oscillatory) 'prefactor'
occurring in the definition of the upper or lower Minkowski content (prior
to taking the corresponding limit as $\varepsilon \to 0^+$); see formula (1.6) on page 9
above.

We leave it as an exercise for the interested reader to specialize to the
above example the tube formulas obtained for generalized Cantor strings in
the lattice case of Section 6.3.1, and to interpret the resulting expressions
as $k \uparrow \infty$ and as $k \downarrow 0$.

Remark 10.12. We note that the notion of fractal lacunarity may also
help connect aspects of our work with earlier physical work of which we
have recently become aware and that was aimed in part at understanding the relationship between lacunarity and turbulence, crack propagation
or fractal growth, among other physical applications. (See, for example,
[BadPo, BallBlu1–3, BessGM, FoTuVa, SmFoSp].)

10.3   Fractality and Self-Similarity

We refine our definition of fractality as follows. The roughness of a fractal
set is first of all characterized by its Minkowski dimension D. Then, either
it has nonreal complex dimensions with real part D, or it only has nonreal
complex dimensions with smaller real part. These two cases correspond to,
respectively, a fractal set that is not Minkowski measurable, and one that
is. (See Chapter 6, especially Sections 6.2 and 6.3.1.) In the first case, we
say that the set is maximally fractal, whereas in the second case, the set is
fractal only in its less-than-D-dimensional features.

From our analysis of self-similar strings in Chapter 2 (Theorem 2.13),
it follows that lattice strings are maximally fractal in this sense, whereas
nonlattice strings are not fractal in dimension D, but are fractal in infinitely
many dimensions less than and arbitrarily close to D.

Conjecturally, the dichotomy lattice vs. nonlattice, and the corresponding characterization of the nature of fractality, applies to a wide variety of situations, including drums with self-similar fractal boundary, drums with self-similar fractal membrane, random fractals as well as 'approximately self-similar sets', such as limit sets of Fuchsian and Kleinian groups or hyperbolic Julia sets [Su, BedKS, Lal2]. This conjecture was first formulated in [Lap3, Conjectures 2–6, pp. 159, 163, 169, 175, 190, 198], to which we refer for a more detailed description of these situations. One obtains, conjecturally, a corresponding lattice–nonlattice dichotomy in the description of the shape and the sound of a self-similar drum, in the nature of random walks and Brownian motion on a self-similar fractal, and in the nature of several function spaces associated with a self-similar fractal.

The following conjecture is very natural in our framework and significantly supplements the geometric aspects of [Lap3, Conjecture 3, p. 163] for self-similar drums. It is the higher-dimensional analogue of Theorem 2.13, which corresponds to self-similar sets in $\mathbb{R}$.

Conjecture 10.13. *The exact analogue of Theorem 2.13 holds for every (strictly) self-similar set*[10] *in $\mathbb{R}^d$ (and thus for any drum with self-similar fractal boundary), satisfying the open set condition.*[11]

In particular, the geometric zeta function has a meromorphic continuation to all of $\mathbb{C}$, with a single (simple) pole at $s = D$, where $D \in (d-1, d)$ is the Minkowski dimension of the set. All the other complex dimensions lie in the closed half-plane $\operatorname{Re} s \leq D$. Furthermore, the complex dimensions of a lattice self-similar set are periodically distributed on finitely many vertical lines, with period equal to the oscillatory period of the set, defined exactly as in Equation (2.27) and Definition 2.10. On the other hand, a nonlattice self-similar set does not have any nonreal complex dimension on the line $\operatorname{Re} s = D$, but it has infinitely many complex dimensions arbitrary close (from the left) to this line. Moreover, its complex dimensions exhibit an almost periodic structure.

For example, consider the snowflake drum, whose boundary, the snowflake curve, consists of three congruent von Koch curves fitted together (see Figure 10.6 and [Fa1, Figure 0.2, p. xv]). The volume $V(\varepsilon)$ of its inner tubular neighborhoods is approximately given by

$$\varepsilon^{2-D} \frac{\sqrt{3}}{4} 4^{-\{x\}} \left(\frac{3}{5} 9^{\{x\}} + 6 \cdot 3^{\{x\}} - 1 \right), \tag{10.3}$$

---

[10]As in [Lap3, Conjecture 3], here and in Section 10.4, we also allow for the set (viewed as the boundary of a self-similar drum) to be composed of a finite union of congruent copies of such a self-similar set, just as for the snowflake drum.

[11]See Remarks 2.2 and 2.17.

Figure 10.6: The snowflake drum.

where $x = \log_3 \varepsilon^{-1}$ and $D = \log_3 4$. By formula (1.13) on page 12, we obtain, with $\mathbf{p} = 2\pi / \log 3$,[12]

$$- \sum_{n=-\infty}^{\infty} \frac{3\sqrt{3} \log 3 \, (4\log 4 + \log 3)\varepsilon^{2-D-in\mathbf{p}}}{8(-\log \frac{9}{4} + 2\pi in)(\log \frac{4}{3} + 2\pi in)(\log 4 + 2\pi in)} \tag{10.4}$$

for the approximate volume of the inner ε-neighborhoods. Hence the complex dimensions of the snowflake curve are $\{D + in\mathbf{p}\}_{n\in\mathbb{Z}}$ and they are simple; see Figure 10.7. This is in agreement with Conjecture 10.13 because the snowflake curve is a lattice self-similar set with Minkowski dimension $D = \log_3 4$ and oscillatory period $\mathbf{p} = 2\pi / \log 3$ (since, in the notation of Theorem 2.13, we have $N = 4$ and $r_1 = r_2 = r_3 = r_4 = 1/3$).

---

[12]Even though it is not accurate for the residues, formula (10.4) is accurate for the complex dimensions.

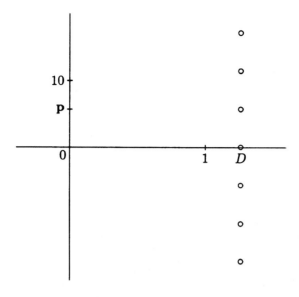

Figure 10.7: The complex dimensions of the snowflake drum. $D = \log_3 4$ and $\mathbf{p} = 2\pi / \log 3$.

Remark 10.14 (Tube formula for self-similar sets). We further conjecture that a tube formula holds for self-similar fractal sets in $\mathbb{R}^d$, with $d \geq 1$, naturally extending the tube formula for self-similar fractal strings in $\mathbb{R}$ obtained in Section 6.3.1, Theorems 6.20 and 6.21. (In particular, for self-similar fractal sprays—studied in Section 5.6.2 and in Section 5.6.1 for the example of the Sierpinski drum—this clearly follows from our explicit formulas exactly as in the case of self-similar strings.) Much as in the proof of Theorems 6.20 and 6.21, it would then follow that a nonlattice self-similar set in $\mathbb{R}^d$ is Minkowski measurable, whereas a lattice self-similar set, such as the Sierpinski gasket or the von Koch snowflake curve (viewed as self-similar fractal boundaries), is not. This statement is exactly the geometric content of [Lap3, Conjecture 3, p. 163] (and, in particular, of [Lap3, Conjecture 2, p. 159] for the special case of snowflake-type curves); see Remark 6.26 above. It is noteworthy that after the present work was completed, the statement concerning the Minkowski measurability of nonlattice self-similar sets was proved by Gatzouras [Gat] by means of the Renewal Theorem (much as was done in [Lap3, Section 4.4.1b] and in [Fa3] for the $d = 1$ case, and earlier in [Lal1] in a related situation, see Remark 6.27 above). Because of the use of the Renewal Theorem, however, the statement concerning the non-Minkowski measurability of lattice self-similar sets does not follow from the result of [Gat], although our methods from Section 6.3.1 would yield it directly.

Remark 10.15 (Approximately self-similar sets). In light of [Lap3, Conjecture 4, p. 175] (motivated in part by work of Lalley in [Lal1–3]), Conjecture 10.13 has a natural counterpart for approximately self-similar fractals (in the sense of [Lap3, §4.5] alluded to above). In that case, the dichotomy lattice vs. nonlattice must be defined by means of Lalley's nonlinear analogue of the Renewal Theorem [Lal2–3].

Remark 10.16. We intend to discuss in more detail the examples of the snowflake drum and of the Devil's staircase, along with several additional examples, in a future work [Lap-vF5].

Remark 10.17. Various extensions of the present theory of complex dimensions are considered in several works in preparation. In short, they deal with the following situations: (i) Certain (Cantor-type) self-similar fractals and the associated elliptic operators [ElLap]; (ii) Random (for example, statistically self-similar) fractals [HamLap]; (iii) Multifractals (non-homogeneous fractals which typically have a continuum of real fractal dimensions [Fa2, Chapter 17; Man3]); see [JafLap].

10.4   The Spectrum of a Fractal Drum

In this section, we consider a bounded open set in $\Omega \subset \mathbb{R}^d$, with a fractal boundary $\partial\Omega$, and equipped with a Laplace operator Δ. The counting function of the frequencies of Δ is denoted by $N_\nu(x)$. We refer to Appendix B for information and references about the case when Ω is a manifold with smooth boundary.

10.4.1   The Weyl–Berry Conjecture

Hermann Weyl's Conjecture (about manifolds with smooth boundary, see Section B.4.1 in Appendix B), has been extended by Michael V. Berry in [Berr1–2]. Berry's intriguing and stimulating conjecture was formulated in terms of the Hausdorff dimension H of $\partial\Omega$. In [BroCa], Brossard and Carmona have disproved this conjecture and suggested that the Hausdorff dimension H should be replaced by the Minkowski dimension D of $\partial\Omega$. Using probabilistic techniques, they have also obtained an analogue of estimate (B.2) (expressed in terms of D) for the spectral partition function $\theta_\nu(t)$ (rather than for $N_\nu(x)$) in the case of the Dirichlet Laplacian.

Later on, using analytical techniques (extending, in particular, those of [Weyl1–2], [CouHi] and [Met]), Lapidus [Lap1] has shown that the counterpart of estimate (B.2) holds for the spectral counting function $N_\nu(x)$ itself. More precisely, provided that $D \in (d-1, d)$ and $\mathcal{M}^*(D; \partial\Omega) < \infty$, the following Weyl asymptotic formula with sharp error term holds pointwise:[13]

$$N_\nu(x) = c_d \operatorname{vol}(M)x^d + O(x^D),\tag{10.5}$$

as $x \to \infty$.[14,15] (Here, the positive constant c_d depends only on d and is given just after formula (B.2) in Appendix B.) Analogous results were obtained in [Lap1] for the Neumann Laplacian (under a suitable assumption on $\partial\Omega$) and more generally, for positive elliptic differential operators (with Dirichlet, Neumann, or mixed boundary conditions); see [Lap1, Theorem 2.1 and Corollaries 2.1–2.2, pp. 479–480]. We mention that the snowflake drum is a natural example of a (self-similar) fractal drum for which estimate (10.5) holds pointwise, either for Dirichlet or Neumann boundary conditions; see [Lap1, Example 5.4, pp. 518–519] along with Section 10.4.2 below.

Afterward, as was briefly discussed in particular in Section 7.1 above, the one-dimensional case (i.e., $d = 1$)—corresponding to arbitrary fractal strings rather than to fractal drums—was studied in great detail in [LapPo1–2] and [LapMa1–2]. Also see the recent work [HeLap1–2] which extends most of the results in [Lap1, LapPo1–2, LapMa1–2] and [Ca1] by using a notion of generalized Minkowski content which is defined by means of some suitable gauge functions other than the traditional power functions in measuring the irregularities of the fractal boundary. Also note that, even when $d = 1$, the work in [HeLap1–2] does not lie within our present framework of complex dimensions because it allows, for instance, for logarithmic singularities in the corresponding zeta functions.

We note that the a-string—introduced in [Lap1] and studied in Section 5.5.1 above (see also [LapPo1–2])—provides a simple example showing that the Hausdorff dimension (or its companion, the packing dimension) cannot be used in this context to measure the roughness of the boundary $\partial\Omega$ of a fractal drum; see [Lap1, Examples 5.1 and 5.1′, pp. 512–515], along with Remark 1.3 in Chapter 1.

Various extensions of the above results and partial results toward a suitable substitute for the Weyl–Berry Conjecture for drums with fractal boundary have now been obtained in a number of papers, including [Lap2–3, LapPo1–3, LapMa1–2, FlVa, Ger, GerSc1–2, Ca1–2, vB, vB-Le, HuaSl, Fl-

---

[13]Recall from Remark 1.3 in Chapter 1 that we always have $d - 1 \le H \le D \le d$; see [Lap1, Corollary 3.2, p. 486]. Further, $\mathcal{M}^*(D; \partial\Omega)$ denotes the upper Minkowski content of $\partial\Omega$, defined as in Remark 1.1.

[14]If $D = d - 1$ (the nonfractal case in the terminology of [Lap1], see also Figure 7.1 on page 165), the result of [Lap1] yields instead the error term $O(x^D \log x)$ obtained previously by G. Métivier in [Met] in that situation and, in the special case of piecewise smooth boundaries, in [CouHi].

[15]We note that the analogue of Weyl's formula (without error term) for nonsmooth domains was first obtained in [BiSo].

LeVa, LeVa, MolVa, vB-Gi, HeLap1–2]. The interested reader will find in those papers, in particular, several examples of monotonic or oscillatory behavior in the asymptotics of $N_\nu(x)$ or of $\theta_\nu(t)$. The theory presented in this book—once suitably extended to higher dimensions (see, in particular, Sections 10.2, 10.3, 10.4.2 and 10.4.3)—should help shed new light on these examples. We note that many of these examples can be viewed as fractal sprays (as in [LapPo3]) and hence already lie within the scope of the present theory; see, for example, Section 5.6 above.

10.4.2   The Spectrum of a Self-Similar Drum

It is conjectured in [Lap3] that the (frequency) spectrum of a drum with self-similar fractal boundary does not have oscillations of order D in the nonlattice case, whereas it has (multiplicatively) periodic oscillations of order D in the lattice case. Here, as before, D denotes the Minkowski dimension of the boundary of the drum. More specifically, as $x \to \infty$, it is conjectured in [Lap3] that in the nonlattice case (respectively, lattice case), the spectral counting function $N_\nu(x)$ has a monotonic (respectively, oscillatory) asymptotic second term of the form a constant times x^D (respectively, $g(\log x)x^D$, for a nonconstant periodic function g of period $2\pi/\mathbf{p}$, the additive generator of the boundary). By a standard Abelian argument ([Si, Theorem 10.2, p. 107] or [Lap1, Appendix A, pp. 521–522]), the same statement would also hold for the spectral partition function $\theta_\nu(t)$ as $t \to 0^+$, with x replaced by t^{-1}. (For strictly self-similar drums, see [Lap3, Conjecture 3, p. 163] along with the important special case of snowflake-type drums discussed in [Lap3, Conjecture 2, p. 159], including the usual snowflake drum. Also see [Lap3, Conjecture 4, p. 175] and Remark 10.15, for the broader class of approximately self-similar drums, in the sense of Section 10.3 above.)

In the special case of self-similar sprays (as defined in Section 5.6.2 and corresponding to the Dirichlet or Neumann Laplacian on a disconnected open set), the results of Section 5.6 above, Theorem 5.24 (for the Sierpinski drum) and especially Theorem 5.27, agree exactly with this conjecture. (See also the earlier references given in Section 5.6, including [FlVa, Ger, GerScl–2, Lap2–3, LeVa, vB-Le].)

Moreover, in the case of the Dirichlet Laplacian on a connected domain, the interesting example of the snowflake drum (see Figure 10.6), a lattice self-similar drum, has been studied in [FlLeVa] from the point of view of the spectral partition function $\theta_\nu(t)$ (or of the closely related notion of heat content). See also the recent extension in [vB-Gi] to a one-parameter family of snowflake-type drums (much as in [Lap3, Conjecture 2, p. 159]). Again, these results agree with [Lap3, Conjectures 2 and 3]. However, we point out two open problems in this situation:

(i) As far as we know, in the lattice case, the counterpart for $\theta_\nu(t)$ of the periodic function g has not been proved to be nonconstant, even for the example of the snowflake drum. We note that for the latter example, the counterpart of g was verified numerically to be nontrivial in [FlLeVa].

(ii) To our knowledge, no such (pointwise) results have been obtained for the spectral counting function $N_\nu(x)$ rather than for $\theta_\nu(t)$. We note that technically, the conjectured pointwise result for $N_\nu(x)$ is significantly harder to establish than for $\theta_\nu(t)$.

Remark 10.18. We refer the interested reader to [LapNeRnGri] (and [Gri-Lap]) for a computer graphics-aided study of the frequency spectrum and of the normal modes of vibration of the snowflake drum (coined 'snowflake harmonics'). Mathematically, the latter are the eigenfunctions of the Dirichlet Laplacian on the snowflake domain; that is, on the bounded domain having the snowflake curve for fractal boundary (see Figure 10.6). We also refer to the earlier paper [LapPan] for a mathematical study of the pointwise behavior of these eigenfunctions (and of their gradient) near the fractal boundary, for a class of simply connected domains including the snowflake domain. The work of [LapNeRnGri] and [LapPan] was motivated in part by intriguing physical work and experiments in [SapGoM] on the vibrations of fractal drums, and was aimed in the long-term at understanding how (and why) fractal shapes—such as coastlines, blood vessels and trees—arise in nature. We note that the numerical data in [LapNeRnGri] (along with those in [SapGoM], [FlLeVa] and [GriLap]) should be useful to investigate the open problems and conjectures stated in the present subsection and in Section 10.4.3 below.

Given the theory developed in this book, in particular Theorem 2.13, it is natural to complement the spectral aspects of [Lap3, Conjecture 3, p. 163] by the following conjecture, which is a partial analogue of Conjecture 10.13 in Section 10.3 for the spectral (rather than for the geometric) complex dimensions of a self-similar drum.[16]

Conjecture 10.19. *Assume, for simplicity, that $d - 1 < D < d$, where D is the Minkowski dimension of the boundary of a (strictly) self-similar drum $\Omega \subset \mathbb{R}^d$ (satisfying the open set condition, see Remarks 2.2 and 2.17). Then D is the only real spectral complex dimension of the drum, other than d itself, and it is simple. (Also see the end of Remark 10.20 below.)*

Moreover, on the vertical line $\operatorname{Re} s = D$, a nonlattice drum does not have any nonreal spectral complex dimensions, whereas a lattice drum has an infinite sequence of spectral complex dimensions, contained in the arithmetic progression $\{D + i n\mathbf{p} : n \in \mathbb{Z}\}$, where $\mathbf{p}$ is the oscillatory period of the drum.

---

[16]By definition, the (visible) spectral dimensions of a fractal drum are the poles (within a given window W) of a meromorphic extension of $\zeta_\nu(s)$, the spectral zeta function of this drum. Henceforth, we omit the adjective 'visible' when referring to spectral complex dimensions and it is implicitly understood that W is a suitable neighborhood of $\{s \colon \operatorname{Re} s \geq D\}$.

In addition, a nonlattice drum has spectral complex dimensions arbitrarily close (from the left) to the line Re $s = D$, *whereas this is not the case for a lattice drum.*

Finally, the spectral complex dimensions of a nonlattice drum are almost periodically distributed (in the sense of Theorem 2.13); that is, they can be approximated by those of a suitable sequence of lattice drums.

We point out that it follows from Conjecture 10.19 and from our explicit formulas (namely, from Theorem 4.20) that the conclusion of [Lap3, Conjectures 2 and 3] holds pointwise for the spectral partition function $\theta_\nu(t)$ and distributionally for the eigenvalue counting function $N_\nu(x)$.

Remark 10.20. As was noted in [Lap2-3], it follows easily from Weyl's asymptotic formula with error term ([Lap1, Theorem 2.1], estimate (10.5) above) that for any fractal drum such that $\mathcal{M}^*(D; \partial\Omega) < \infty$ and $D < d$, the spectral zeta function $\zeta_\nu(s)$ has a meromorphic extension to the open half-plane Re $s > D$, with a single (simple) pole at $s = d$, either for Dirichlet or Neumann boundary conditions.[17] Therefore, all the spectral complex dimensions of the fractal drum other than d itself lie in the closed half-plane Re $s \le D$. This is the case, in particular, for the self-similar drums considered in Conjecture 10.19 (under the hypothesis of footnote 17).

Remark 10.21. In view of [Lap3, Conjecture 4, p. 175] and Remark 10.15 above, an entirely analogous conjecture can be made about the spectrum of approximately self-similar fractal drums. (For a discussion of the case of drums with self-similar fractal membrane, see [Lap6, Section 8] which would complete the rigorous results of [KiLap] regarding an analogue of Weyl's formula for Laplacians on self-similar fractals.)

Remark 10.22. A conjecture more precise than Conjecture 10.19 could be made for the spectrum of self-similar drums. It would be much closer to the statement of Conjecture 10.13 for the geometry of self-similar drums. At this point, however, we prefer to refrain from formulating it until a better understanding of a suitable counterpart of the spectral operator has been obtained in the context of self-similar fractal drums. (Recall that the spectral operator discussed in this book relates the geometric and the spectral complex dimensions of a fractal string or more generally, of a fractal spray; see Section 5.3.2 and Definition 5.22 in Section 5.6.) Further insight into this difficult problem may be gained by examining the question raised in the next subsection, regarding the relationship between the spectrum and the dynamics of a fractal drum.

---

[17]Here and in Conjecture 10.19 above, only for the Neumann Laplacian, we assume as in [Lap1] or [Lap3, Conjecture 3] that Ω is a quasidisc, which is the case, for instance, for the snowflake domain or its natural generalizations; see [Lap1, Section 4.2.B and Example 5.4] along with [LapPan].

10.4.3 Spectrum and Periodic Orbits

In [Gut2, Section 16.5 and Chapter 17], it is explained how the spectrum of the Laplacian on a Riemannian manifold is related to the periodic orbits of a particle moving in the manifold. By a trace formula, reminiscent of the Selberg Trace Formula, the spectral partition function—or rather, its quantum-mechanical analogue, a suitable distributional trace of the unitary group $\left\{e^{it\Delta}\right\}_{t\in\mathbb{R}}$, where Δ is the Laplacian on the manifold—is expressed as a sum over the periodic orbits. This relationship—which is rather surprising at first and is usually known in the physics literature as the Gutzwiller Trace Formula [Gut1-2]—follows from a heuristic application of the method of stationary phase to a Feynman path integral.[18]

A sample of physical and mathematical works related to the Gutzwiller Trace Formula includes [Gut1-2, BallanBlo, BalaVor, Berr3, Vor, BerrHow, BraBh] and [Col, Chaz, DuGu].

As an example, we consider fractal strings. In this case, all the quantities involved can be worked out explicitly. As in Gutzwiller [Gut2, Section 16.5, Formula (16.13)], we show how counting frequencies can be transformed into counting periodic orbits,[19] using the Poisson Summation Formula.

Given $t > 0$, let

$$z_{\nu,\mathcal{B}}(t) = \sum_{k\in\mathbb{Z}\setminus\{0\}} e^{-\pi k^2 t} = 2\sum_{k=1}^{\infty} e^{-\pi k^2 t}$$

be the spectral partition function for the (normalized) squared frequencies (i.e., eigenvalues) of the Bernoulli string. (See footnote 4 on page 229 of Appendix B.) Then the squared frequencies of the fractal string $\mathcal{L} = \{l_j\}_{j=1}^{\infty}$ are counted by[20]

$$z_{\nu,\mathcal{L}}(t) = 2\sum_{j=1}^{\infty}\sum_{k=1}^{\infty} e^{-\pi k^2 l_j^{-2} t}.$$

Since, by the Poisson Summation Formula (see [Ti, Section 2.3], [Pat, Theorem 2.2] or [Sch1, Eq. (VII.7.5)]), we have

$$z_{\nu,\mathcal{B}}(t) + 1 = \frac{1}{\sqrt{t}}\sum_{k\in\mathbb{Z}} e^{-\pi k^2 t^{-1}},$$

---

[18]We wish to thank Michael V. Berry for a conversation about this subject [Berr4].

[19]Note that here we view the Bernoulli string as the unit circle, instead of as the unit interval $(0,1)$. The periodic orbits of a particle are the orbits of this particle around the circle. Hence we have the basic periodic orbit corresponding to going around once, and repetitions thereof.

[20]Recall that the (normalized, nonzero) frequencies of the Laplacian on a circle of length l_j are $k \cdot l_j^{-1}$, with $k \in \mathbb{Z}\setminus\{0\}$. Here, we use a normalization such that the square of these frequencies is equal to $\pi k^2 l_j^{-2}$ (instead of $k^2 l_j^{-2}$), in order to obtain elegant formulas. Further, following the convention used, in particular, in Chapter 1 and in Appendix B, we exclude here the zero frequency.

we find that

$$
z_{\nu,\mathcal{L}}(t) = \sum_{j=1}^{\infty} \left(\frac{l_j}{\sqrt{t}} - 1 + \frac{l_j}{\sqrt{t}} \sum_{k \in \mathbb{Z};\, k \neq 0} e^{-\pi k^2 l_j^2 t^{-1}} \right)
$$

$$
= \frac{L}{\sqrt{t}} + \sum_{j=1}^{\infty} \left(-1 + \frac{l_j}{\sqrt{t}} \sum_{k \in \mathbb{Z};\, k \neq 0} e^{-\pi k^2 l_j^2 t^{-1}} \right), \tag{10.6}
$$

where $L = \sum_{j=1}^{\infty} l_j$ is the total length of the string. We recognize the analogue of the Weyl term, $L/\sqrt{t}$, and the double sum in (10.6) extends over the periodic orbits (of lengths $k \cdot l_j$, $k = 1, 2, \dots$) of a particle on the string. The convergence of this series is subtle, and can be checked by noting that

$$
\frac{l_j}{\sqrt{t}} \sum_{k \in \mathbb{Z};\, k \neq 0} e^{-\pi k^2 l_j^2 t^{-1}} = \int_{-\infty}^{\infty} e^{-\pi x^2}\, dx + O\left(l_j\right),
$$

as $j \to \infty$.

For the frequency counting function itself, we obtain a similar result. Note that the Fourier transform of the characteristic function of $(0, x)$ is

$$
\frac{1 - e^{-2\pi i x y}}{2\pi i y}.
$$

Thus we find, again by means of the Poisson Summation Formula,

$$
\sum_{1 \leq k \leq x} 1 = x - \frac{1}{2} + \sum_{k=1}^{\infty} \frac{\sin 2\pi x k}{\pi k}.
$$

Hence

$$
N_{\nu}(x) = \sum_{j=1}^{\infty} \sum_{1 \leq k \leq x l_j} 1 = Lx - \sum_{j=1}^{\infty} \left(\frac{1}{2} - \sum_{k=1}^{\infty} \frac{\sin 2\pi x l_j k}{\pi k} \right). \tag{10.7}
$$

We recover the Weyl term, Lx, and a sum over all the periodic orbits. That this series converges is again a subtle matter.

For higher-dimensional fractal sets, a similar method could possibly be applied, although this is still a remote prospect, as we briefly discuss below. Note that when the boundary of a bounded open set is fractal, as is the case for the snowflake drum, the determination of the periodic orbits becomes problematic, since a particle bounces off the boundary in unpredictable directions. The bouncing of a point-particle is not even well defined.

Remark 10.23 (Self-similar fractal drums and fractal billiards).   Motivated in part by results or comments from [Berr1–3, Lal2–3, PaPol; BedKS, esp. Sections 6 and 8] and the relevant references therein, including several of those mentioned at the beginning of this subsection, the first author has formulated in [Lap3, Conjecture 6, p. 198] a 'metaconjecture' underlying many of the conjectures from Part II of [Lap3] briefly discussed in Sections 10.3 and 10.4.2, both for (approximately) self-similar drums with fractal membrane and with fractal boundary (as in the present situation). Roughly speaking, it suggests that there should exist a suitable dynamical system associated with a given self-similar drum (viewed, say, as a billiard table) such that an analogue of the Selberg (or, more generally, the Gutzwiller) Trace Formula holds in this context. Accordingly, the lattice–nonlattice dichotomy in the spectrum of self-similar drums could be understood as follows: In the nonlattice case, there would exist an invariant ergodic measure with respect to which the periodic orbits of the fractal billiard are equidistributed, whereas in the lattice case, the periodic orbits would be concentrated and their oscillatory behavior (of order D) would be described by their distribution with respect to a suitable measure. Therefore, conjecturally, this would explain dynamically the absence or the presence of oscillations of order D in the spectrum of self-similar drums, in the nonlattice or lattice case, respectively.

As is clear from the discussion in this subsection, in the present case of drums with fractal boundary, we are still far from being able to formulate such a conjecture precisely, let alone to prove or disprove it, even for the snowflake drum. Nevertheless, this is a challenging problem that appears to be worth investigating further in the future.

10.5   The Complex Dimensions as Geometric Invariants

We close this chapter with a section of a very speculative nature, regarding a possible cohomological interpretation of the complex dimensions of a fractal. In Section 6.1.1, we pointed out an analogy between the explicit formula for the volume of the tubular neighborhoods of a fractal string and formula (6.13) for the volume of the tubular neighborhoods of a submanifold M of Euclidean space. Similarly, in Appendix B, Remark B.4 and Theorem B.5, it is explained that the principal parts of the spectral zeta function associated with the Laplacian on a Riemannian manifold M have a geometric interpretation and can be expressed in terms of invariants of the cotangent bundle on M; see the formulas (B.7) and (B.8).

Thus both the poles and the principal parts of the spectral zeta function of a manifold have a geometric interpretation in terms of the de Rham complex of the manifold. In part because of this analogy, we want to raise the question of whether a theory of 'complex cohomology' of fractal strings, and more generally, of fractal drums, could exist. This theory would be based entirely on the geometric and spectral zeta functions of the fractal.

We propose the following dictionary for translating homological properties in terms of zeta functions. This is motivated in part by analogy with the two situations mentioned above. It is also suggested by the situation of algebraic varieties over a finite field, along with the associated zeta functions (see Section 9.4) and étale cohomology theory (in particular, the Weil Conjectures [Wei3, 7], [ParSh1, Chapter 4, Sections 1.1–1.3]). This situation bears some similarity with Cantor strings and other lattice strings, especially in the case of curves over a finite field.

homological	property of the zeta function
Poincaré duality	functional equation
grading by integers $0, 1, \ldots, d$	lines of poles at $\operatorname{Re} s = 0, 1/2, \ldots, d/2$
eigenvalues of Frobenius in dimension j on a variety over the finite field $\mathbb{F}_q$	q^ω, where ω is a complex dimension with $\operatorname{Re} \omega = j/2$
eigenvalues of the infinitesimal generator of a dynamical system on a manifold	ω, the complex dimensions
Lefschetz fixed point formula	coefficients of the zeta function
functoriality	the zeta function divides another one

Sometimes a stratification into different dimensions will be possible, such as for manifolds, and also for lattice strings. But for nonlattice strings, a stratification will be impossible. It is therefore a problem to know how such a theory would be set up algebraically, since there will not be a chain complex equipped with the usual boundary maps. Instead, the cycles in such a theory could be represented by Dirichlet series with analytic continuation up to a certain line $\operatorname{Re} s = \Theta$, whereas the boundaries would be represented by Dirichlet series that converge at $s = \Theta$ (i.e., with abcissa of convergence less than Θ). As a consequence, a homology class would be represented by a Dirichlet series with analytic continuation up to $\operatorname{Re} s = \Theta$, modulo Dirichlet series that converge there. Thus, only the information contained in the poles with their principal parts remains. In particular, only the asymptotic behavior of the coefficients of the Dirichlet series will be important in representing a complex cohomology class.

Appendix A
Zeta Functions in Number Theory

In this appendix, we collect some basic facts about zeta functions in number theory to which our theory of explicit formulas can be applied. We refer to [Lan, Chapters VIII and XII–XV], [ParSh1, Chapter 4] and [ParSh2, Chapter 1, §6, Chapter 2, §1.13] for more complete information and proofs.

Let K be an algebraic number field of degree d over $\mathbb{Q}$, and let $\mathcal{O}$ be the ring of integers of K. The *norm* of an ideal $\mathfrak{a}$ of $\mathcal{O}$ is defined as the number of elements of the ring $\mathcal{O}/\mathfrak{a}$:

$$N\mathfrak{a} = \#\mathcal{O}/\mathfrak{a}. \tag{A.1}$$

The norm is multiplicative: $N(\mathfrak{a}\mathfrak{b}) = N\mathfrak{a}N\mathfrak{b}$. Furthermore, an ideal has a unique factorization into prime ideals,

$$\mathfrak{a} = \mathfrak{p}_1^{a_1} \dots \mathfrak{p}_k^{a_k}, \tag{A.2}$$

where $\mathfrak{p}_1, \dots \mathfrak{p}_k$ are prime ideals.

A.1 The Dedekind Zeta Function

We denote by r_1 the number of real embeddings $K \to \mathbb{R}$ and by r_2 the number of pairs of complex conjugate embeddings $K \to \mathbb{C}$. Thus $r_1 + 2r_2 = d$. Further, w stands for the number of roots of unity contained in K. Associated with K, we need the *discriminant* $\mathrm{disc}(K)$, the *class number* h, and the *regulator* $\mathfrak{R}$. We refer to [Lan, p. 64 and p. 109] for the definition of these terms.

The *Dedekind zeta function* of K is defined for $\mathrm{Re}\, s > 1$ by

$$\zeta_K(s) = \sum_{\mathfrak{a}} (N\mathfrak{a})^{-s} = \sum_{n=1}^{\infty} A_n n^{-s}. \tag{A.3}$$

Here, $\mathfrak{a}$ runs over the ideals of $\mathcal{O}$, and, in the second expression, A_n denotes the number of ideals of $\mathcal{O}$ of norm n. This function has a meromorphic continuation to the whole complex plane, with a unique (simple) pole at $s = 1$, with residue

$$\frac{2^{r_1} (2\pi)^{r_2} h \mathfrak{R}}{w \sqrt{\mathrm{disc}(K)}}. \tag{A.4}$$

(See [Lan, Theorem 5 and Corollary, p. 161].)

From unique factorization and the multiplicativity of the norm, we deduce the Euler product for $\zeta_K(s)$: For $\mathrm{Re}\, s > 1$,

$$\zeta_K(s) = \prod_{\mathfrak{p}} \frac{1}{1 - (N\mathfrak{p})^{-s}}, \tag{A.5}$$

where $\mathfrak{p}$ runs over the prime ideals of $\mathcal{O}$.

Example A.1 (The case $K = \mathbb{Q}$). Ideals of $\mathbb{Z}$ are generated by positive integers, and the norm of the ideal $n\mathbb{Z}$ is n, for $n \geq 1$. Hence,

$$\zeta_{\mathbb{Q}}(s) = \sum_{n=1}^{\infty} n^{-s} = \zeta(s), \tag{A.6}$$

so that the Dedekind zeta function of $\mathbb{Q}$ is the Riemann zeta function. The Euler product for $\zeta_{\mathbb{Q}}(s)$ is given by

$$\zeta_{\mathbb{Q}}(s) = \prod_{p} \frac{1}{1 - p^{-s}}, \tag{A.7}$$

where p runs over the rational prime numbers.

A.2 Characters and Hecke *L*-series

Let χ be an ideal-character of $\mathcal{O}$, belonging to the cycle $\mathfrak{c}$ (see [Lan, Chapter VIII, §3 and Chapter VI, §1] for a complete explanation of the terms). The *L-series* associated with χ is defined by

$$L_{\mathfrak{c}}(s, \chi) = \sum_{(\mathfrak{a}, \mathfrak{c})=1} \chi(\mathfrak{a}) (N\mathfrak{a})^{-s}. \tag{A.8}$$

By the multiplicativity of the norm, this function has an Euler product

$$L_{\mathfrak{c}}(s,\chi) = \prod_{\mathfrak{p} \nmid \mathfrak{c}} \frac{1}{1 - \chi(\mathfrak{p})\,(N\mathfrak{p})^{-s}}. \tag{A.9}$$

This zeta function can be completed with factors corresponding to the divisors of $\mathfrak{c}$, to obtain a function $L(s,\chi)$, called the *Hecke L-series* associated with χ. It is related to the Dedekind zeta function as follows: Let L be the class field associated with an ideal class group of $\mathcal{O}$, and let ζ_L be the Dedekind zeta function of L. Let χ run over the characters of the ideal class group. Then,

$$\zeta_L(s) = \prod_{\chi} L(s,\chi). \tag{A.10}$$

Example A.2 (The case $K = \mathbb{Q}$). A multiplicative function χ on the positive integers gives rise to a *Dirichlet L-series*

$$L(s,\chi) = \sum_{n=1}^{\infty} \chi(n)n^{-s}. \tag{A.11}$$

The value $\chi(n)$ only depends on the class of n modulo a certain positive integer c. The minimal such c is called the *conductor* of χ.

A.3 Completion of *L*-Series, Functional Equation

The most fundamental property of the Dedekind zeta function, and of the *L*-series, is that it can be completed to a function that is symmetric about $s = \frac{1}{2}$. Let $\Gamma(s)$ be the gamma function. Let $\zeta_{\mathbb{R}}(s) = \pi^{-s/2}\Gamma(s/2)$ and $\zeta_{\mathbb{C}}(s) = (2\pi)^{-s}\Gamma(s)$. Denote by $\mathrm{disc}(K)$ the discriminant of K. Then the function

$$\xi_K(s) = \mathrm{disc}(K)^{s/2}\big(\zeta_{\mathbb{R}}(s)\big)^{r_1}\big(\zeta_{\mathbb{C}}(s)\big)^{r_2}\zeta_K(s) \tag{A.12}$$

has a meromorphic continuation to the whole complex plane, with poles located only at $s = 1$ and $s = 0$, and it satisfies the functional equation

$$\xi_K(1-s) = \xi_K(s). \tag{A.13}$$

We deduce from this key fact that the function

$$\psi_K(\sigma) := \limsup_{t\to\infty} \frac{\log|\zeta_K(\sigma + it)|}{\log t}, \tag{A.14}$$

defined for $\sigma \in \mathbb{R}$, is given by the following simple formula for $\sigma \notin (0, 1)$:

$$\psi_K(\sigma) = \begin{cases} 0, & \text{for } \sigma \geq 1, \\ \dfrac{d}{2}(1 - 2\sigma), & \text{for } \sigma \leq 0. \end{cases}$$

It is known that ψ_K is convex on the real line. (The Lindelöf Hypothesis says that $\psi_K(1/2) = 0$.) We deduce the following property (see [Lan, Chapter XIII, §5]): For every real number σ and $\varepsilon > 0$, there exists a constant C, depending on σ and ε, such that for all real numbers t with $|t| > 1$,

$$|\zeta_K(\sigma + it)| \leq C|t|^{\psi_K(\sigma) + \varepsilon}. \tag{A.15}$$

Thus $\zeta_K(s)$ satisfies the hypotheses ($\mathbf{H_1}$) and ($\mathbf{H_2}$) of Chapter 4.

The formalism required to prove for general L-series the functional equation (A.13) and the estimate (A.15) about the growth along vertical lines was developed in Tate's thesis [Ta]. We refer to [Lan, Chapter XIV] for the corresponding results.

Remark A.3. Our theory also applies to the more general L-series associated with nonabelian representations of $\mathbb{Q}$, such as those considered in [RudSar]. (The abelian case corresponds to the Hecke L-series discussed above.) These L-series can also be completed at infinity, and they satisfy a functional equation, relating the zeta function associated with a given representation with the zeta function associated with the contragredient representation. Moreover, they have an Euler product representation, much like that of $L(s, \chi)$, except that the p-th Euler factor may be a polynomial in p^{-s} of degree larger than one.

We note that these zeta functions are called primitive L-series in [Rud-Sar]. According to the Langlands Conjectures, they are the building blocks of the most general L-series occurring in number theory. See, for example, [Gel, KatSar, RudSar].

A.4   Epstein Zeta Functions

A natural generalization of the Riemann zeta function is provided by the so-called Epstein zeta functions [Ep]. (See, for example, [Te, Section 1.4] for detailed information about these functions.) Let $q = q(x)$ be a positive definite quadratic form on $\mathbb{R}^d$, with $d \geq 1$. Then the associated *Epstein zeta function* is defined by[1]

$$\zeta_q(s) = \sum_{\mathbf{n} = (n_1, \ldots, n_d) \in \mathbb{Z}^d \setminus \{0\}} q(\mathbf{n})^{-s/2}. \tag{A.16}$$

---

[1]Our convention is the same as in [Lap2, §4], for instance, but is different from the traditional one used in [Te].

It can be shown that $\zeta_q(s)$ has a meromorphic continuation to all of $\mathbb{C}$, with a simple pole at $s = d$ having residue $\pi^{d/2}(\det q)^{-1/2}\Gamma(d/2)^{-1}$. Further, $\zeta_q(s)$ satisfies a functional equation analogous to that satisfied by $\zeta(s)$; namely,

$$\pi^{-s/2}\Gamma(s/2)\,\zeta_q(s) = (\det q)^{-1/2}\pi^{-(d-s)/2}\Gamma((d-s)/2)\,\zeta_{q^{-1}}(d-s),$$
$$\text{(A.17)}$$

where q^{-1} is the positive definite quadratic form associated with the inverse of the matrix of q. (See [Te, Theorem 1, p. 59].)

In the important special case when $q(x) = x_1^2 + \cdots + x_d^2$, the corresponding Epstein zeta function is denoted $\zeta_d(s)$ in Section 1.4, Equation (1.42), and can be viewed as a natural higher-dimensional analogue of the Riemann zeta function $\zeta(s)$. Indeed, we have $\zeta_1(s) = 2\zeta(s)$. (See, for example, [Lap2, §4] and the end of Section 1.4 above.) Since in this case, q is associated with the d-dimensional identity matrix, we have $q^{-1} = q$ and $\det q = 1$, so that (A.17) takes the form of a true functional equation relating $\zeta_d(s)$ and $\zeta_d(d-s)$. Moreover, $\zeta_d(s)$ has an Euler product like that of $\zeta(s)$ if and only if $d = 1, 2, 4$ or 8, corresponding to the real numbers, the complex numbers, the quaternions and the octaves, respectively.

Remark A.4. More generally, one can also consider the Epstein-like zeta functions considered in [Es1]. Such Dirichlet series are associated with suitable homogeneous polynomials of degree greater than or equal to one.

Remark A.5. For all the above examples of zeta functions in Sections A.1–A.4, hypotheses $(\mathbf{H_1})$ and $(\mathbf{H'_2})$ (Equations (4.16) and (4.18), page 80) are satisfied with a window W equal to all of $\mathbb{C}$. Moreover, for the example from Section A.5 below, one can take W to be a right half-plane of the form $\operatorname{Re} s \geq \sigma_0$, for a suitable $\sigma_0 > 0$.

A.5   Other Zeta Functions in Number Theory

The flexibility of our theory of explicit formulas with an error term allows us to apply it to other zeta functions that do not necessarily satisfy a functional equation.

As an example, we mention the zeta function

$$\mathcal{P}(s) = \sum_p (\log p)\, p^{-s},$$

which was studied by M. van Frankenhuysen in [vF2, §3.9] in connection with the ABC conjecture. To obtain information about this function, we consider the logarithmic derivative of the Euler product of the Riemann zeta function,

$$-\frac{\zeta'(s)}{\zeta(s)} = \sum_{p,m} (\log p)\, p^{-ms}.$$

This function has simple poles at $s = 1$ and at each zero of $\zeta(s)$. By Möbius inversion,

$$\begin{aligned}
\mathcal{P}(s) &= -\frac{\zeta'(s)}{\zeta(s)} + \frac{\zeta'(2s)}{\zeta(2s)} + \frac{\zeta'(3s)}{\zeta(3s)} + \frac{\zeta'(5s)}{\zeta(5s)} - \frac{\zeta'(6s)}{\zeta(6s)} + \cdots \\
&= \sum_{n=1}^{\infty} \mu(n)\left(-\frac{\zeta'(ns)}{\zeta(ns)}\right),
\end{aligned}$$

where $\mu(n)$ is the Möbius function, defined on the positive integers by

$$\mu(n) = \begin{cases} 0, & \text{if } n \text{ is not square-free,} \\ (-1)^k, & \text{if } n = p_1 \ldots p_k \text{ is square-free.} \end{cases}$$

It follows that the poles of $\mathcal{P}(s)$ accumulate on the line $\operatorname{Re} s = 0$. Hence this line is a natural boundary for the analytic continuation of $\mathcal{P}(s)$. But on a line $\operatorname{Re} s = \sigma > 0$ that does not meet any of the poles of $\mathcal{P}(s)$, this function is bounded by a constant times $\log|t|$, for $t = \operatorname{Im} s > 2$. Hence this function satisfies hypotheses ($\mathbf{H_1}$) and ($\mathbf{H_2}$) (Equations (4.16) and (4.17), page 80) with $W = \{s \colon \operatorname{Re} s \geq \sigma_0\}$, where $\sigma_0 > 0$ is suitably chosen. For example, if the Riemann Hypothesis is true, one can take any positive value for σ_0 other than $1/n$, $n = 1, 2, \ldots$.

Appendix B
Zeta Functions of Laplacians and Spectral Asymptotics

In this appendix, we provide a brief overview of some of the results from spectral geometry that are relevant to the study of the spectral zeta function associated with a Laplacian Δ on a smooth compact Riemannian manifold M. For the simplicity of exposition, we focus on the case when M is a closed manifold (i.e., is without boundary). However, as is briefly explained at the end of the appendix, all the results stated for closed manifolds are known to have a suitable counterpart for the case of a compact manifold with boundary. An important special case of the latter situation is that when M is a smooth bounded open set in Euclidean space $\mathbb{R}^d$ and $\Delta = -\sum_{k=1}^{d} \partial^2/\partial x_k^2$ is the associated Dirichlet or Neumann Laplacian.

By necessity of concision, our presentation is somewhat sketchy and imprecise. For a much more detailed treatment of these matters, we refer the interested reader to some of the articles or books cited below, including [Min1–2, MinPl, Kac, McKSin, Sel, BergGM, AtPSin, Gi, Höl1–3, Gru, AndLap1–2], along with the relevant references therein.

B.1 Weyl's Asymptotic Formula

Let M be a closed, d-dimensional, smooth, compact and connected Riemannian manifold. We assume throughout that the closed manifold M is equipped with a fixed Riemannian metric g. Let Δ be the (positive) Laplacian (or Laplace–Beltrami operator) on M associated with g.[1] It is well known that Δ has a discrete (frequency) spectrum, written in increasing order according to multiplicity:

$$0 < f_1 \leq f_2 \leq \cdots \leq f_j \leq \cdots,$$

where $f_j \to +\infty$ as $j \to \infty$. Here, by convention, the frequencies of Δ are defined as the square root of its eigenvalues λ_j.[2,3]

Next, let $N_\nu(x) = N_{\nu,M}(x)$ be the associated *spectral counting function* (or counting function of the frequencies):

$$N_\nu(x) = \# \{j \geq 1 : f_j \leq x\}, \quad \text{for } x > 0. \tag{B.1}$$

Then Weyl's classical asymptotic formula [Wey1–2] states that

$$N_\nu(x) = c_d \operatorname{vol}(M)x^d + o\left(x^d\right), \tag{B.2}$$

as $x \to \infty$, where $c_d = (2\pi)^{-d}\mathcal{B}_d$ and $\mathcal{B}_d$ denotes the volume of the unit ball in $\mathbb{R}^d$. Recall that $\mathcal{B}_d = \pi^{d/2}/\Gamma(d/2 + 1)$, where $\Gamma = \Gamma(s)$ is the usual gamma function. Further, $\operatorname{vol}(M)$ denotes the Riemannian volume of M.

The leading term in (B.2),

$$W(x) = W_M(x) = (2\pi)^{-d}\mathcal{B}_d \operatorname{vol}(M)x^d, \tag{B.3}$$

is often referred to as the 'Weyl term' in the literature.

Remark B.1. We note that Weyl's formula (for Euclidean domains, say) plays an important role in mathematical physics and can be given interesting physical interpretations; see, for example, [CouHi, Kac, ReSi3, Si], along with [BaltHi].

Remark B.2. Weyl's original result has been improved in various ways. One extension consists in giving a (sharp) remainder estimate for Weyl's asymptotic law, of the form

$$N_\nu(x) = c_d \operatorname{vol}(M)x^d + O\left(x^{d-1}\right), \tag{B.4}$$

as $x \to \infty$. This result is due to Hörmander [Höl] in the case of closed manifolds, and to Seeley [Se4–5] (for $d \leq 3$ or to Pham The Lai [Ph]

---

[1]Using Einstein's summation convention, Δ is given in local coordinates by

$$\Delta = -\frac{1}{\sqrt{\det g}} \frac{\partial}{\partial x_\alpha} g^{\alpha\beta} \sqrt{\det g} \frac{\partial}{\partial x_\beta},$$

where $g = (g_{\alpha\beta})_{\alpha,\beta=1}^d$ and $g^{-1} = (g^{\alpha\beta})_{\alpha,\beta=1}^d$.

[2]We have used a slightly different normalization in the rest of this book; see, for example, Section 1.3 and footnote 1 of the introduction.

[3]Throughout this discussion, we ignore the zero eigenvalue of the Neumann Laplacian. Alternatively, we can replace Δ by $\Delta + \alpha$, for some positive constant α.

(for $d \geq 4$) in the case of manifolds with boundary (for example, for the Dirichlet or Neumann Laplacian on a smooth bounded open set in $\mathbb{R}^d$). We refer the interested reader to [Hö2–3] for a detailed exposition of these results.

B.2   Heat Asymptotic Expansion

Now, let us denote by $z_\nu(t) = z_{\nu,M}(t)$ the trace of the heat semigroup $\{e^{-t\Delta} : t \geq 0\}$ generated by Δ.[4] Thus, $z_\nu(t)$ is given by

$$z_\nu(t) = \text{Trace}\left(e^{t\Delta}\right)$$
$$= \sum_{j=1}^{\infty} e^{-t\lambda_j}, \tag{B.5}$$

for every $t > 0$.

A well known Tauberian argument shows that Weyl's formula (B.2) is equivalent to the following asymptotic formula for $z_\nu(t)$ (see, for example, [Kac] or [Si]):

$$z_\nu(t) = e_d \, \text{vol}(M) t^{-d/2} + o\left(t^{-d/2}\right), \tag{B.6}$$

as $t \to 0^+$, where $e_d = \Gamma(d/2 + 1)c_d$. Using the above expression for c_d and $\mathcal{B}_d$, one finds $e_d = (4\pi)^{-d/2}$.

Remark B.3. The fact that (B.2) implies (B.6) is immediate and follows from a simple Abelian argument; see, e.g., [Si, Theorem 10.2, p. 107] or [Lap1, Appendix A, pp. 521–522]. However, the converse relies on Karamata's Tauberian Theorem [Si, Theorem 10.3, p. 108] (which is closely related to the Wiener–Ikehara Tauberian Theorem [Pos, Section 27, pp. 109–112; Sh, Theorem 14.1, p. 115]). In addition, even the existence of an error term in (B.6) does not imply the corresponding Weyl formula with error term in (B.2).[5]

More generally, a key result due in its original form to Minakshisundaram and Pleijel [MinPl] (building, in particular, on work of the first of these

---

[4]For convenience, we use here for $z_\nu(t)$ the standard convention encountered in the literature on spectral geometry; that is, we work with the trace of $e^{-t\Delta}$ rather than of $e^{-t\sqrt{\Delta}}$. The latter choice would correspond to the spectral partition function

$$\theta_\nu(t) = \text{Trace}\left(e^{-t\sqrt{\Delta}}\right) = \sum_{j=1}^{\infty} e^{-tf_j} = \int_0^{\infty} e^{-tx} dN_\nu(x),$$

as we defined it in Section 5.2.3 above; see, for example, Equation (5.25).

[5]We stress that in contrast to much of the rest of this book, all the asymptotic formulas in this appendix are interpreted pointwise.

authors [Min1–2] in a closely related context) states that $z_\nu(t)$ has the following asymptotic expansion (in the sense of Poincaré):[6]

$$z_\nu(t) \sim \sum_{k \geq 0} \alpha_k t^{-(d-k)/2}, \tag{B.7}$$

as $t \to 0^+$, where the coefficients $\alpha_k = \alpha_k(M)$ are integrals with respect to the Riemannian volume measure of M of suitable local geometric invariants of M. Namely, for each $k = 0, 1, 2, \ldots$,

$$\alpha_k(M) = \int_M \alpha_k(y, M) \, d\operatorname{vol}_M(y), \tag{B.8}$$

where the function $\alpha_k(\cdot, M)$ can be expressed as a (locally invariant) polynomial of (suitable contractions of) the Riemann curvature tensor of M and of its covariant derivatives. In that sense, it is a local invariant of M.

Moreover, in the present situation, one can show that

$$\alpha_k(M) = 0, \quad \text{if } k \text{ is odd.} \tag{B.9}$$

Remark B.4. For example, α_0 is equal to $e_d \operatorname{vol}(M)$, while α_2 is proportional to the integral over M of the scalar curvature of M.[7] In general, the explicit computation of the coefficients α_k is difficult but a large amount of information is now available, particularly in the present case of closed manifolds. We refer to Gilkey's book [Gi, Sections 1.7, 1.10, 4.8, and 4.9] for a detailed treatment of this matter. Further information about heat asymptotic expansions and related issues can be found, for instance, in the papers by McKean and Singer [McKSin] or Atiyah, Patodi and Singer [AtPSin], and in [AndLap1–2] or in the first unnumbered subsection of [JohLap, Section 20.2.B], along with the relevant references therein. Also see [BergGM] for many interesting examples of spectra of Laplacians on Riemannian manifolds.

---

[6]This asymptotic formula can be interpreted as follows: For each fixed integer $k_0 \geq 0$,

$$z_\nu(t) = \sum_{k \leq k_0} \alpha_k t^{-(d-k)/2} + O\left(t^{(k_0+1-d)/2}\right),$$

as $t \to 0^+$.

[7]By application of the Gauss–Bonnet Formula (as extended by S.-S. Chern [Chern1–2] to every dimension d), it follows from the latter statement that the Euler characteristic of M is 'audible' (i.e., can be recovered from the spectrum of M); see, e.g., [McKSin, pp. 44–45]. (Recall that the Euler characteristic of M vanishes when d is odd.)

B.3   The Spectral Zeta Function and Its Poles

Let us next introduce the *spectral zeta function* of the Laplacian on M (or simply, the zeta function of Δ) $\zeta_\nu(s) = \zeta_{\nu,M}(s)$:[8]

$$\zeta_\nu(s) = \text{Trace}\left(\Delta^{-s/2}\right)$$
$$= \sum_{j=1}^{\infty} f_j^{-s}. \tag{B.10}$$

Note that in view of (B.5), we have the following relation between $\zeta_\nu(s)$ and $z_\nu(t)$:

$$\zeta_\nu(2s) = \frac{1}{\Gamma(s)} \int_0^\infty z_\nu(t) t^{s-1} \, dt. \tag{B.11}$$

Hence, by Weyl's asymptotic formula (B.2) (or equivalently, by (B.6)), $\zeta_\nu(s)$ extends holomorphically to the open right half-plane $\text{Re}\,s > d$. Further, according to (B.7), $\zeta_\nu(s)$ has a simple pole at $s = d$. It follows that the abcissa of convergence of the Dirichlet series $\zeta_\nu(s) = \sum_{j=1}^{\infty} f_j^s$ is equal to d, the dimension of the manifold M.

More generally, the asymptotic expansion (B.7) combined with relation (B.11) above yields the following key theorem (see [MinPl] and, for instance, [Gi, Section 1.10, especially Lemma 1.10.1, p. 79]):

Theorem B.5. *The spectral zeta function $\zeta_\nu(s)$ of a closed Riemannian manifold M admits a meromorphic extension to the whole complex plane, with simple poles located at d and at a (subset of) the points $d - 2, d - 4$, $d - 6, \ldots$. Further, for $k = 0, 1, 2, \ldots$, the residue at $s = d - k$ is equal to $2\alpha_k(M)/\Gamma((d - k)/2)$, where $\alpha_k = \alpha_k(M)$ is the k-th coefficient in the 'heat asymptotic expansion' (B.7).*

More precisely, $\zeta_\nu(s)$ is holomorphic except for simple poles located at

$$\begin{cases} s = d - 2q, \ q = 0, 1, 2, \ldots, & \text{if } d \text{ is odd,} \\ s = d, d - 2, d - 4, \ldots, 4, 2, & \text{if } d \text{ is even.} \end{cases}$$

Remark B.6. From our present point of view, the first part of Theorem B.5 is the most important one. It implies that all the poles of the spectral zeta function of a smooth manifold are located on the real axis, in contrast to what happens for 'fractal manifolds', as is illustrated in the main body of this book.

---

[8]In the usual terminology, $\zeta_\nu(s)$ is the zeta function of $\sqrt{\Delta}$, because, according to our present conventions, the frequencies f_j of Δ are given by $f_j = \sqrt{\lambda_j}$, where the λ_j's are the eigenvalues of Δ, written in nondecreasing order. The reader should keep this in mind when comparing our formulas with those in [Gi], for example.

Remark B.7. Strictly speaking, although $s = d$ is always a (simple) pole of $\zeta_\nu(s)$, as was explained above, the other points mentioned in Theorem B.5 may not be poles of $\zeta_\nu(s)$, because the associated residue may happen to vanish. For example, if $M = \mathbb{T}^d = \mathbb{R}^d/\mathbb{Z}^d$ is the standard flat d-dimensional torus (i.e., the unit cube $[0,1]^d$ with its faces identified, as at the end of Section 1.4), then $\zeta_\nu(s) = \zeta_{\nu,M}(s)$ is the normalized Epstein zeta function associated with the standard quadratic form $q_d(x) = x_1^2 + \cdots + x_d^2$ for $x = (x_1, \ldots, x_d) \in \mathbb{R}^d$; namely,

$$\zeta_\nu(s) = \zeta_d(s) = \sum_{(n_1,\ldots,n_d)\in\mathbb{Z}^d\setminus\{0\}} \left(n_1^2 + \cdots + n_d^2\right)^{-s/2}$$

as in Equation (1.42) above.[9] Therefore, for any $d \geq 1$, $s = d$ is the only pole of $\zeta_\nu(s) = \zeta_d(s)$, and it is simple. (See Appendix A, Section A.4, or [Te, Section 1.4].) In order to reconcile this fact with the statement of Theorem B.5, it suffices to note that $M = \mathbb{T}^d$ has zero Euler characteristic and vanishing curvature. An entirely analogous comment can be made about a general Epstein zeta function $\zeta_q(s)$ considered in Section A.4 of Appendix A (or in [Te, Section 1.4]), which can be viewed as the spectral zeta function of the Laplacian on a flat torus $M = \mathbb{R}^d/\Lambda$, where Λ is a lattice of $\mathbb{R}^d$ with associated positive definite quadratic form $q = q(x)$. (See Equation (A.16).)

B.4 Extensions

Various extensions of the above results are known in spectral geometry. We mention only a few, which are most relevant to our situation or that may help clarify certain issues:

(i) Formulas (B.2), (B.7) and Theorem B.5 apply to the more general situation of a (positive) elliptic differential operator $\mathcal{P}$ (instead of the Laplacian Δ). If $\mathcal{P}$ is of order $m > 0$, say, then we define the j-th frequency of $\mathcal{P}$ by $f_j = \lambda_j^{1/m}$, where λ_j is the j-th eigenvalue of $\mathcal{P}$, written in nondecreasing order according to multiplicity. With this convention, the exponent of x in (B.2) remains equal to d, while the exponent of t^{-1} in (B.6) and (B.7) is now equal to d/m and $(d-k)/m$, respectively. Further, the poles of $\zeta_\nu(s)$ also remain the same as in Theorem B.5. On the other hand, in (B.2) and in (B.3), the constant $(2\pi)^{-d}\mathcal{B}_d$ will be replaced by $(2\pi)^{-d}$ times a volume in phase space (i.e., in the cotangent bundle of M) determined by the principal symbol of $\mathcal{P}$. Moreover, with the obvious change in notation, in the analogue of (B.7) and (B.8), the local invariants $\alpha_k(\cdot, \mathcal{P})$

---

[9]For convenience, we are using here the normalized eigenvalues of Δ on $\mathbb{T}^d$.

are now expressed as (locally invariant) polynomials of the total symbol of $\mathcal{P}$ and of its covariant derivatives.

(ii) Let us now assume that M is a (smooth, compact) manifold with boundary. For elliptic boundary value problems on M (and, in particular, for the prototypical cases of the Dirichlet and Neumann Laplacians on a smooth bounded open set of d-dimensional Euclidean space $\mathbb{R}^d$), the analogue of Weyl's asymptotic formula (B.2) and of the Minakshisundaram–Pleijel heat asymptotic expansion (B.7) still holds. It takes the same form as above, except that in the counterpart of (B.7), the coefficients α_k (or the corresponding local invariants) are more complicated to compute.[10] In addition, a suitable counterpart of Theorem B.5 also holds; see [MinPl] and [McKSin]. In particular, the poles of ζ_ν are all simple and located on the real axis. Perhaps the most complete treatment of these questions in the case of manifolds with boundary can be found in Grubb's book [Gru], which also deals with the more general case of elliptic pseudodifferential boundary value problems on M.[11] Besides the earlier papers [Min1–2, MinPl] (which study slightly different notions of spectral zeta functions of Laplacians, motivated by the work of Carleman [Car]), other useful references in this setting include the aforementioned paper by McKean and Singer [McKSin], along with the classical paper by Mark Kac [Kac] entitled *Can one hear the shape of a drum?*, which gives some related results on certain planar domains.

In a seminal paper, entitled *Complex powers of elliptic operators*, Seeley [Se1] has used modern analytical tools to study spectral zeta functions. In turn, Seeley's paper (along with its sequel for boundary value problems [Se2–3]) has stimulated a number of further developments related to the zeta functions of elliptic pseudodifferential operators. (See, for example, [Sh, Chapter II] and [Gru].)

B.4.1 Monotonic Second Term

Under the assumptions of Remark B.2 for a manifold with smooth boundary, it need not be the case that $N_\nu(x)$ admits (pointwise) an asymptotic second term as $x \to \infty$. (Contrast this statement with the fact that $z_\nu(t)$ has an asymptotic expansion of every order as $t \to 0^+$; see formula (B.7).) Knowing when $N_\nu(x)$ admits a monotonic asymptotic second term (i.e., of the form a nonzero constant times x^{d-1}) is the object of Hermann Weyl's Conjecture [Weyl1–2]. In a beautiful work, Ivrii [Ivr1–2] has partially solved

---

[10]In fact, to our knowledge, no explicit algorithm is known to calculate every α_k in this case, although a great deal of information is available.

[11]For a general pseudodifferential operator $\mathcal{P}$ on M, the heat asymptotic expansion may contain logarithmic terms, corresponding to the singularities of the symbol of $\mathcal{P}$. This is not the case, however, for an elliptic differential operator, and hence for a Laplacian on M. (See Corollary 4.27, page 388 and the comment on page 390 in [Gru].)

this conjecture. More specifically, for example for the Dirichlet or Neumann Laplacian, respectively, he shows that on a manifold M with boundary ∂M, we have (with the obvious notation for the volume of M and ∂M),

$$N_\nu(x) = c_d \operatorname{vol}_d(M) x^d \mp g_{d-1} \operatorname{vol}_{d-1}(\partial M) x^{d-1} + o\left(x^{d-1}\right), \qquad \text{(B.12)}$$

as $x \to \infty$, provided a suitable condition is satisfied.[12] (Here, the positive constant g_{d-1} is explicitly known in terms of $d-1$, the dimension of the smooth boundary ∂M.) Positive results toward Weyl's Conjecture were also obtained by Melrose [Mel1–2] for manifolds with concave boundary. We refer the interested reader to volumes III and IV of Hörmander's treatise [Hö3] as well as to Ivrii's recent book [Ivr3] for further information about this subject.

Finally, we note that situations where $N_\nu(x)$ has an oscillatory behavior (beyond the Weyl term) have been analyzed, in particular, by Duistermaat and Guillemin [DuGu] in function of the concentration of periodic geodesics (or, more generally, of bicharacteristics). Also see the beginning of Section 10.4.3 for a sample of related mathematical and physical works, including the papers by Colin de Verdière [Col] and Chazarain [Chaz].

---

[12]Roughly speaking, this condition says that the set of multiply reflected periodic geodesics of M forms a set of measure zero, with respect to Liouville measure in phase space (i.e., in the cotangent bundle of M). This condition (which is sufficient but not necessary) is known to be generic among smooth Euclidean domains, but is difficult to verify for any concrete example.

References

[Ahl]　L. V. Ahlfors, *Complex Analysis*, 3d. ed., McGraw-Hill, London, 1985.

[AndLap1]　S. I. Andersson and M. L. Lapidus (eds.), *Progress in Inverse Spectral Geometry*, Trends in Mathematics, vol. 1, Birkhäuser-Verlag, Basel and Boston, 1997.

[AndLap2]　S. I. Andersson and M. L. Lapidus, *Spectral Geometry: An introduction and background material for this volume*, in: [AndLap1, pp. 1–14].

[AtPSin]　M. Atiyah, V. K. Patodi and I. M. Singer, Spectral asymmetry and Riemannian geometry, I, *Math. Proc. Cambridge Philos. Soc.* **77** (1975), 43–69; II, *ibid.* **78** (1975), 405–432; III, *ibid.* **79** (1976), 71–99.

[BadPo]　R. Badii and A. Politi, Intrinsic oscillations in measuring the fractal dimension, *Phys. Lett.* A **104** (1984), 303–305.

[BalaVor]　N. L. Balazs and A. Voros, Chaos on the pseudosphere, *Phys. Rep.*, No. 3, **143** (1986), 109–240.

[BallBlu1]　R. C. Ball and R. Blumenfeld, Universal scaling of the stress field at the vicinity of a wedge crack in two dimensions and oscillatory self-similar corrections to scaling, *Phys. Rev. Lett.* **65** (1990), 1784–1787.

[BallBlu2] R. C. Ball and R. Blumenfeld, Sidebranch selection in fractal growth, *Europhys. Lett.* **16** (1991), 47–52.

[BallBlu3] R. C. Ball and R. Blumenfeld, Probe for morphology and hierarchical corrections in scale-invariant structures, *Phys. Rev. E* **47** (1993), 2298–3002.

[BallanBlo] R. Ballan and C. Bloch, Solution of the Schrödinger equation in terms of classical paths, *Ann. Physics* **85** (1974), 514–546.

[BaltHi] H. P. Baltes and E. R. Hilf, *Spectra of Finite Systems*, B. I.-Wissenschaftsverlag, Vienna, 1976.

[Ban] T. Banchoff, Critical points and curvature for embedded polyhedra, *J. Differential Geom.* **1** (1967), 245–256. II, in: *Differential Geometry* (Proc. Special Year, Maryland), Progress in Math., vol. 32, Birkhäuser, Boston, 1983, pp. 34–55.

[Bar] K. Barner, On A. Weil's explicit formula, *J. Reine Angew. Math.* **323** (1981), 139–152.

[Bea] A. F. Beardon, *Iteration of Rational Functions*, Springer-Verlag, Berlin, 1991.

[BedFi] T. Bedford and A. M. Fisher, Analogues of the Lebesgue density theorem for fractal sets of reals and integers, *Proc. London Math. Soc.* (3) **64** (1992), 95–124.

[BedKS] T. Bedford, M. Keane and C. Series (eds.), *Ergodic Theory, Symbolic Dynamics and Hyperbolic Spaces*, Oxford Univ. Press, Oxford, 1991.

[Bér] P. Bérard, Spectres et groupes cristallographiques I: Domaines euclidiens, *Invent. Math.* **58** (1980), 179–199.

[BergGM] M. Berger, P. Gauduchon and E. Mazet, *Le Spectre d'une Variété Riemannienne*, Lecture Notes in Math., vol. 194, Springer-Verlag, Berlin, 1971.

[BergGo] M. Berger and B. Gostiaux, *Differential Geometry: Manifolds, Curves and Surfaces*, English transl., Springer-Verlag, Berlin, 1988.

[Berr1] M. V. Berry, *Distribution of modes in fractal resonators*, in: Structural Stability in Physics (W. Güttinger and H. Eikemeier, eds.), Springer-Verlag, Berlin, 1979, pp. 51–53.

[Berr2] M. V. Berry, *Some geometric aspects of wave motion: Wavefront dislocations, diffraction catastrophes, diffractals*, in: Geometry of the Laplace Operator, Proc. Sympos. Pure Math., vol. 36, Amer. Math. Soc., Providence, R. I., 1980, pp. 13–38.

[Berr3] M. V. Berry, The Bakerian lecture, 1987: Quantum chaology, *Proc. Roy. Soc. London Ser. A* **413** (1987), 183–198.

[Berr4] M. V. Berry, *Private communication*, January 1999.

[BerrHow] M. V. Berry and C. J. Howls, High orders of the Weyl expansion for quantum billiards: resurgence of periodic orbits, and the Stokes phenomenon, *Proc. Roy. Soc. London Ser. A* **447** (1994), 527–555.

[BesTa] A. S. Besicovitch and S. J. Taylor, On the complementary intervals of a linear closed set of zero Lebesgue measure, *J. London Math. Soc.* **29** (1954), 449–459.

[BessGM] D. Bessis, J. S. Geronimo and P. Moussa, Mellin transforms associated with Julia sets and physical applications, *J. Statist. Phys.* **34** (1984), 75–110.

[BiSo] M. S. Birman and M. Z. Solomyak, Spectral asymptotics of non-smooth elliptic operators, I, *Trans. Moscow Math. Soc.* **27** (1972), 3–52; II, *ibid.* **28** (1973), 3–34.

[Bl] W. Blaschke, *Integralgeometrie*, Chelsea, New York, 1949.

[BogKe] E. B. Bogomolny and J. P. Keating, Random matrix theory and the Riemann zeros I: Three-and-four-point correlations, *Nonlinearity* **8** (1995), 1115–1131.

[Bohr] H. Bohr, *Almost Periodic Functions*, Chelsea, New York, 1951.

[Bom] E. Bombieri, *Counting points on curves over finite fields (d'après S. A. Stepanov)*, Séminaire Bourbaki, 25ème année 1972/73, no. 430, Lecture Notes in Math. Vol. 383, Springer-Verlag, New York, 1974, pp. 234–241.

[Bou] G. Bouligand, Ensembles impropres et nombre dimensionnel, *Bull. Sci. Math.* (2) **52** (1928), 320–344 and 361–376.

[Bow] R. Bowen, Symbolic dynamics for hyperbolic flows, *Amer. J. Math.* **95** (1973), 429–460.

[BraBh] M. Brack and R. K. Bhaduri, *Semiclassical Physics*, Frontiers in Physics, Addison-Wesley, Reading, 1997.

[Bré] H. Brézis, *Analyse Fonctionnelle: Théorie et Applications*, Masson, Paris, 1983.

[BroCa] J. Brossard and R. Carmona, Can one hear the dimension of a fractal?, *Commun. Math. Phys.* **104** (1986), 103–122.

[Ca1]   A. M. Caetano, Some domains where the eigenvalues of the Dirichlet Laplacian have non-power second term asymptotic estimates, *J. London Math. Soc.* (2) **43** (1991), 431–450.

[Ca2]   A. M. Caetano, On the search for the asymptotic behaviour of the eigenvalues of the Dirichlet Laplacian for bounded irregular domains, *Intern. J. Appl. Sci. Comput.* **2** (1995), 261–287.

[Car]   T. Carleman, Propriétés asymptotiques des fonctions fondamentales des membranes vibrantes, *Scand. Math. Congress* (1934), 34–44.

[Chaz]   J. Chazarain, Formule de Poisson pour les variétés riemanniennes, *Invent. Math.* **24** (1974), 65–82.

[CheeMüS1]   J. Cheeger, W. Müller and R. Schrader, On the curvature of piecewise flat manifolds, *Commun. Math. Phys.* **92** (1984), 405–454.

[CheeMüS2]   J. Cheeger, W. Müller and R. Schrader, Kinematic and tube formulas for piecewise linear spaces, *Indiana Univ. Math. J.* **35** (1986), 737–754.

[Chern1]   S.-S. Chern, A simple intrinsic proof of the Gauss–Bonnet formula for closed Riemannian manifolds, *Ann. of Math.* **45** (1944), 747–752.

[Chern2]   S.-S. Chern, On the curvature integrals in a Riemannian manifold, *Ann. of Math.* **46** (1945), 674–684.

[Chern3]   S.-S. Chern, On the kinematic formula in integral geometry, *J. of Math. and Mech.* **16** (1966), 101–118.

[Coh]   D. L. Cohn, *Measure Theory*, Birkhäuser, Boston, 1980.

[Col]   Y. Colin de Verdière, Spectre du laplacien et longueur des géodésiques périodiques, I et II, *Compositio Math.* **27** (1973), 83–106 and 159–184.

[CouHi]   R. Courant and D. Hilbert, *Methods of Mathematical Physics*, vol. I, English transl., Interscience, New York, 1953.

[Cr]   H. Cramér, Studien über die Nullstellen der Riemannschen Zetafunktion, *Math. Z.* **4** (1919), 104–130.

[Da]   H. Davenport, *Multiplicative Number Theory*, 2nd ed., Springer-Verlag, New York, 1980.

[Del]   J. Delsarte, Formules de Poisson avec reste, *J. Anal. Math.* **17** (1966), 419–431.

[dV1] C.-J. de la Vallée Poussin, Recherches analytiques sur la théorie des nombres; Première partie: La fonction $\zeta(s)$ de Riemann et les nombres premiers en général, *Ann. Soc. Sci. Bruxelles Sér. I* **20** (1896), 183–256.

[dV2] C.-J. de la Vallée Poussin, Sur la fonction $\zeta(s)$ de Riemann et le nombre des nombres premiers inférieurs à une limite donnée, *Mém. Couronnés et Autres Mém. Publ. Acad. Roy. Sci., des Lettres Beaux-Arts Belg.* **59** (1899–1900).

[Den1] C. Deninger, Local L-factors of motives and regularized determinants, *Invent. Math.* **107** (1992), 135–150.

[Den2] C. Deninger, Lefschetz trace formulas and explicit formulas in analytic number theory, *J. Reine Angew. Math.* **441** (1993), 1–15.

[Den3] C. Deninger, *Evidence for a cohomological approach to analytic number theory*, in: Proc. First European Congress of Mathematics (A. Joseph *et al.*, eds.), vol. I, Paris, July 1992, Birkhäuser-Verlag, Basel, 1994, pp. 491–510.

[DenSc] C. Deninger and M. Schröter, A distributional theoretic proof of Guinand's functional equation for Cramér's V-function and generalizations, *J. London Math. Soc.* **52** (1995), 48–60.

[DoFr] J. D. Dollard and C. N. Friedman, *Product Integration, with Application to Differential Equations*, Encyclopedia of Mathematics and Its Applications, vol. 10, Addison-Wesley, Reading, 1979.

[DuGu] J. J. Duistermaat and V. Guillemin, The spectrum of positive elliptic operators and periodic bicharacteristics, *Invent. Math.* **29** (1975), 39–79.

[EdEv] D. E. Edmunds and W. D. Evans, *Spectral Theory of Differential Operators*, Oxford Mathematical Monographs, Oxford University Press, Oxford, 1987.

[Edw] H. M. Edwards, *Riemann's Zeta Function*, Academic Press, New York, 1974.

[ElLap] G. Elek and M. L. Lapidus, Elliptic operators and their zeta functions on self-similar fractals, in preparation.

[Ep] P. Epstein, Zur Theorie allgemeiner Zetafunktionen, I, *Math. Ann.* **56** (1903), 614–644; II, *ibid.* **63** (1907), 205–216.

[Es1] D. Essouabri, Singularités des séries de Dirichlet associées à des polynômes de plusieurs variables et applications en théorie analytique des nombres, *Ann. Inst. Fourier (Grenoble)* **47** (1996), 429–484.

[Es2]   D. Essouabri, *Private communication*, June 1996.

[EvPeVo] C. J. G. Evertsz, H.-O. Peitgen and R. F. Voss (eds.), *Fractal Geometry and Analysis: The Mandelbrot Festschrift*, World Scientific, Singapore, 1996.

[Fa1]   K. J. Falconer, *The Geometry of Fractal Sets*, Cambridge Univ. Press, Cambridge, 1985.

[Fa2]   K. J. Falconer, *Fractal Geometry: Mathematical Foundations and Applications*, Wiley, Chichester, 1990.

[Fa3]   K. J. Falconer, On the Minkowski measurability of fractals, *Proc. Amer. Math. Soc.* **123** (1995), 1115–1124.

[Fed1]   H. Federer, Curvature measures, *Trans. Amer. Math. Soc.* **93** (1959), 418–491.

[Fed2]   H. Federer, *Geometric Measure Theory*, Springer-Verlag, New York, 1969.

[Fel]   W. Feller, *An Introduction to Probability Theory and its Applications*, vol. II, Wiley, New York, 1966.

[FlLeVa] J. Fleckinger, M. Levitin and D. Vassiliev, Heat equation on the triadic von Koch snowflake, *Proc. London Math. Soc.* (3) **71** (1995), 372–396.

[FlVa]   J. Fleckinger and D. Vassiliev, An example of a two-term asymptotics for the "counting function" of a fractal drum, *Trans. Amer. Math. Soc.* **337** (1993), 99–116.

[FoTuVa] J.-D. Fournier, G. Turchetti and S. Vaienti, Singularity spectrum of generalized energy integrals, *Phys. Lett. A* **140** (1989), 331–335.

[Fu1]   J. H. G. Fu, Tubular neighborhoods in Euclidean spaces, *Duke Math. J.* **52** (1985), 1025–1046.

[Fu2]   J. H. G. Fu, Curvature measures of subanalytic sets, *Amer. J. Math.* **116** (1994), 819–880.

[Gab]   O. Gabber, *Private communication*, June 1997.

[Gat]   D. Gatzouras, Lacunarity of self-similar sets and stochastically self-similar sets, 1999; to appear in *Trans. Amer. Math. Soc.*

[Gel]   S. Gelbart, An elementary introduction to the Langlands program, *Bull. Amer. Math. Soc.* (N. S.) **10** (1984), 177–219.

[Ger]   J. Gerling, *Untersuchungen zur Theorie von Weyl–Berry–Lapidus*, Graduate Thesis (Diplomarbeit), Dept. of Physics, Universität Osnabrück, Germany, May 1992.

[GerSc1] J. Gerling and H.-J. Schmidt, Self-similar drums and generalized Weierstrass functions, *Physica A* **191** (1992), 536–539.

[GerSc2] J. Gerling and H.-J. Schmidt, *Three-term asymptotic of the spectrum of self-similar fractal drums*, preprint, 1996.

[Gi]   P. B. Gilkey, *Invariance Theory, the Heat Equation, and the Atiyah–Singer Index Theorem*, 2nd ed., Publish or Perish, Wilmington, 1984. (New rev. and enl. ed. in "Studies in Advanced Mathematics", CRC Press, Boca Raton, 1995.)

[Gra]   A. Gray, *Tubes*, Addison-Wesley, Reading, 1990.

[GriLap] C. A. Griffith and M. L. Lapidus, *Computer graphics and the eigenfunctions for the Koch snowflake drum*, in: [AndLap1, pp. 95–109].

[Gru]   G. Grubb, *Functional Calculus of Pseudodifferential Boundary Problems*, 2nd ed. (of the 1986 ed.), Progress in Mathematics, vol. 65, Birkhäuser, Boston, 1996.

[Gui1] H. P. Guinand, A summation formula in the theory of prime numbers, *Proc. London Math. Soc.* (2) **50** (1948), 107–119.

[Gui2] H. P. Guinand, Fourier reciprocities and the Riemann zeta-function, *Proc. London Math. Soc.* (2) **51** (1950), 401–414.

[Gut1] M. C. Gutzwiller, Periodic orbits and classical quantization conditions, *J. Math. Phys.* **12** (1971), 343–358.

[Gut2] M. C. Gutzwiller, *Chaos in Classical and Quantum Mechanics*, Interdisciplinary Applied Mathematics, vol. 1, Springer-Verlag, Berlin, 1990.

[Had1] J. Hadamard, Étude sur les propriétés des fonctions entières et en particulier d'une fonction considérée par Riemann, *J. Math. Pures Appl.* (4) **9** (1893), 171–215. (Reprinted in [Had3, pp. 103–147].)

[Had2] J. Hadamard, Sur la distribution des zéros de la fonction $\zeta(s)$ et ses conséquences arithmétiques, *Bull. Soc. Math. France* **24** (1896), 199–220. (Reprinted in [Had3, pp. 189–210].)

[Had3] J. Hadamard, *Oeuvres de Jacques Hadamard*, Tome I, Editions du Centre National de la Recherche Scientifique, Paris, 1968.

[HamLap] B. M. Hambly and M. L. Lapidus, Complex dimensions of random fractals, in preparation.

[Haran] S. Haran, Riesz potentials and explicit sums in arithmetic, *Invent. Math.* **101** (1990), 696–703.

[HardW] G. H. Hardy and E. M. Wright, *An Introduction to the Theory of Numbers*, fourth edition, Oxford, 1960.

[HeLap1] C. Q. He and M. L. Lapidus, Generalized Minkowski content and the vibrations of fractal drums and strings, *Mathematical Research Letters* **3** (1996), 31–40.

[HeLap2] C. Q. He and M. L. Lapidus, Generalized Minkowski content, spectrum of fractal drums, fractal strings and the Riemann zeta-function, *Memoirs Amer. Math. Soc.*, No. 608, **127** (1997), 1–97.

[Hö1] L. Hörmander, The spectral function of an elliptic operator, *Acta Math.* **121** (1968), 193–218.

[Hö2] L. Hörmander, *The Analysis of Linear Partial Differential Operators*, vol. I, *Distribution Theory and Fourier Analysis*, 2nd ed. (of the 1983 ed.), Springer-Verlag, Berlin, 1990.

[Hö3] L. Hörmander, *The Analysis of Linear Partial Differential Operators*, vols. II–IV, Springer-Verlag, Berlin, 1983 & 1985.

[HuaSl] C. Hua and B. D. Sleeman, Fractal drums and the n-dimensional modified Weyl–Berry conjecture, *Commun. Math. Phys.* **168** (1995), 581–607.

[HurWa] W. Hurewicz and H. Wallman, *Dimension Theory*, Princeton Univ. Press, Princeton, 1941.

[Hut] J. E. Hutchinson, Fractals and self-similarity, *Indiana Univ. Math. J.* **30** (1981), 713–747.

[In] A. E. Ingham, *The Distribution of Prime Numbers*, 2nd ed. (reprinted from the 1932 ed.), Cambridge University Press, Cambridge, 1992.

[Ivi] A. Ivić, *The Riemann Zeta-Function: The Theory of the Riemann Zeta-Function with Applications*, Wiley, New York, 1985.

[Ivr1] V. Ja. Ivrii, Second term of the spectral asymptotic expansion of the Laplace–Beltrami operator on manifolds with boundary, *Functional Anal. Appl.* **14** (1980), 98–106.

[Ivr2]   V. Ja. Ivrii, *Precise Spectral Asymptotics for Elliptic Operators Acting in Fiberings over Manifolds with Boundary*, Lecture Notes in Math., vol. 1100, Springer-Verlag, Berlin, 1984.

[Ivr3]   V. Ja. Ivrii, *Microlocal Analysis and Precise Spectral Asymptotics*, Springer-Verlag, Berlin, 1998.

[JafLap] S. Jaffard and M. L. Lapidus, Complex dimensions of multifractals, in preparation.

[JohLap] G. W. Johnson and M. L. Lapidus, *The Feynman Integral and Feynman's Operational Calculus*, Oxford Mathematical Monographs, Oxford Univ. Press, Oxford, January 2000, in press. (ISBN 0-19-853574-0; 774 pages.)

[JorLan1] J. Jorgenson and S. Lang, On Cramer's theorem for general Euler products with functional equation, *Math. Ann.* **297** (1993), 383–416.

[JorLan2] J. Jorgenson and S. Lang, *Basic Analysis of Regularized Series and Products*, Lecture Notes in Math., vol. 1564, Springer-Verlag, New York, 1993.

[JorLan3] J. Jorgenson and S. Lang, *Explicit Formulas for Regularized Products and Series*, Lecture Notes in Math., vol. 1593, Springer-Verlag, New York, 1994, pp. 1–134.

[Kac]   M. Kac, Can one hear the shape of a drum?, *Amer. Math. Monthly* (Slaught Memorial Papers, No. 11) (4) **73** (1966), 1–23.

[KahSa] J.-P. Kahane and R. Salem, *Ensembles Parfaits et Séries Trigonométriques*, Hermann, Paris, 1963.

[KatSar] N. M. Katz and P. Sarnak, Zeroes of zeta functions and symmetry, *Bull. Amer. Math. Soc.* (N. S.) **36** (1999), 1–26.

[Ke]   J. P. Keating, *The Riemann zeta-function and quantum chaology*, in: Quantum Chaos (G. Casati, I. Guarneri and U. Smilanski, eds.), North-Holland, Amsterdam, 1993, pp. 145–185.

[KiLap] J. Kigami and M. L. Lapidus, Weyl's problem for the spectral distribution of Laplacians on p.c.f. self-similar fractals, *Commun. Math. Phys.* **158** (1993), 93–125.

[Ko]   O. Kowalski, Additive volume invariants of Riemannian manifolds, *Acta Math.* **145** (1980), 205–225.

[Lal1]   S. P. Lalley, Packing and covering functions of some self-similar fractals, *Indiana Univ. Math. J.* **37** (1988), 699–709.

[Lal2] S. P. Lalley, Renewal theorems in symbolic dynamics, with applications to geodesic flows, noneuclidean tessellations and their fractal limits, *Acta Math.* **163** (1989), 1–55.

[Lal3] S. P. Lalley, *Probabilistic counting methods in certain counting problems of ergodic theory*, in: [BedKS, pp. 223–258].

[Lan]   S. Lang, *Algebraic Number Theory*, 3d. ed. (of the 1970 ed.), Springer-Verlag, New York, 1994.

[LanCh] S. Lang and W. Cherry, *Topics in Nevanlinna Theory*, Lecture Notes in Math., vol. 1433, Springer-Verlag, New York, 1990.

[Lap1] M. L. Lapidus, Fractal drum, inverse spectral problems for elliptic operators and a partial resolution of the Weyl–Berry conjecture, *Trans. Amer. Math. Soc.* **325** (1991), 465–529.

[Lap2] M. L. Lapidus, *Spectral and fractal geometry: From the Weyl–Berry conjecture for the vibrations of fractal drums to the Riemann zeta-function*, in: Differential Equations and Mathematical Physics (C. Bennewitz, ed.), Proc. Fourth UAB Intern. Conf. (Birmingham, March 1990), Academic Press, New York, 1992, pp. 151–182.

[Lap3] M. L. Lapidus, *Vibrations of fractal drums, the Riemann hypothesis, waves in fractal media, and the Weyl–Berry conjecture*, in: Ordinary and Partial Differential Equations (B. D. Sleeman and R. J. Jarvis, eds.), vol. IV, Proc. Twelfth Intern. Conf. (Dundee, Scotland, UK, June 1992), Pitman Research Notes in Math. Series, vol. 289, Longman Scientific and Technical, London, 1993, pp. 126–209.

[Lap4] M. L. Lapidus, Fractals and vibrations: Can you hear the shape of a fractal drum?, *Fractals* **3**, No. 4 (1995), 725–736. (Special issue in honor of Benoît B. Mandelbrot's 70th birthday. Reprinted in [EvPeVo, pp. 321–332].)

[Lap5] M. L. Lapidus, Analysis on fractals, Laplacians on self-similar sets, noncommutative geometry and spectral dimensions, *Topological Methods in Nonlinear Analysis*, **4** (1994), 137–195.

[Lap6] M. L. Lapidus, *Towards a noncommutative fractal geometry? Laplacians and volume measures on fractals*, in: Harmonic Analysis and Nonlinear Differential Equations, Contemporary Mathematics, vol. 208, American Math. Soc., Providence, R. I., 1997, pp. 211–252.

[LapFl] M. L. Lapidus and J. Fleckinger-Pellé, Tambour fractal: vers une résolution de la conjecture de Weyl–Berry pour les valeurs propres du laplacien, *C. R. Acad. Sci. Paris Sér. I Math.* **306** (1988), 171–175.

[LapMa1] M. L. Lapidus and H. Maier, Hypothèse de Riemann, cordes fractales vibrantes et conjecture de Weyl–Berry modifiée, *C. R. Acad. Sci. Paris Sér. I Math.* **313** (1991), 19–24.

[LapMa2] M. L. Lapidus and H. Maier, The Riemann hypothesis and inverse spectral problems for fractal strings, *J. London Math. Soc.* (2) **52** (1995), 15–34.

[LapNeRnGri] M. L. Lapidus, J. W. Neuberger, R. J. Renka and C. A. Griffith, Snowflake harmonics and computer graphics: Numerical computation of spectra on fractal domains, *Intern. J. Bifurcation & Chaos* **6** (1996), 1185–1210.

[LapPan] M. L. Lapidus and M. M. H. Pang, Eigenfunctions of the Koch snowflake drum, *Commun. Math. Phys.* **172** (1995), 359–376.

[LapPo1] M. L. Lapidus and C. Pomerance, Fonction zêta de Riemann et conjecture de Weyl–Berry pour les tambours fractals, *C. R. Acad. Sci. Paris Sér. I Math.* **310** (1990), 343–348.

[LapPo2] M. L. Lapidus and C. Pomerance, The Riemann zeta-function and the one-dimensional Weyl–Berry conjecture for fractal drums, *Proc. London Math. Soc.* (3) **66** (1993), 41–69.

[LapPo3] M. L. Lapidus and C. Pomerance, Counterexamples to the modified Weyl–Berry conjecture on fractal drums, *Math. Proc. Cambridge Philos. Soc.* **119** (1996), 167–178.

[Lap-vF1] M. L. Lapidus and M. van Frankenhuysen, *Complex dimensions of fractal strings and explicit formulas for geometric and spectral zeta-functions*, Preprint, IHES/M/97/34, Institut des Hautes Études Scientifiques, Bures-sur-Yvette, France, April 1997.

[Lap-vF2] M. L. Lapidus and M. van Frankenhuysen, *Complex dimensions and oscillatory phenomena, with applications to the geometry of fractal strings and to the critical zeros of zeta-functions*, Preprint, IHES/M/97/38, Institut des Hautes Études Scientifiques, Bures-sur-Yvette, France, May 1997.

[Lap-vF3] M. L. Lapidus and M. van Frankenhuysen, *Complex dimensions of fractal strings and oscillatory phenomena in fractal geometry and arithmetic*, in: Spectral Problems in Geometry and Arithmetic (T. Branson, ed.), Contemporary Mathematics, vol. 237, Amer. Math. Soc., Providence, R. I., 1999, pp. 87–105. (Also; Preprint, IHES/M/97/85, Institut des Hautes Études Scientifiques, Bures-sur-Yvette, France, November 1997.)

246   References

[Lap-vF4] M. L. Lapidus and M. van Frankenhuysen, (i) *Complex dimensions of fractal strings and oscillatory phenomena*; (ii) *Zeta-functions and explicit formulas for the geometry and spectrum of fractal strings*, Abstracts #918-35-537 and 918-35-539, Abstracts Amer. Math. Soc. **18**, No. 1 (1997), pp. 82–83. (Presented by M. L. L. and M. v. F., respectively, at the Annual Meeting of the American Mathematical Society, San Diego, Calif., Jan. 11, 1997, Special Session on "Analysis, Diffusions and PDEs on Fractals"; AMS meeting 918, event code AMS SS M1.)

[Lap-vF5] M. L. Lapidus and M. van Frankenhuysen, *Fractality and complex dimensions*, in preparation.

[LeVa] M. Levitin and D. Vassiliev, Spectral asymptotics, renewal theorem, and the Berry conjecture for a class of fractals, *Proc. London Math. Soc.* (3) **72** (1996), 188–214.

[Man1] B. B. Mandelbrot, *The Fractal Geometry of Nature*, rev. and enl. ed. (of the 1977 ed.), W. H. Freeman, New York, 1983.

[Man2] B. B. Mandelbrot, *Measures of fractal lacunarity: Minkowski content and alternatives*, in: Fractal Geometry and Stochastics (C. Bandt, S. Graf and M. Zähle, eds.), Progress in Probability, vol. 37, Birkhäuser-Verlag, Basel, 1995, pp. 15–42.

[Man3] B. B. Mandelbrot, *Multifractals and 1/f Noise (Wild Self-Affinity in Physics)*, Springer-Verlag, New York, 1998.

[MarVu] O. Martio and M. Vuorinen, Whitney cubes, *p*-capacity, and Minkowski content, *Exposition Math.* **5** (1987), 17–40.

[Mat] P. Mattila, *Geometry of Sets and Measures in Euclidean Spaces (Fractals and Rectifiability)*, Cambridge Univ. Press, Cambridge, 1995.

[Maz] V. G. Maz'ja, *Sobolev Spaces*, Springer-Verlag, Berlin, 1985.

[McKSin] H. P. McKean and I. M. Singer, Curvatures and the eigenvalues of the Laplacian, *J. Differential Geom.* **1** (1967), 43–69.

[Mel1] R. B. Melrose, *Weyl's conjecture for manifolds with concave boundary*, in: Geometry of the Laplace Operator, Proc. Sympos. Pure Math., vol. 36, Amer. Math. Soc., Providence, R. I., 1980, pp. 254–274.

[Mel2] R. B. Melrose, *The trace of the wave group*, Contemporary Mathematics, vol. 27, Amer. Math. Soc., Providence, R. I., 1984, pp. 127–167.

[Met]   G. Métivier, Valeurs propres de problèmes aux limites elliptiques irréguliers, *Bull. Soc. Math. France, Mém.* **51–52** (1977), 125–219.

[Mil]   J. Milnor, *Euler characteristic and finitely additive Steiner measure*, in: John Milnor: Collected Papers, vol. 1, Geometry, Publish or Perish, Houston, 1994, pp. 213–234. (Previously unpublished.)

[Min1]   S. Minakshisundaram, A generalization of Epstein zeta-functions, *Canad. J. Math.* **1** (1949), 320–329.

[Min2]   S. Minakshisundaram, Eigenfunctions on Riemannian manifolds, *J. Indian Math. Soc.* **17** (1953), 158–165.

[MinPl]   S. Minakshisundaram and Å. Pleijel, Some properties of the eigenfunctions of the Laplace-operator on Riemannian manifolds, *Canad. J. Math.* **1** (1949), 242–256.

[Mink]   H. Minkowski, *Theorie der konvexen Körper, insbesondere Begründung ihres Oberflächenbegriffs*, in: Gesammelte Abhandlungen von Hermann Minkowski (part II, Chapter XXV), Chelsea, New York, 1967, pp. 131–229. (Originally reprinted in: Gesamm. Abh., vol. II, Leipzig, 1911.)

[MolVa]   S. Molchanov and B. Vainberg, On spectral asymptotics for domains with fractal boundaries, *Commun. Math. Phys.* **183** (1997), 85–117.

[Mor]   P. A. P. Moran, Additive functions of intervals and Hausdorff measure, *Math. Proc. Cambridge Philos. Soc.* **42** (1946), 15–23.

[Od1]   A. M. Odlyzko, On the distribution of spacings between zeros of the zeta-function, *Math. Comp.* **48** (1987), 273–308.

[Od2]   A. M. Odlyzko, *The 10^{20}-th zero of the Riemann zeta-function and 175 millions of its neighbors*, preprint, AT&T Bell Labs, Murray Hill, 1991 (book to appear).

[Od-tR]   A. M. Odlyzko and H. J. J. te Riele, Disproof of the Mertens conjecture, *J. Reine Angew. Math.* **357** (1985), 138–160.

[PaPol]   W. Parry and M. Pollicott, *Zeta Functions and the Periodic Orbit Structure of Hyperbolic Dynamics*, Astérisque, vols. 187–188, Soc. Math. France, Paris, 1990.

[ParSh1]   A. N. Parshin and I. R. Shafarevich (eds.), *Number Theory*, vol. I, *Introduction to Number Theory*, Encyclopedia of Mathematical Sciences, vol. 49, Springer-Verlag, Berlin, 1995. (Written by Yu. I. Manin and A. A. Panchishkin.)

[ParSh2] A. N. Parshin and I. R. Shafarevich (eds.), *Number Theory*, vol. II, *Algebraic Number Fields*, Encyclopedia of Mathematical Sciences, vol. 62, Springer-Verlag, Berlin, 1992. (Written by H. Koch.)

[Pat] S. J. Patterson, *An Introduction to the Theory of the Riemann Zeta-Function*, Cambridge Univ. Press, Cambridge, 1988.

[Ph] Pham The Lai, Meilleures estimations asymptotiques des restes de la fonction spectrale et des valeurs propres relatifs au laplacien, *Math. Scand.* **48** (1981), 5–38.

[Pi] M. A. Pinsky, The eigenvalues of an equilateral triangle, *SIAM J. Math. Anal.* **11** (1980), 819–827.

[Pos] A. G. Postnikov, *Tauberian Theory and Its Applications*, Proc. Steklov Inst. of Math., vol. 144, No. 2, American Math. Soc., Providence, R. I., 1980.

[Pu1] C. R. Putnam, On the non-periodicity of the zeros of the Riemann zeta-function, *Amer. J. Math.* **76** (1954), 97–99.

[Pu2] C. R. Putnam, Remarks on periodic sequences and the Riemann zeta-function, *Amer. J. Math.* **76** (1954), 828–830.

[ReSi1] M. Reed and B. Simon, *Methods of Modern Mathematical Physics*, vol. I, *Functional Analysis*, rev. and enl. ed. (of the 1975 ed.), Academic Press, New York, 1980.

[ReSi2] M. Reed and B. Simon, *Methods of Modern Mathematical Physics*, vol. II, *Fourier Analysis, Self-Adjointness*, Academic Press, New York, 1975.

[ReSi3] M. Reed and B. Simon, *Methods of Modern Mathematical Physics*, vol. IV, *Analysis of Operators*, Academic Press, New York, 1979.

[Rie1] B. Riemann, *Ueber die Anzahl der Primzahlen unter einer gegebenen Grösse*, Monatsb. der Berliner Akad., 1858/60, pp. 671–680. (Reprinted in [Rie2, pp. 145–155]; English transl. in [Edw, Appendix, pp. 299–305].)

[Rie2] B. Riemann, *Gesammelte Mathematische Werke*, Teubner, Leipzig, 1892, No. VII. (Reprinted by Dover Books, New York, 1953.)

[Rog] C. A. Rogers, *Hausdorff Measures*, Cambridge Univ. Press, Cambridge, 1970.

[Roq] P. Roquette, Arithmetischer Beweis der Riemannschen Vermutung in Kongruenzfunktionenkörpern beliebigen Geslechts, *J. Reine Angew. Math.* **191** (1953), 199–252.

[Ru1]   W. Rudin, *Fourier Analysis on Groups*, Interscience Publishers, John Wiley & Sons, New York, 1962.

[Ru2]   W. Rudin, *Real and Complex Analysis*, 3rd ed., McGraw-Hill, New York, 1987.

[RudSar] Z. Rudnick and P. Sarnak, Zeros of principal *L*-functions and random matrix theory, *Duke Math. J.* **81** (1996), 269–322.

[Rue]   D. Ruelle, Generalized zeta-functions for Axiom A basic sets, *Bull. Amer. Math. Soc.* **82** (1976), 153–156.

[SapGoM] B. Sapoval, Th. Gobron and A. Margolina, Vibrations of fractal drums, *Phys. Rev. Lett.* **67** (1991), 2974–2977.

[ScSo] M. Schröter and C. Soulé, *On a result of Deninger concerning Riemann's zeta-function*, in: Motives, Proc. Sympos. Pure Math., vol. 55, American Math. Soc., Providence, R. I., 1994, pp. 745–747.

[Sch1] L. Schwartz, *Théorie des Distributions*, rev. and enl. ed. (of the 1951 ed.), Hermann, Paris, 1966.

[Sch2] L. Schwartz, *Méthodes Mathématiques pour les Sciences Physiques*, Hermann, Paris, 1961.

[Se1]   R. T. Seeley, *Complex powers of elliptic operators*, in: Proc. Symp. Pure Math., vol. 10, American Math. Soc., Providence, R. I., 1967, pp. 288–307.

[Se2]   R. T. Seeley, The resolvent of an elliptic boundary problem, *Amer. J. Math.* **91** (1969), 889–920.

[Se3]   R. T. Seeley, Analytic extensions of the trace associated with elliptic boundary problems, *Amer. J. Math.* **91** (1969), 963–983.

[Se4]   R. T. Seeley, A sharp asymptotic remainder estimate for the eigenvalues of the Laplacian in a domain of $\mathbb{R}^3$, *Adv. in Math.* **29** (1978), 244–269.

[Se5]   R. T. Seeley, An estimate near the boundary for the spectral counting function of the Laplace operator, *Amer. J. Math.* **102** (1980), 869–902.

[Ser]   J.-P. Serre, *A Course in Arithmetic*, English transl., Springer-Verlag, Berlin, 1973.

[Sh]   M. A. Shubin, *Pseudodifferential Operators and Spectral Theory*, Springer-Verlag, Berlin, 1987.

250   References

[Si]   B. Simon, *Functional Integration and Quantum Physics*, Academic Press, New York, 1979.

[SmFoSp]   L. A. Smith, J.-D. Fournier and E. A. Spiegel, Lacunarity and intermittency in fluid turbulence, *Phys. Lett. A* **114** (1986), 465–468.

[Stein]   J. Steiner, *Über parallele Flächen*, Monatsb. preuss. Akad. Wiss., Berlin, 1840, pp. 114–118. (Reprinted in: Gesamm. Werke vol. II, pp. 173–176.)

[Step]   S. A. Stepanov, On the number of points of a hyperelliptic curve over a finite prime field, *Izv. Akad. Nauk SSSR, Ser. Mat.* **33** (1969), 1103–1114.

[Str1]   R. S. Strichartz, Fourier asymptotics of fractal measures, *J. Functional Anal.* **89** (1990), 154–187.

[Str2]   R. S. Strichartz, Self-similar measures and their Fourier transforms, I, *Indiana Univ. Math. J.* **39** (1990), 797–817; II, *Trans. Amer. Math. Soc.* **336** (1993), 335–361; III, *Indiana Univ. Math. J.* **42** (1993), 367–411.

[Su]   D. Sullivan, Entropy, Hausdorff measures old and new, and limit sets of geometrically finite Kleinian groups, *Acta Math.* **153** (1984), 259–277.

[Ta]   J. T. Tate, *Fourier Analysis in Number Fields and Hecke's Zeta-Functions*, Ph.D. Dissertation, Princeton University, Princeton, N. J., 1950. (Reprinted in: *Algebraic Number Theory*, J. W. S. Cassels and A. Fröhlich (eds.), Academic Press, New York, 1967, pp. 305–347.)

[Te]   A. Terras, *Harmonic Analysis on Symmetric Spaces and Applications*, vol. I, Springer-Verlag, New York, 1985.

[Ti]   E. C. Titchmarsh, *The Theory of the Riemann Zeta-Function*, 2nd ed. (revised by D. R. Heath-Brown), Oxford Univ. Press, Oxford, 1986.

[Tr1]   C. Tricot, Douze définitions de la densité logarithmique, *C. R. Acad. Sci. Paris Sér. I Math.* **293** (1981), 549–552.

[Tr2]   C. Tricot, Two definitions of fractional dimension, *Math. Proc. Cambridge Philos. Soc.* **91** (1982), 57–74.

[Tr3]   C. Tricot, *Curves and Fractal Dimensions*, Springer-Verlag, New York, 1995.

[Va]   V. S. Varadarajan, Some remarks on the analytic proof of the Prime
        Number Theorem, *Nieuw Archief voor Wiskunde* **16** (1998), 153–
        160.

[vB]   M. van den Berg, Heat content and Brownian motion for some re-
        gions with a fractal boundary, *Probab. Theory and Related Fields*
        100 (1994), 439–456.

[vB-Gi] M. van den Berg and P. B. Gilkey, A comparison estimate for the
        heat equation with an application to the heat of the *s*-adic von Koch
        snowflake, *Bull. London Math. Soc.*, in press.

[vB-Le] M. van den Berg and M. Levitin, Functions of Weierstrass type
        and spectral asymptotics for iterated sets, *Quart. J. Math. Oxford*
        (2) **47** (1996), 493–509.

[vF1]   M. van Frankenhuysen, *Counting the number of points on an alge-
        braic curve*, appeared in: *Algoritmen in de algebra, a seminar on
        algebraic algorithms*, Nijmegen, 1993 (ed. A. H. M. Levelt).

[vF2]   M. van Frankenhuysen, *Hyperbolic Spaces and the abc Conjecture*,
        Ph.D. Dissertation (Proefschrift), Katholieke Universiteit Nijme-
        gen, Netherlands, 1995.

[vL-vdG] J. H. van Lint and G. van der Geer, *Introduction to Cod-
        ing Theory and Algebraic Geometry*, DMV Seminar, Band 12,
        Birkhäuser, Boston, 1988.

[vL-tR-W] J. van de Lune, H. J. J. te Riele and D. Winter, On the zeros
        of the Riemann zeta function in the critical strip, IV, *Math. Comp.*
        46 (1986), 667–681.

[vM1]   H. von Mangoldt, *Auszug aus einer Arbeit unter dem Titel: Zu Rie-
        mann's Abhandlung 'Über die Anzahl der Primzahlen unter einer
        gegebenen Grösse'*, Sitzungsberichte preuss. Akad. Wiss., Berlin,
        1894, pp. 883–896.

[vM2]   H. von Mangoldt, Zu Riemann's Abhandlung 'Über die Anzahl der
        Primzahlen unter einer gegebenen Grösse', *J. Reine Angew. Math.*
        114 (1895), 255–305.

[Vor]   A. Voros, Spectral functions, special functions and the Selberg zeta
        function, *Commun. Math. Phys.* **110** (1987), 439–465.

[Wa]   M. Watkins, Arithmetic progressions of zeros of Dirichlet *L*-func-
        tions, Preprint, 1998.

[Weil] A. Weil, On the Riemann hypothesis in function-fields, *Proc. Nat.
        Acad. Sci. U.S.A.* **27** (1941), 345–347. (Reprinted in [Wei7, vol. I,
        pp. 277–279].)

[Wei2] A. Weil, *Sur les courbes algébriques et les variétés qui s'en déduisent*, Pub. Inst. Math. Strasbourg VII (1948), pp. 1–85. (Reprinted in: *Courbes algébriques et variétés abéliennes*, Hermann, Paris, 1971.)

[Wei3] A. Weil, Number of solutions of equations in finite fields, *Bull. Amer. Math. Soc.* **55** (1949), 497–508. (Reprinted in [Wei7, vol. I, pp. 399–410].)

[Wei4] A. Weil, *Sur les "formules explicites" de la théorie des nombres premiers*, Comm. Sém. Math. Lund, Université de Lund, Tome supplémentaire (dédié à Marcel Riesz), (1952), pp. 252–265. (Reprinted in [Wei7, vol. II, pp. 48–61].)

[Wei5] A. Weil, *Fonction zêta et distributions*, Séminaire Bourbaki, 18ème année, 1965/66, no. 312, Juin 1966, pp. 1–9. (Reprinted in [Wei7, vol. III, pp. 158–163].)

[Wei6] A. Weil, Sur les formules explicites de la théorie des nombres, *Izv. Mat. Nauk (Ser. Mat.)* **36** (1972), 3–18; English transl. in: *Math. USSR, Izv.* **6** (1973), 1–17. (Reprinted in [Wei7, vol. III, pp. 249–264].)

[Wei7] A. Weil, *André Weil: Oeuvres Scientifiques* (Collected Papers), vols. I, II and III, 2nd ed. (with corrected printing), Springer-Verlag, Berlin and New York, 1980.

[Wey1] H. Weyl, Über die Abhängigkeit der Eigenschwingungen einer Membran von deren Begrenzung, *J. Reine Angew. Math.* **141** (1912), 1–11. (Reprinted in [Wey4, vol. I, pp. 431–441].)

[Wey2] H. Weyl, Das asymptotische Verteilungsgesetz der Eigenwerte linearer partieller Differentialgleichungen, *Math. Ann.* **71** (1912), 441–479. (Reprinted in [Wey4, vol. I, pp. 393–430].)

[Wey3] H. Weyl, On the volume of tubes, *Amer. J. Math.* **61** (1939), 461–472. (Reprinted in [Wey4, vol. III, pp. 658–669].)

[Wey4] H. Weyl, *Hermann Weyl: Gesammelte Abhandlungen* (Collected Works), vols. I and III, Springer-Verlag, Berlin and New York, 1968.

[Wid]   D. Widder, *The Laplace Transform*, Princeton Univ. Press, Princeton, 1946.

[Zag]   D. Zagier, *Private communication*, 1994.

[Zyg]   A. Zygmund, *Trigonometric Series* I and II, 2nd ed., Cambridge Univ. Press, Cambridge, 1959.

Conventions

$f(x) = O(g(x))$	$f(x)/g(x)$ is bounded
$f(x) = o(g(x))$	$f(x)/g(x)$ tends to 0
$f(x) \sim g(x)$	$f(x)/g(x)$ tends to 1
$f(x) \ll g(x)$	same meaning as $f(x) = O(g(x))$
$\approx$	approximately equal to
$d \mid n$	d divides n
$A \backslash B$	the set of points in A that do not lie in B
$\#A$	the cardinality of the finite set A
$\mathbf{N}$	the set of nonnegative integers $0, 1, 2, 3, \ldots$
$\mathbf{N}^* = \mathbf{N} \backslash \{0\}$	the set of positive integers $1, 2, 3, \ldots$
$\mathbf{Z}$	the set of integers $\ldots, -3, -2, -1, 0, 1, 2, 3, \ldots$
$\mathbf{R}, \mathbf{C}$ and $\mathbf{Q}$	the set of real, complex and rational numbers, respectively
$\mathbf{R}_+^*$	the multiplicative group of positive real numbers
i	the square root of -1
$s = \sigma + it$	s is a complex number with $\sigma = \operatorname{Re} s$ and $t = \operatorname{Im} s$
$\log x$	the natural logarithm of x
$\log_a x$	$\log x / \log a$, the logarithm of x in base a

Symbol Index

Index

List of Figures

Acknowledgements

We would like to thank the Institut des Hautes Etudes Scientifiques (IHES), of which we were members while much of this research was performed. The work of Michel L. Lapidus was supported by the National Science Foundation under grants DMS-9207098 and DMS-9623002, and that of Machiel van Frankenhuysen by the Marie Curie Fellowship ERBFMBICT960829 of the European Community.

The first author would also like to thank Alain Connes, Rudolf H. Riedi, and Christophe Soulé for helpful conversations and/or references during the preliminary phase of this work. In addition, the authors are grateful to Gabor Elek and Jim Stafney as well as to several anonymous referees, for their helpful comments on the preliminary versions of this book.

Finally, we are grateful to Ann Kostant, Executive Editor of Mathematics and Physics at Birkhäuser Boston, for her enthusiasm and constant encouragement, as well as for her guidance in preparing the manuscript for publication.

Last but not least, the first author would like to thank his wife, Odile, and his children, Julie and Michaël, for supporting him through long periods of sleepless nights, either at the Résidence de l'Ormaille of the IHES in Bures-sur-Yvette or at their home in Riverside, California, while the theory presented in this research monograph was being developed or the book was in the process of being written.

Part of this work was presented by the authors at the Special Session on "Analysis, Diffusions and PDEs on Fractals" held during the Annual Meeting of the American Mathematical Society (San Diego, January 1997), and at the Special Session on "Dynamical, Spectral and Arithmetic Zeta-Functions" held during the Annual Meeting of the American Mathematical Society (San Antonio, January 1999).[1] It was also presented by the first author in invited talks at the CBMS-NSF Conference on "Spectral Problems in Geometry and Arithmetic" (Iowa City, August 1997) and at the Conference on "Recent Progress in Noncommutative Geometry" (Lisbon, Portugal, September 1997), as well as in the Programs on "Spectral Geometry" (June–July 1998) and on "Number Theory and Physics" (September 1998), both held at the Erwin Schroedinger International Institute for Mathematical Physics in Vienna, Austria. In addition, it was presented by the first author at the Basic Research Institute in the Mathematical Sciences (BRIMS) in Bristol, UK, in April 1999 and in the Program on "Mathematics and Applications of Fractals" (March–April 1999) held at the Isaac Newton Institute for Mathematical Sciences of the University of Cambridge, England.

---

[1]Abstracts #918-35-537 and 539, Abstracts Amer. Math. Soc. **18** No. 1 (1997), 82–83, and Abstracts #939-58-84 and 85, *ibid.* **20** No. 1 (1999), 126–127.

CPSIA information can be obtained at www.ICGtesting.com
Printed in the USA
LVOW102131171012

303372LV00002B/20/P